D1564106

Conditioning Agents for Hair and Skin

COSMETIC SCIENCE AND TECHNOLOGY

Series Editor

ERIC JUNGERMANN

Jungermann Associates, Inc.
Phoenix, Arizona

1. Cosmetic and Drug Preservation: Principles and Practice, *edited by Jon J. Kabara*
2. The Cosmetic Industry: Scientific and Regulatory Foundations, *edited by Norman F. Estrin*
3. Cosmetic Product Testing: A Modern Psychophysical Approach, *Howard R. Moskowitz*
4. Cosmetic Analysis: Selective Methods and Techniques, *edited by P. Boré*
5. Cosmetic Safety: A Primer for Cosmetic Scientists, *edited by James H. Whittam*
6. Oral Hygiene Products and Practice, *Morton Pader*
7. Antiperspirants and Deodorants, *edited by Karl Laden and Carl B. Felger*
8. Clinical Safety and Efficacy Testing of Cosmetics, *edited by William C. Waggoner*
9. Methods for Cutaneous Investigation, *edited by Robert L. Rietschel and Thomas S. Spencer*
10. Sunscreens: Development, Evaluation, and Regulatory Aspects, *edited by Nicholas J. Lowe and Nadim A. Shaath*
11. Glycerine: A Key Cosmetic Ingredient, *edited by Eric Jungermann and Norman O. V. Sonntag*
12. Handbook of Cosmetic Microbiology, *Donald S. Orth*
13. Rheological Properties of Cosmetics and Toiletries, *edited by Dennis Laba*
14. Consumer Testing and Evaluation of Personal Care Products, *Howard R. Moskowitz*
15. Sunscreens: Development, Evaluation, and Regulatory Aspects. Second Edition, Revised and Expanded, *edited by Nicholas J. Lowe, Nadim A. Shaath, and Madhu A. Pathak*
16. Preservative-Free and Self-Preserving Cosmetics and Drugs: Principles and Practice, *edited by Jon J. Kabara and Donald S. Orth*
17. Hair and Hair Care, *edited by Dale H. Johnson*
18. Cosmetic Claims Substantiation, *edited by Louise B. Aust*
19. Novel Cosmetic Delivery Systems, *edited by Shlomo Magdassi and Elka Touitou*
20. Antiperspirants and Deodorants: Second Edition, Revised and Expanded, *edited by Karl Laden*

21. Conditioning Agents for Hair and Skin, *edited by Randy Schueller and Perry Romanowski*

ADDITIONAL VOLUMES IN PREPARATION

Principles of Polymer Science and Technology in Cosmetics and Personal Care, *edited by E. Desmond Goddard and James V. Gruber*

Botanicals in Cosmetics, *edited by Larry Smith*

Conditioning Agents for Hair and Skin

edited by
Randy Schueller
Perry Romanowski
Alberto-Culver Company
Melrose Park, Illinois

MARCEL DEKKER, INC.

NEW YORK • BASEL

ISBN: 0-8247-1921-2

This book is printed on acid-free paper.

Headquarters
Marcel Dekker, Inc.
270 Madison Avenue, New York, NY 10016
tel: 212-696-9000; fax: 212-685-4540

Eastern Hemisphere Distribution
Marcel Dekker AG
Hutgasse 4, Postfach 812, CH-4001 Basel, Switzerland
tel: 44-61-261-8482; fax: 44-61-261-8896

World Wide Web
http://www.dekker.com

The publisher offers discounts on this book when ordered in bulk quantities. For more information, write to Special Sales/Professional Marketing at the headquarters address above.

Current printing (last digit):
10 9 8 7 6 5 4 3 2 1

PRINTED IN THE UNITED STATES OF AMERICA

About the Series

The Cosmetic Science and Technology series was conceived to permit discussion of a broad range of current knowledge and theories of cosmetic science and technology. The series is made up of books written either by a single author or by a number of contributors to an edited volume. Authorities from industry, academia, and the government have participated in writing these books.

The aim of this series is to cover the many facets of cosmetic science and technology. Topics are drawn from a wide spectrum of disciplines ranging from chemistry, physics, biochemistry, and analytical and consumer evaluations to safety, efficacy, toxicity, and regulatory questions. Organic, inorganic, physical, and polymer chemistry, emulsion technology, microbiology, dermatology, toxicology, and related fields all play a role in cosmetic science.

There is little commonality in the scientific methods, processes, or formulations required for the wide variety of cosmetic and toiletries manufactured. Products range from hair care, oral care, and skin care preparations to lipsticks, nail polishes and extenders, deodorants, body powders, and aerosols to such products as antiperspirants, dandruff and acne treatments, antimicrobial soaps, and sunscreens.

Cosmetics and toiletries represent a highly diversified field with many subsections of science and "art." Indeed, even in these days of high technology, "art" and intuition continue to play an important part in the development of formulations, their evaluation, and the selection of raw materials. There is a move toward more sophisticated scientific methodologies in the fields of claim substantiation, safety testing, product evaluation, and chemical analyses and a better understanding of the properties of skin and hair.

Emphasis in this series is placed on reporting the current status of cosmetic technology and science in addition to historical reviews. The series has grown, dealing with the constantly changing technologies and trends in the cosmetic industry, including globalization. Several of the books have been translated into Japanese and Chinese. Contributions range from highly sophisticated and scientific treatises, to primers, descriptions of practical applications, and pragmatic presentations. Authors are encouraged to present their own concepts as well as established theories. Contributors have been asked not to shy away from fields that are still in a state of transition, or to hesitate to present detailed discussions of their own work. Altogether, we intend to develop in this series a collection of critical surveys and ideas covering diverse phases of the cosmetic industry.

Conditioning Agents of Hair and Skin is the twenty-first book published in the Cosmetic Science and Technology series. The book includes detailed discussions of the biology of skin, the largest component of the human body, and hair. Not only are skin and hair responsible for our appearance, but they also provide important protective properties. Two other key areas covered in the book are the chemicals used as conditioners in hair and skin care products and the methods and new techniques for determining the efficacy of these products and their ability to deliver conditioning to skin and hair. In addition to providing "cosmetic" effects, the conditioning agents must be functional, provide hair treatment, and make positive contributions to the health of the skin.

I want to thank all the contributors for taking part in this project and particularly the editors, Randy Schueller and Perry Romanowski, for developing the concept of this book and contributing a chapter. Special recognition is also due to Sandra Beberman and the editorial staff at Marcel Dekker, Inc. In addition, I would like to thank my wife, Eva, without whose constant support and editorial help I would never have undertaken this project.

Eric Jungermann, Ph.D.

Preface

Biological surfaces, such as hair and skin, are vulnerable to damage from a variety of external sources. Such damage can make hair rough and unmanageable and look dull, while skin can become dry, scaly, and itchy. Cosmetic products are used to counteract this damage and to make skin and hair look and feel better; in other words, the products put these surfaces in better condition. Hair conditioning products are primarily intended to make wet hair easier to detangle and comb and make dry hair smoother, shinier, and more manageable. Skin conditioning products are primarily intended to moisturize while providing protection from the drying effects of sun, wind, and harsh detergents. The functional raw materials responsible for the conditioning ability of these products are the focus of this book.

Conditioning Agents for Hair and Skin was prepared by cosmetic formulators, for cosmetic formulators. Our objective is to provide information that is useful to anyone involved in formulating personal care products, from the novice chemist to the seasoned veteran. For the beginning chemist, we aim to provide a solid foundation of technical knowledge. For the seasoned formulator, we detail the latest state-of-the-art ingredients and testing procedures used in their evaluation.

The book is structured to give a complete review of the subject. The first chapter serves as a general introduction. We define conditioning and provide an overview of the types of materials and formulations used to achieve these effects. In addition, we discuss how to evaluate the performance of conditioning products. Chapters 2 and 3 review the biological and physicochemical aspects of hair and skin in order to provide an understanding of what conditioning agents should accomplish.

The next several chapters comprise the bulk of the text; they deal with the individual conditioning agents used in hair and skin care products. The materials we have included were chosen because of their different conditioning effects on biological surfaces. This section begins with a review of petrolatum, an occlusive material that provides conditioning by sealing moisture in skin. Next is a chapter about humectants, materials that attract and bind water to skin and hair. Conditioning agents that impart emolliency are described next in a chapter about esters and oils. The rest of the chapters in this section deal with ingredients that have an electrostatic affinity for biological surfaces. These include proteins and classic quaternized ammonium compounds as well as pseudo-cationic surfactants, like amine oxides. A discussion of cationic polymers is included as well. Two chapters in this section discuss novel silicone-derived materials and how they function in formulations. After discussing chemical raw materials we felt it important to place this information in perspective by including a review of factors to consider when formulating these materials into finished products. Finally, we end the book by examining the testing methods currently available for evaluation of conditioning formulations.

We believe this approach will give the beginner an excellent overview of the subject while providing the veteran chemist with new insights into raw materials, formulations, and testing. As in many works of this type, there is some overlap of material between chapters and, while this may seem redundant, it is important to recognize the multifunctionality of many conditioning agents. As with any work of this nature, the state of the art is constantly changing and we welcome comments from readers to be considered for incorporation into future editions.

Randy Schueller
Perry Romanowski

Contents

Contributors

Eric S. Abrutyn Senior Product Development Leader, The Andrew Jergens Company, Cincinnati, Ohio

John Carson Technical Director, Croda, Inc., Edison, New Jersey

Zoe Diana Draelos Clinical Associate Professor, Department of Dermatology, Wake Forest University School of Medicine, Winston-Salen, and Dermatology Consulting Services, High Point, North Carolina

Peter M. Elias Professor of Dermatology, University of California, San Francisco, and Veterans Affairs Medical Center, San Francisco, California

David T. Floyd Technical Director, Goldschmidt Chemical Corporation, Hopewell, Virginia

Kevin F. Gallagher Croda, Inc., Edison, New Jersey

Bruce W. Gesslein Technical Manager, Specialty Chemicals Division, Ajinomoto U.S.A., Inc., Teaneck, New Jersey

Burghard H. Grüning Senior Research Manager, Goldschmidt Chemical Corporation, Hopewell, Virginia

Bernard Idson Professor, College of Pharmacy, University of Texas at Austin, Austin, Texas

Janusz Jachowicz Science Fellow, International Specialty Products, Wayne, New Jersey

Matthew F. Jurczyk Marketing Manager, Goldschmidt Chemical Corporation, Hopewell, Virginia

David L. Miller President, CuDerm/Bionet, Inc., Dallas, Texas

David S. Morrison Senior Research Associate, Penreco, The Woodlands, Texas

Gary A. Neudahl Technical Services Manager, Costec, Inc., Palatine, Illinois

Anthony J. O'Lenick, Jr. President, Lambent Technologies, Norcross, Georgia

Ronald L. Rizer Chief Investigator and Vice President of Operations, Thomas J. Stephens & Associates, Inc., Carrollton, Texas

Perry Romanowski Senior Research Chemist, Research and Development, Alberto-Culver Company, Melrose Park, Illinois

Randy Schueller Group Leader, Research and Development, Alberto-Culver Company, Melrose Park, Illinois

Monya L. Sigler Senior Investigator, Thomas J. Stephens & Associates, Inc., Carrollton, Texas

Mort Westman President, Westman Associates, Inc., Oak Brook, Illinois

1

Introduction to Conditioning Agents for Hair and Skin

Randy Schueller and Perry Romanowski
Alberto-Culver Company, Melrose Park, Illinois

I. WHAT IS "CONDITIONING"?

A. Definition of Conditioning

The purpose of this book is to educate the reader about the materials used to condition hair and skin. To best accomplish this, the book is divided into three sections: Part One reviews the biological reasons hair and skin need to be conditioned, Part Two discusses the chemical conditioning agents that are available to the formulator, and Part Three provides strategies for formulating and evaluating products containing these agents. Before beginning the discussion, we must first define what we mean when we talk about conditioning. We suggest that the reader view conditioning not as a single element or property, but as the combined effect of many influencing factors. Conditioning can mean many things depending on the circumstances, and therefore any functional definition of the term must be flexible enough to encompass multiple meanings. For the purposes of this book we offer the following definition:

> A product can be said to have conditioning properties if it improves the quality of the surface to which it is applied, particularly if this improvement involves the correction or prevention of some aspects of surface damage.

With this definition as a framework with which to work, we can turn the discussion to specific elements to consider when attempting to condition hair and skin.

B. Components of Conditioning

Just as conditioning is not a single entity, the condition of hair and skin cannot be described by a single variable. Instead, the condition of the surface of hair and skin is determined by a series of properties, which can be considered to be the components of conditioning. Viewed as a whole, these components determine the overall condition of hair and skin. We have attempted to group these components into categories, but this is difficult because the categories overlap and interact to a large degree. For example, when discussing hair conditioning properties, one may say that softness is one beneficial conditioning characteristic and that shine is another. But don't both of these properties really depend on the smoothness of the surface of the hair? Therefore, should one assume that surface smoothness is the ultimate aspect of conditioning, the cosmetic equivalent of quantum physics' quark? No, because surface analysis reveals only *one* aspect of the hair's condition and cannot fully explain other contributing factors, such as moisture content, electrical conductivity, and lipid composition. To obtain a complete assessment of the hair's condition, these other factors must be taken into consideration. In our definition of conditioning, improvement to the substrate (hair or skin) is measured by evaluating several physical, chemical, and perceptual factors. These factors can be sorted into the following categories.

Structural factors (how hair and skin are constructed)
Compositional elements (what physiochemical components hair and skin contain)
Visual appearance (how hair and skin look to the observer)
Tactile perception (how hair and skin feel to the observer)
Physical properties (how hair and skin behave)

These factors will be discussed separately as they relate to hair and skin.

1. Structural Factors

a. Hair. Clearly, the structure of hair is an important consideration when assessing its condition. Not surprisingly, Robbins (1, p. 212) notes that hair is damaged by the breakdown or removal of structural elements. The specific characteristics of these elements are the focus of Chapter 2 and will not be discussed in detail here. In this introductory chapter we will simply note that hair consists of protein that is organized into the cuticle (the outer layer of overlapping scales) and the cortex (the inner region consisting of bundled protein fibers). It is the cuticle that is of primary interest when discussing conditioning products. We know from examining electron micrographs that when the cuticular scales are raised, chipped, or broken, the hair's surface is perceptibly rougher. Even relatively simple actions, such as combing, can cause lifting of these scales, which in turn leaves the cuticle even more vulnerable to subsequent damage. With continued physical abrasion these cuticular defects can become cracks or fissures which penetrate into the cortex. Sufficient degradation of the cortex will ultimately lead to breakage and loss of the hair shaft. Therefore, maintaining a smooth cuticular surface is essential to the condition of hair.

b. Skin. Skin is also composed of protein but, unlike hair, it is a living organ. Chapters 2 and 3 discuss the biology of skin structures in detail. The skin's outer layer, the stratum corneum, is analogous to the hair's cuticle in that it consists of flat, hardened, keratinized cells. As with hair, severe perturbations in the outer protective layer of skin can result in increasing degrees of damage. This damage is especially prevalent at low temperature/humidity conditions, which can cause skin to become relatively inflexible and inelastic. The resulting cracks and fissures in the stratum corneum can cleave the epidermis and lead to bleeding, inflammation, and infection (2). For both skin and hair, a contiguous surface structure is key to ensuring the good condition of the substrate.

2. Compositional Elements

a. Hair. In addition to the proteinaceous structure of hair, there are additional elements of composition to consider. Lipids, which are essential to maintaining the pliability of hair, are present both in the free form and bound in cell structures. The free lipids are primarily the oily secretions of the sebaceous glands which provide surface gloss and lubrication to the hair; however, they may also detract from its condition by making the hair appear greasy or weighed down. The structural lipids are part of the intercellular complex and are important because they help cement together the protein structures, particularly in keeping the cuticle cells. It is worth noting that the intercellular spaces, where these lipids reside, may be important pathways to deliver conditioning agents inside the hair (1, p. 44). When these lipids are stripped from hair, the result is that the hair becomes more brittle and prone to damage.

Water is another important component in hair. Water also helps maintain pliability, which keeps the hair shafts from fracturing. Generally, the moisture content in hair will equilibrate to the external humidity. However, when hair is dried by heating, the water level does not equilibrate to preheating levels without submersion or rewetting. This phenomenon, known as histeresis (1, p. 77) can occur as a result of the heating processes commonly used in hair care (e.g., blow dryers and curling irons).

b. Skin. Skin, being a living organ, has a more complex structure than hair, but it also depends on a mixture of lipids and water to maintain its optimum condition. In fact, epidermal lipids account for about 10–12% of skin's dry tissue weight. These lipids consist of phospholipids, sphingolipids, free and esterified fatty acids, and free and esterified cholesterol (2). For the adventurous reader, an in-depth biochemical review of skin lipid composition can be found in Chapter 3.

Water content in skin is essential to its health. As skin ages, it loses some of its natural ability to retain water, which in turn negatively affects its condition. Skin conditioning products can be extremely helpful in combating problems resulting from lack of moisture. In addition to moisture and lipids, skin also contains materials which act as natural humectants (e.g., sodium hyaluronate and sodium pyrollidone carboxylate). Such materials are commonly included in formulations designed to moisturize skin.

Another aspects of skin's condition related to compositional elements is pH. The skin's surface is slightly acid, with an inherent pH between 4 and 6. This value varies depending on the area of the body, however, and areas with higher moisture tend to have higher pH value (this includes the axilla, inguinal regions, and in between fingers and toes) (3). This "acid mantle," as it is referred to, assists the body in warding off infection. Therefore, proper pH is an indicator that skin is in good condition.

It is important that the chemist recognize the role these compositional elements (lipids, moisture, and acid/base groups) play in determining the condition of hair and skin. Conditioning formulations should strive to maintain the proper balance of these elements.

3. Appearance

a. Hair. Appearance is another significant indicator of the condition of hair and skin. Shine is one visual aspect that is strongly associated with the "health" of hair. The term "health" is somewhat paradoxical, considering that hair is not composed of living tissue. Nonetheless, one major marketer has been quite successful with the claim that its products result in "hair so healthy it shines." Regardless of the claims, shine is really the end result of several of the processes/factors described in the preceding sections (i.e., smoothness of surface structure and the presence of the appropriate levels of lipids and moisture). Although definitions vary somewhat, shine, gloss, and luster can be measured both instrumentally and subjectively to provide an indication of the condition of hair. On a similar note, color is also associated with the "health" of hair; i.e., hair that has vibrant color is assumed to be in better condition. In reality this may not be the case, since hair color which is artificially induced can actually be damaging to the hair.

b. Skin. While shine or gloss per se may not necessarily be a desirable goal for skin conditioners, it may be an advantage for a facial moisturizer to leave the skin with a "healthy glow." We note that "healthy" appearance claims make more sense when directed to skin, since that is a living organ. Another visual indicator of the quality of skin condition is its color. Color is an important indicator because dry skin can take on a whitish or grayish cast as light is scattered by the loose scales of the stratum corneum; in blacks this gives skin an "ashy" appearance (2). Severely dry skin may have a reddened or cracked appearance, which serves as a visual cue to its poor condition.

Appearance is a key factor of hair and skin condition that should not be taken lightly. By addressing problems with the substrate's structural and compositional elements, it is likely that its appearance will improve as well. Furthermore, there may be additional visual cues the formulator may be able to impart which will reinforce the impression of such a healthy condition (artificial hair color might be one such example).

4. Tactile Perception

a. Hair. Tactile sensations are extremely important barometers of hair and skin condition. For hair, conditioning agents can augment the soft feel that a smooth cuticle and proper moisture/lipid balance produces. Even when hair has been significantly damaged, i.e., when the cuticle scales are raised or abraded, conditioning agents can ameliorate the rough feel that normally accompanies this condition. In this regard, conditioning agents can give signals on wet hair as well as dry. For example, some conditioning agents (e.g., volatile silicones) are designed to provide transitory lubrication while hair is wet. Hopefully, in addition to providing a temporary improvement in the feel of the hair, the formulator will be able to provide some form of long-term aid (such as providing substantive lubrication) which will lessen the likelihood that further damage will occur during subsequent styling processes.

b. Skin. The sensations of touch are particularly critical when evaluating the condition of skin. This is somewhat ironic considering that the sensory mechanisms for touch reside in the skin itself. Skin that feels rough or dry to the touch is, almost by definition, in poor condition. This rough feel has been quantified by both instrumental and subjective methods. Comparative studies have been conducted on skin roughness using a variety of methods such as image analysis, as well as in-vivo skin testing techniques (4). Relieving perceived dryness is a "must" for skin conditioning products. In fact, Idson refers to skin conditioning as "dry skin relief" which is achieved primarily by adding materials to maintain or restore the elasticity or flexibility of the horny layer. This approach requires the surface of the skin be hydrated with water (or water-miscible agents) or that it is lubricated and occluded with water-insoluble agents (2).

Consumers' perception of the dry feel of their skin is also an important consideration, although one might question whether such self-diagnosis is relevant to true skin conditioning. We would respond to this issue in two ways. First, in the cosmetic industry, often perception is reality. If consumers feel that a certain product is providing conditioning properties, no matter how technically accurate this perception is, they are likely to be satisfied with that product. Second, numerous scientific studies have correlated consumer perception with objective measurements. For example, one study found good correlation between consumers' assessment of skin condition and actual measurement of sebum control. Another found a high correlation between perceived skin oiliness and skin surface lipid quantity. Finally, a good correlation also exists between perceived smoothness/roughness and measured transepidermal water loss (TEWL) and skin surface morphology (5).

In addition to surface texture-related tactile sensations, itching is another sensation of interest. Technically known as pruritus, a strong itching sensation is especially common as the aging skin loses some of its hygroscopic nature (2). By improving general skin condition, conditioning products can ease the factors which cause itching. Although these tactile sensations are perhaps the most subjective indicators of skin condition, they are also perhaps the most meaningful to consumers.

5. *Physical Properties*

a. Hair. Ultimately, deterioration of the hair shaft will be measurable in terms of specific physical properties. These physical properties represent another aspect of the hair's condition, and they reveal aspects that are not necessarily reflected by the components discussed previously. As we have noted, degeneration of the hair's condition can be assessed by examining its surface structure, measuring depletion of lipids or moisture, or even simply touching or observing its surface. But none of these evaluations can determine how easily the hair will break or how its frictional properties have been affected. By objectively measuring the actual physical properties of hair, additional insight can be gained into its condition. Properties of interest include tensile strength, frictional characteristics, and assembly properties of bulk hair.

Tensile strength is a measurement of how the strength of hair and how prone it is to breakage. A number of instruments are used commercially to evaluate the tensile properties of hair (1, p. 299). The frictional characteristics of the hair's surface are a good measure of how much damage it has endured. Frictional forces can be measured instrumentally and are affected by a variety of factors including the diameter of the hair, ambient humidity, and the degree to which the hair has been damaged (1, p. 335). The bulk properties of assembled hair fibers can also provide further insight into the hairs' condition. These include measurements of body or volume, manageability, and combing properties (1, p. 359). Electrical conductivity measurements can quantify the degree to which hair will experience "flyaway," which is the result of a buildup of static electric charges.

b. Skin. The condition of skin can similarly be evaluated using physical measurements. Tensile strength is not as relevant as it is in the case of hair, but elasticity measurements are commonly employed as a barometer of skin condition. A general reduction in skin elasticity is thought to be one of the primary causes of wrinkles, and therefore is a major indicator of skin condition (6). One instrumental method for quantifying skin elasticity employs a hand-held probe which measures the suction force required to life a 2-mm area of skin (7). As with hair, friction properties of skin can provide information on its condition. Friction is a function of specific desquamation, where the upper layers of the skin are defatted and dehydrated. When this happens, the cells may stick together and come off in clumps. An object moving across this roughened, dry surface will experience increased friction (2). Frictional forces can be measured and correlated with morphology. Finally, we mention that electrical conductivity is an important physical measurement in assessing the moisture level of skin, which in turn is crucial in evaluating the skin's condition. A number of commercial instruments are commonly used for this purpose.

When these five components (structure, compositional elements, appearance, tactile sensations, and physical properties) are looked at together, they form a multidimensional picture of the condition of the hair or skin. Of course, depending on the nature of the surface and the type of conditioning effect being sought, one or more of them may be more important than the others. It is important for the chemist to realize that their conditioning formulations should address the appropriate areas. The specific performance objectives of the product being formulated should be carefully compared to the components of conditioning as outlined in this chapter to suggest formulation approaches the chemist can use to ensure the product is successful. Refer to Chapter 12 for an expanded discussion of the philosophy of formulating conditioning products.

II. CONDITIONING QUESTIONS—WHO, WHAT, WHERE, AND WHY

For our definition of conditioning to be useful to the chemist during the formula development process, additional questions must be asked to expand upon what is really meant by conditioning. When cosmetic chemists deal with the challenge of formulating a product intended to condition hair or skin, they must first have a firm grasp on what is required of that product. They must understand how the product should look, smell, and feel; how expensive the formula should be, the type of packaging it will be in, and a myriad of other details. But first and foremost, they must understand what kind of conditioning it is expected to deliver. To know this, the chemist needs to know the answer to several other questions first.

A. Toward Whom Is the Conditioning Benefit Targeted?

Even with an understanding of these components of conditioning, formulating chemists must understand the context in which conditioning applies. The first question is: "Who is asking the question?" This may sound a bit flippant, but it really is an important point, because the level of knowledge, sophistication, and experience of the inquiring party will help determine the answer. Idson points out that dry skin, for example, "means different things depending whether one is a sufferer, a dermatologist, a cosmetic manufacturer, or a geneticist" (2). For example, a consumer's awareness of conditioning may be limited to recognizing that a rough patch of skin, e.g., on the elbow, is softened. It is possible that simple application of a conditioning oil may be enough to satisfy this expectation. The marketing brand manager in charge of the project may have similar expectations. On the other hand, a chemist or dermatologist may expect conditioning agents to measurably improve the moisture content of skin. In this case, more effective moisturizing agents such as occlusives or humectants may be required. It is important to recognize that the term "conditioning" can have different meanings to different audiences. This point is echoed by Professor Dikstein, who notes that in seeking a moisturizing product, the consumer may really be looking for "a product that replenishes skin oils." This may be the case even though technically such a product may not actually reduce dryness or increase moisture content. The subjective element should not be overlooked (5).

B. What Kind of Conditioning Is the Product Intended to Do?

As noted earlier, numerous terms are used by professionals and lay people to denote conditioning. Conditioning skin products can be moisturizers, toners, exfolliants, etc., while hair care products can be moisturizers, detanglers, antistatic, hydrators, and so forth. The intended purpose of the product will determine the type of formula that is appropriate. All conditioning products must address some, or all, of the five components of conditioning outlined above. The key is for the chemist to understand what kind of conditioning is required and to ensure that the formulation addresses the appropriate conditioning needs.

C. Where Is the Product to Be Used?

Another key question to ask is: "Where is the conditioning effect to take place?" Is it targeted toward hair or skin? The requirements of these two substrates are quite different (these differences are reviewed in detail in Chapters 2 and 3). Even more specific, but less obvious, is the question: "What *part* of the skin or hair is being conditioned?" The conditioning requirements of rough elbows are quite different than those of the delicate skin around the eyes. Likewise, frayed or split ends of hair may have different requirements than the oilier hair close to the scalp. The chemist needs to understand the particular target area for the product before being able to select appropriate conditioning agents.

D. Why Is Conditioning of This Product Required?

This leads us to the next question, which is: "Why is conditioning being included as part of this product profile?" This is a very important question given that conditioning may not be the primary purpose of the product. For example, a hand and body lotion has different conditioning requirements than a conditioning nail polish remover. While both are claimed to be "conditioning," the expectations are quite different. The requirements for the nail polish remover may be satisfied as long as the nail and the surrounding tissue is somewhat refatted and protected from the lipid-stripping action of the solvents that the product may employ to dissolve the polish. The hand and body lotion may need to significantly increase the moisture content of the skin. Conditioning claims are often made as a secondary benefit; understanding the role of conditioning agents in the product will help the chemist decide on appropriate agents to include.

The questions discussed above may be best answered in the form of a comprehensive description of the product. This "product profile" should clearly spell out the expectations for the product. The use of product profiles in formulating conditioning products is discussed in Chapter 12. Once the chemist knows the who, what, where, and why of conditioning, he or she can begin to make meaningful choices as to which conditioning agents are best suited for a given product.

III. FORMAT OF THE BOOK

As mentioned at the beginning of this chapter, this book is organized into three parts in order to present a thorough and practical look at the subject of conditioning agents. These parts discuss the physiological reasons why skin and hair need to be conditioned, the chemical compounds that can provide conditioning, and methods for developing and testing conditioning products.

The text begins with a description of the biology of skin and hair and discusses the physiological factors that create the need for conditioning. Chapter 2 provides a general look at the structure of both hair and skin and introduces some of the strategies employed for conditioning these surfaces. It also expands on the definition of conditioning, providing an appropriate background for the rest of the book. Chapter 3 examines the composition of skin, particularly biological lipids, in detail. It also describes the various components that are naturally present in skin and suggests mechanisms by which they are involved in conditioning skin.

The second part of this volume discusses the specific ingredients which are used as conditioning agents. The ingredients have been chosen as representatives of the different classes of materials actually used as conditioning agents in personal care formulations. Each ingredient chapter provides a general background on the materials in question, including a definition, description of chemical and physical properties, brief history, method of manufacture, and basic function. Additionally, information is provided about the way these materials interact with hair and skin surfaces and information regarding their use in formulations. Finally, each chapter speculates on the future of these raw materials and what derivative materials may be developed.

The first three chapters in Part Two deal with conditioning agents which are designed primarily to improve the condition of skin, though hair applications are also considered. Chapter 4 discusses occlusive agents, which are materials that condition by creating a moisture barrier between the biological surfaces and air. Petrolatum is used as the classic example of an occlusive agent. Humectants, which are materials that condition by binding moisture, are the subject of Chapter 5. These conditioning agents include materials such as glycerin and propylene glycol. Chapter 6 examines the chemical and physical properties of emollients. These hydrophobic materials are used to impart a specific feel to hair and skin.

Chapter 7 looks at proteins and how they can provide conditioning benefits to both skin and hair. Data are given which show real conditioning benefits from the use of these materials. The next two chapters discuss the growing area of silicone conditioning agents. The reactions to produce long-chain siloxane polymers and applications for these materials in personal care products are provided in Chapter 8. Chapter 9 examines the latest in cutting-edge silicone derivative technology.

Chapter 10 discusses conditioning through the use of cationic surfactants. These materials bond ionically to skin and hair to improve their condition. This section concludes with Chapter 11, which is about the various types of polymers used for conditioning. Much research is going on in this area of polymers, and it is anticipated that these materials will continue to find expanded use in formulations.

The third part of the book deals with the formulation and evaluation of personal care products designed to condition hair and skin. The practical aspects of formulating conditioning products is discussed in Chapter 12. It presents a practical method for designing a product from the idea stage all the way through to production. The last two chapters provide a look at methods for testing the effectiveness of conditioning products. Chapter 13 describes classic procedures for testing hair, along with the latest evaluation methods. Chapter 14 examines test methodologies for skin. It describes various methods of evaluating skin condition including both in-vivo and in-vitro testing.

REFERENCES

1. Robbins CR. Chemical and Physical Behavior of Human Hair. 3d ed. New York: Springer-Verlag, 1994:212.
2. Idson B. Dry skin moisturizing and emolliency. Cosmet Toilet 1992 (July); 107(7): 70.
3. Robbins CR. Chemical and Physical Behavior of Human Hair. 3d ed. New York: Springer-Verlag, 1994:44.
4. Ibid., p. 77.
5. Idson, B. Dry skin moisturizing and emolliency. Cosmet Toilet 1992 (July); 107(7): 71.
6. Yosipovitch G, Maibach H. Skin surface pH: a protective acid mantle. Cosmet Toilet 1996 (Dec.); 111(12):101.
7. Idson, B. Dry skin moisturizing and emolliency. Cosmet Toilet 1992 (July); 107(7): 70.
8. Schrader K, Bielfeldt S. Comparative studies of skin roughness measurements by image analysis and several in vivo skin testing measurements. J Soc Cosmet Chem 1991; 42(6):385.
9. Idson B. Dry skin moisturizing and emolliency. Cosmet Toilet 1992 (July); 107(7): 69.
10. Ayala L, Dikstein S. Objective measurement and self-assessment of skin care treatments. Cosmet Toilet 1996 (June); 111(6):91.
11. Idson, B. Dry skin moisturizing and emolliency. Cosmet Toilet 1992 (July): 107(7): 69.
12. Robbins CR. Chemical and Physical Behavior of Human Hair. 3d ed. New York: Springer-Verlag, 1994:299.
13. Ibid., p. 335.
14. Ibid., p. 359.
15. Imokawa G, Takema Y. Fine wrinkle formation: etiology and prevention. Cosmet Toilet 1993 (Dec.); 108(12):65.
16. Takema Y, Yorimoto K. The relationship between age related changes in the physical properties of wrinkles in human facial skin. J Soc Cosmet Chem 1995; 46:163–173.
17. Idson B. Dry skin moisturizing and emolliency. Cosmet Toilet 1992 (July); 107(7): 70.
18. Ibid.
19. Ayala L, Dikstein S. Objective measurement and self-assessment of skin care treatments. Cosmet Toilet 1996 (June); 111(6):96.

2
Biology of the Hair and Skin

Zoe Diana Draelos
Wake Forest University School of Medicine, Winston-Salem, and Dermatology Consulting Services, High Point, North Carolina

I. INTRODUCTION

A. The Need to Condition

The hair and skin are responsible for the entire visual appearance of both men and women. All visible major body surfaces are covered with some type of hair or skin, important for healthy functioning of the body and creation of an identifiable image of cosmetic value. An understanding of the biology of hair and skin is key to developing products designed to maintain the optimum cosmetic appearance of the individual. This chapter focuses on those aspects of hair and skin physiology that affect the use and development of conditioning agents for the hair and skin.

The hair and skin are different from other body organs in that the hair and skin create the milieu suitable for optimum functioning of other body organs and systems while sustaining contact with the external environment. Conditioning agents are necessary because humans live in an outdoor global environment, with wide variations in humidity and temperature, and have created indoor environments that are not always optimal for hair and skin functioning. Thus, conditioning agents are the means by which people can endure tremendous climatic variability without an increase in skin disease processes or a decrease in the cosmetic value of the hair or skin.

Conditioning of the hair and skin must be a continuous process, however, since both of these substances are in a cycle of constant renewal and shedding. Unlike vital organs, such as the heart, liver, or kidneys, in which limited cellular renewal can occur, the skin completely replaces its outermost layer on a biweekly basis, while hair growth occurs at a rate of 0.35 mm/day (1). This leads to a key difference between hair and skin. Skin is basically a living organ with a few sloughed cells on the surface, while hair is basically a dead organ with a few live cells deep within the skin. Thus, conditioning agents for skin can affect the homeostatic processes of growth and repair by supplementing the body's own genetically inherited renewal mechanisms. This contrasts with conditioning agents for hair, which have no effect on growth and cannot affect cellular repair. Hair conditioners can only temporarily improve the cosmetic appearance of damaged hair and must be reapplied as removal occurs. With these basic concepts in mind, the discussion can now turn to the details of the structure and biology of the hair and skin.

B. Historical Perspective

The hair and skin are unique in that they are body structures readily accessible for scientific observation, yet much remains to be understood regarding their growth and regulation. The first article detailing the development of the skin was published by Albert von Kolliker, a Swiss anatomy professor in Wurzburg, in 1847. The first dermatologist to study the subject was P. G. Unna of Hamburg in 1876 (2). A student in Unna's clinic, Martin Engman, Professor of Dermatology, Washington University, St. Louis, became interested in the embryology and development of the hair follicle. His work was furthered by C. H. Danforth, Mildred Trotter, and L. D. Cady, who published the foundation work on hair formation in 1925 (3). Further work on the development of the hair follicle continued with Pinkus (1958) (4), Sengel (1976) (5), and Spearman (1977) (6,7). New observational methods of evaluating the hair and skin have led to the current understanding of the biology of the hair and skin to be discussed.

II. BIOLOGY OF THE SKIN

The skin comprises the largest organ of the body, representing an amazing structure that encases, contours, and conforms to the organism within. It provides a physical and chemical barrier while protecting against environmental insults. It contains numerous transducers sending a continuous array of sensory information to the brain for processing. Lastly, the skin is responsible for external appearance, creating a unique and recognizable image identifiable to others.

A. Development and Anatomy (Figure 1)

The skin develops at 30 to 40 days gestation and consists of two primary layers: the epidermis and dermis (8,9). The epidermis represents the external layer of skin and is composed mainly of keratinocytes, named for their primary proteinaceous component, known as keratin. The keratinocytes are held together by cellular attachments or desmosomes. The epidermis also contains melanocytes, which produce a pigment known as melanin, and specialized immune cells, known as Langerhans cells. It is divided into several distinct functional layers: the basal cell layer (stratum germinativum), from which all new cells are derived; the spinous layer (stratum spinosum); the granular cell layer (stratum granulosum); and the horny cell layer (stratum corneum) (10). The stratum corneum is the outermost layer of the epidermis, comprised of 15 to 20 cell layers, functioning to protect the underlying tissues and nerves from damage. Products aimed at conditioning the skin must have significant impact on the stratum corneum.

The zone between the epidermis and the underlying dermis is referred to as the basement membrane zone or the epidermal–dermal junction (11). Below this lies the dermis, which is a moderately dense fibroelastic connective tissue composed of collagen fibers, elastic fibers, and an interfibrillar gel of glycosaminoglycans. Collagen is synthesized by the dermal fibroblasts and is responsible for the strength and 77% of the fat free dry weight of the skin. Interspersed between the collagen bundles is a network of elastic fibers that allow the skin to resist and recover following mechanical deformation (12).

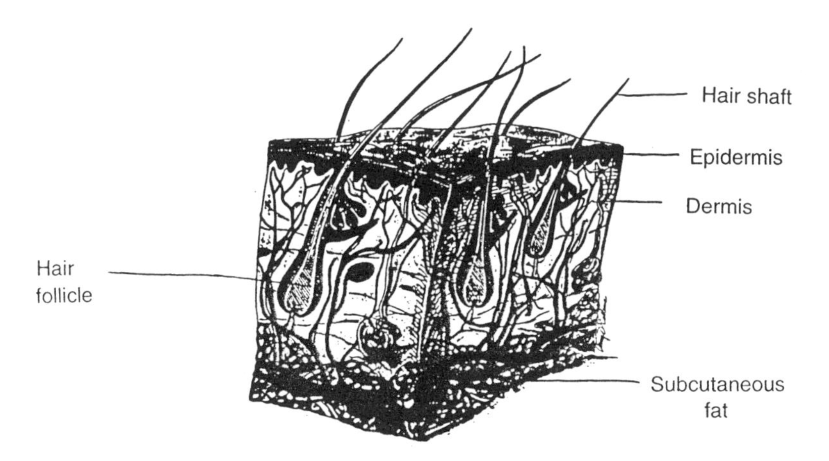

Figure 1 Structure of the skin, demonstrating the relationship between the epidermis, dermis, and follicular units.

The dermis is divided into two parts: the papillary dermis and the reticular dermis (13). The papillary dermis contains smaller collagen and elastic fibers than the reticular dermis, but encloses the extensive circulatory, lymphatic, and nervous system of the skin. It also contains fibronectins, which function as adhesive proteins to attach fibroblasts to collagen, and abundant glycosaminoglycans, also known as ground substance, which are anionic polysaccharides. The major glycosaminoglycans are hyaluronic acid, chondroitin sulfate, dermatan sulfate, and heparin sulfate. They function to maintain adequate water homeostasis within the skin and influence flow resistance to solutes. In essence they are the natural moisturizing substances of the skin, necessary to prevent cutaneous dehydration. They comprise only 0.2% of the dry weight of the skin, but are capable of binding water in volumes up to 1000 times their own.

The reticular dermis is formed of dense collagenous and elastic connective tissue and does not contain many cells or blood vessels or glycosaminoglycans. It overlays the subcutaneous fat, which provides padding for cutaneous mechanical insults and acts as a reservoir for extra caloric intake.

B. Growth Cycle

The skin is in a constant state of renewal as the basal cells divide and replenish aging cells, which are sloughed off from the stratum corneum at the surface. Thus, skin has a tremendous ability to regenerate. Acute injuries to the epidermis alone can be completely repaired without scarring, but dermal injuries usually result in the production of scar tissue which is functionally and cosmetically inferior to the uninjured skin.

The transit time of cells from the lower stratum corneum to desquamation is approximately 14 days, meaning that a new skin surface is present every 2 weeks. Products that are designed to affect the skin surface must be continually applied, since the skin is in a constant state of flux. Complete, full-thickness renewal of the skin from the basal cell layer to the stratum corneum requires 45 to 75 days, meaning that the epidermis completely replaces itself every 2 months (14). Thus products that are designed to affect skin growth must be applied for at least 2 months for a visual impact.

C. Collagen Composition

Collagen forms the major component of the skin and is synthesized as three polypeptide chains by the fibroblasts. The three chains form a triple helix of approximately 1000 amino acids in length, with glycine serving as every third unit accompanied by abundant hydroxyproline and hydroxylysine. Crosslinking between the chains occurs at the sides, not at the ends, accounting for the great strength of the collagen fibers (15). At present, 14 different types of collagen have been identified, each with a different function within the skin and body.

D. Cutaneous Structures Relevant to Skin Condition

Three important structures warranting further comment contribute to the condition of the skin: the stratum corneum, the stratum spinosum, and the sebaceous gland. A more detailed discussion of the numerous structures within the skin is beyond the scope of this chapter. (Further information can be obtained from the excellent text by Lowell A. Goldsmith entitled *Physiology, Biochemistry, and Molecular Biology of the Skin*, 2nd edition, New York, Oxford University Press, 1991.)

The stratum corneum is the outermost layer of the epidermis. It is responsible for protection and is affected by conditioning substances. It is formed by corneocytes which are highly resistant to degradation due to the crosslinking of a soluble protein precursor involucrin, which is acted upon by epidermal transglutaminase. The final product is a 65% insoluble cysteine-rich disulfide crosslinked protein (16). The stratum corneum presents the first barrier to percutaneous absorption, but is subject to degradation through hydration, removal through tape stripping, protein denaturation through the application of solvents, and alteration of the lipid structure (17).

The stratum spinosum is important because it is responsible for production of submicroscopic lamellar granules approximately 100 by 300 nm in diameter. These lamellar granules, also called membrane-coating granules or Odland bodies, are seen at the outer surface of the flattening cells that form the stratum granulosum and are eventually extruded into the extracellular compartment between the lowermost cells of the stratum corneum. Their internal structure is highly ordered into sheets which hold sterols, lipids, lipases, glycosidases, and acid phosphatase (18). The lipases are able to convert polar lipids into nonpolar lipids which coalesce to form crystalline sheets within the intercellular spaces between the corneocytes, thus forming a waterproof barrier to percutaneous absorption (19). The acid phosphatases, on the other hand, are necessary to aid in desquamation of old corneocytes by dissolving the intercellular cement. Thus, the lamellar granules are extremely important in the condition of the external surface of the skin (20).

The last important structure to discuss relevant to the condition of the skin is the sebaceous gland, which is responsible for production of sebum. The sebaceous glands are largest on the face, scalp, mid-back, and mid-chest, producing secretions released into the sebaceous duct which connects the gland to the follicular canal. Sebaceous secretions are produced in response to hormonal stimuli. Sebum has many functions on the skin: it functions as a moisturizer, it enhances barrier properties, and it may act as an antifungal and antibacterial. It is composed of triglycerides (60%), free fatty acids (20%), and squalene (10%). Sebum, in combination with perspiration from the eccrine glands and environmental dirt, coat the external surface of the skin.

E. Epidermal Lipids

Three intercellular lipids are implicated in epidermal barrier function: sphingolipids, free sterols, and free fatty acids (21). In addition, it is thought that the lamellar bodies, discussed previously (Odland bodies, membrane-coating granules, cementsomes), containing sphingolipids, free sterols, and phospholipids, play a key role in barrier function and are essential to trap water and prevent excessive water loss (22,23). The lipids are necessary for barrier function, since solvent extraction of these chemicals leads to xerosis, directly proportional to the amount of lipid removed (24). The major lipid by weight found in the stratum corneum is ceramide, which becomes sphingolipid if glycosylated via the primary alcohol of sphingosine (25). Ceramides possess the majority of the long-chain fatty acids and linoleic acid in the skin.

Other lipids present in the stratum corneum include cholesterol sulfate, free sterols, free fatty acids, triglycerides, sterol wax/esters, squalene, and *n*-alkanes (26). Cholesterol sulfate comprises only 2–3% of the total epidermal lipids, but is important in corneocyte desquamation (27). It appears that corneocyte desquamation is mediated through the desulfatation of cholesterol sulfate (28). Fatty acids are also important, since it has been demonstrated that barrier function can be restored by topical or systemic administration of linoleic acid-rich oils in essential fatty acid-deficient rats (29).

F. Physical Properties

The skin is a tremendously distensible material that adapts quickly to change. This is necessary to accommodate the many alterations in shape required for movement and growth of the human body. Thus, the skin becomes more elastic over time as it is subjected to increasing tensile and torsional forces. Elastic tissue accounts for only 5% of the dermal connective tissue by weight and thus cannot contribute significantly to the viscoelastic properties of the skin. Rather, it is thought that the collagen fibers slip over one another and become aligned through a change in bonding or molecular realignment (30). The tensile strength of collagen is tremendous, with a single 1-mm fiber able to withstand a static load of up to 20 kg.

However, the ability of the skin to respond to mechanical change decreases with age. Some of this change is due to ultraviolet light exposure (extrinsic aging), while other change is due to the natural effects of aging (intrinsic aging) and the effects of long-term oxidative damage on the skin. Biochemical changes noted include decreased glycosaminoglycan content, leading to thinning of the dermis, which predisposes to easy bruisability, wrinkling, and dehydration of the skin. The collagen fibers also change as soluble collagens decrease while insoluble collagens increase (31).

This tremendous ability of the skin to renew itself and respond to change contrasts with hair, a nonliving tissue that cannot adapt or repair damage.

III. BIOLOGY OF THE HAIR

Hair represents a structure that has lost much of its functional significance through evolutionary change, yet it is important to aid in the transduction of sensory information and to create gender identity. The value of hair, however, cannot be underestimated from a social and emotional standpoint, since the appearance of the scalp hair is an intimate part of the perception of self.

A. Development and Anatomy

Hair follicles are formed early in development of the fetus, with eyebrow, upper lip, and chin follicles present at week 9 and the full complement of follicles present by week 22. At this time, the total body number of 5 million follicles is present, with 1 million on the head, of which 100,000 are on the scalp (32). No additional follicles are formed during life. As body size increases, the number of hair follicles per unit area decreases. For example, the average density of hair follicles in the newborn is 1135 per square centimeter, drops to 795 per square centimeter at the end of the first year, and decreases to 615 per square centimeter by the end of the third decade. Continued hair follicle density decreases occur on the scalp with balding (33).

The hair grows from follicles which resemble stocking-like invaginations of the epithelium enclosing an area of dermis known as the dermal paillae (Figure 2). The area of active cell division, the living area of the hair, is formed around the dermal papillae and is known as the bulb, where cell division occurs every 23 to 72 hr (34). The follicles slope into the dermis at varying angles, depending on body location and individual variation, and reside at varying levels between the lower dermis and the subcutaneous fat. In general, larger hairs come from more deeply placed follicles than do finer hairs (35). An arrector pili muscle attaches to the midsection of the follicle wall and ends at the junction between the epidermis and the dermis. In some body areas, a sebaceous gland (oil gland) and an apocrine gland (scent gland) attach above the muscle and open into the follicle. The point at which the arrector pili muscle attaches is known as the hair "bulge" and is considered to be the site where new matrix cells are formed and the hair growth cycle initiated. It takes approximately 3 weeks for a newly formed hair to appear at the scalp surface (36).

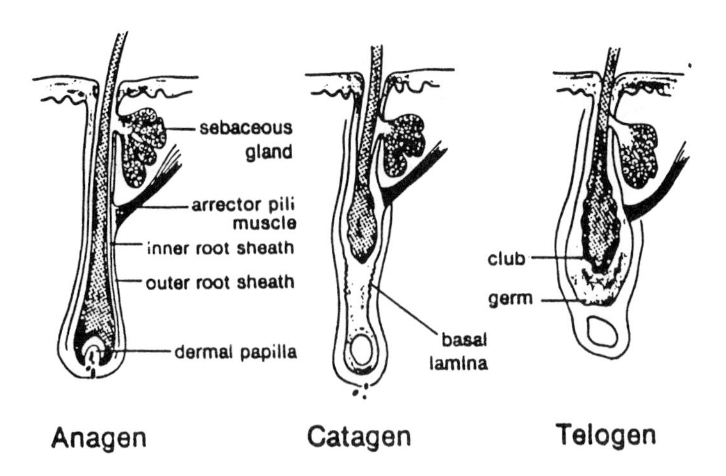

Figure 2 Schematic drawing demonstrating the relationship of the hair follicle to the skin at various stages of growth.

The sebaceous gland is important to the maintenance of the grown hair shaft, as it produces sebum, a natural conditioning agent. Approximately 400 to 900 glands per square centimeter are located on the scalp and represent the largest glands on the body (37). Sebum, composed of free fatty acids and neutral fats, is produced in increased amounts after puberty in males and females. In the female, sebum production declines with age, but this effect is less prominent in males.

B. Growth Cycle

Hair growth continues on a cyclical basis, with each hair growing to genetic and age-determined length, remaining in the follicle for a short period of time without growth, and eventual shedding followed by regrowth (Figure 2) (38). The growth phase, known as anagen, lasts approximately 1000 days, and the transitional phase, or catagen, about 2 weeks (39). The resting phase, or telogen, lasts approximately 100 days. Scalp hair is characterized by a relatively long anagen and a relatively short telogen, with a ratio of anagen to telogen hairs of 90 to 10 (40). Only 1% or less of the follicles are in catagen at any given time. Thus, the healthy individual loses 100 hairs per day. It is estimated that each follicle completes this cycle 10 to 20 times over a lifetime, but the activity of each follicle is independent.

The mechanism signaling the progression from one phase to the next is unknown, but the duration of anagen determines the maximum length to which the hair can be grown. Hair growth can be affected by physical factors (severe illness, surgery, weight change, pregnancy, hormonal alterations, thyroid anomalies, dermatological disease) and emotional factors, but is unaffected by physical alterations limited to the hair shaft (shaving, curling, combing, dyeing, etc.). Plucking of the hairs from resting follicles can stimulate growth, however (41).

C. Composition

Hair is composed of keratin, a group of insoluble cystine-containing helicoidal protein complexes. The hair is made up of an amorphous matrix that is high in sulfur proteins and in which the keratin fibers are embedded. These protein complexes, which form 65–95% of the hair by weight, are extraordinarily resistant to degradation and are thus termed "hard," as opposed to the "soft" keratins that compose the skin (42). Under X-ray crystallography, the hair fiber helix has an alpha diffraction pattern, which changes to a beta diffraction pattern as the hair is stretched and the helix is pulled into a straight chain.

D. Structure (Figure 3)

The hair is composed of closely attached, keratinized fusiform cells arranged in a cohesive fiber (43). The greatest mass of the hair shaft is the cortex, with some shafts also possessing a medulla. The cortex consists of closely packed spindle-shaped cells with their boundaries separated by a narrow gap which contains a proteinaceous intercellular lamella that is thought to cement the cells together (44). Thus, the cortex contributes to the mechanical properties of the hair shaft. The medulla is formed from a protein known as trichohyalin. The function of the medulla remains unknown; however, it contains glycogen (a form of glucose) and melanosomes (pigment packages). In some areas the medulla cells appear to dehydrate and air-filled spaces are left behind. Thicker hairs, such as scalp hairs, are more likely to contain a medulla than are finer body hairs (45).

Surrounding this structure is a protective layer of overlapping, keratinized scales known as the cuticle, which can account for up to 10% of the hair fiber by weight (46). The free edges of the cuticle are directed outward, with the proximal edges resting against the cortex. The cuticular scales are arranged much like roofing shingles to provide 5 to 10 overlapping cell layers, each 350 to 450 nm thick, to protect the hair shaft along its entire length. The cell structure of the cuticle is composed of three major layers: the A layer, the exocuticle, and the endocuticle (Figure 4). It is the clear A layer, which is high in sulfur-containing proteins, that protects the hair from chemical, physical, and environmental insults (47). From a hair conditioning standpoint, an intact cuticle is essential to the cosmetic value of the hair, and the

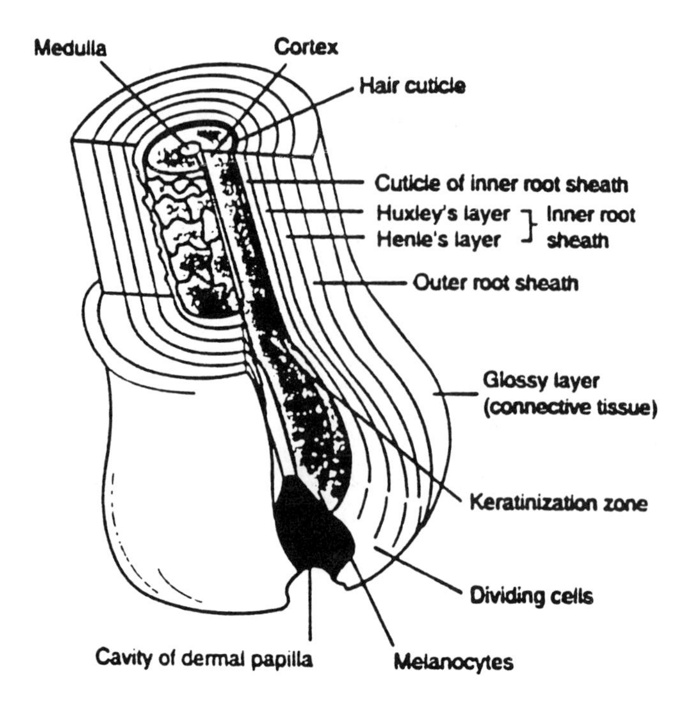

Figure 3 Layers of the hair shaft.

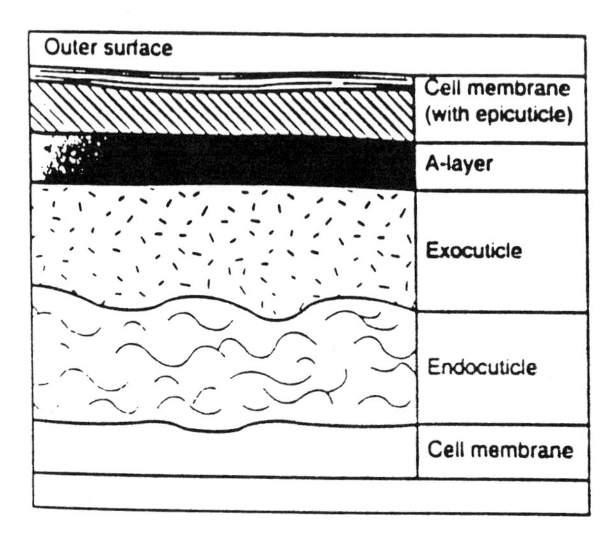

Figure 4 Layers of the cuticle.

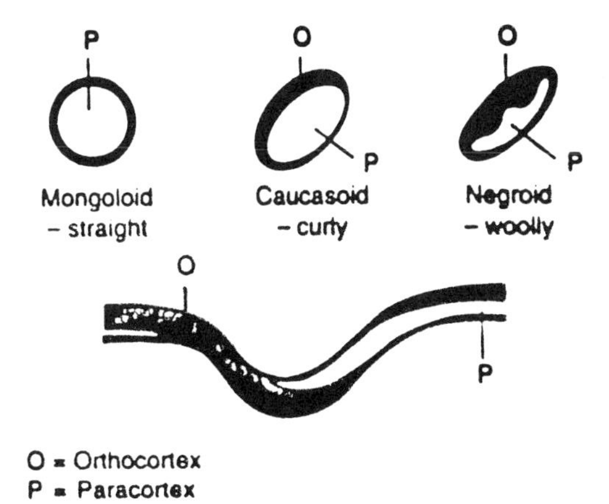

Figure 5 The cross-sectional shape of the hair shaft determines its appearance, physical properties, and conditioning needs.

goal of hair conditioning products is to enhance and restore order to this hair shaft layer.

E. Shape

The degree of curl found in a hair shaft is related directly to its cross-sectional shape, which determines its cosmetic appearance and conditioning needs (Figure 5). Caucasoid hair has an elliptical cross section accounting for a slight curl, while Mongoloid hair has a circular cross section leading to straight hair (48). Negroid hair is identical to Caucasoid and Mongoloid hair in its amino acid content, but has a slightly larger diameter, lower water content, and, most important, a flattened elliptical cross-sectional shape (49). It is the asymmetry of this cross section that accounts for the irregular kinky appearance of Black hair. Hair that is wavy or loosely kinked has a cross-sectional shape in between a circle and flattened ellipse.

The cross-sectional shape of the hair fiber accounts for more than the degree of curl; It also determines the amount of shine and the ability of sebum to coat the hair shaft (50). Straight hair possesses more shine than kinky hair due to its smooth surface, allowing maximum light reflection and ease of sebum movement from the scalp down the hair shaft. The irregularly kinked hair shafts appear duller, even though they may have an intact cuticle, due to rough surface and difficulty encountered in sebum transport from the scalp, even though Negroid hair tends to produce more sebum.

The shape of the hair shaft also determines grooming ease. Straight hair is easiest to groom, since combing friction is low and the hair is easy to arrange in a fashionable style. Kinky hair, on the other hand, demonstrates increased grooming friction, resulting in increased hair shaft breakage. Kinky hair also does not easily conform to a predetermined hair style, unless the shafts are short.

F. Physical Properties

The physical properties of the hair shaft are related to its geometric shape and the organization of its constituents. The cortex is largely responsible for the strength of the hair shaft, but must have an intact cuticle to resist externally applied mechanical stresses. The most important mechanical property of hair is its elasticity, allowing stretching deformation and a return to normal condition. Hair can be stretched to 30% of its original length in water and experience no damage, but irreversible changes occur when hairs are stretched to between 30% and 70% of their original length. Stretching to 80% of original length generally results in hair shaft fracture (51).

The water content of the hair shaft is important to its physical and cosmetic properties. The porosity of the hair shaft is about 20%, allowing a weight increase of 12–18% when soaked in water. The absorption rate is very rapid, with 75% of the maximum absorbable water entering the hair shaft within 4 min (52). Water absorption causes hair shaft swelling, making wetting the first step in cosmetic hair procedures that require entry into the hair shaft. Wetting and subsequent drying of the hair shaft in a predetermined position is also basic to hair styling.

Friction effects on the hair shaft are also an important consideration. It has been shown that wet straight hair has a higher combing friction than dry straight hair (53); however, wet hairs are excellent conductors of electricity while dry hairs conduct electricity poorly. Thus, static electricity preferentially affects dry hair, since the ions are not moving. The presence of static electricity creates a manageability problem known as "flyaways," since the individual hair shafts repel one another. Static electricity can be reduced by decreasing hair friction, combing hair under cooler conditions, or decreasing the resistance of hair fibers by increasing hair moisture.

The preceding discussion has focused on the specifics of the biology of the skin and hair. These ideas now must be expanded to understand how hair and skin biology affect their conditioning needs.

IV. SKIN DISEASE AND CONDITIONING

A. Physiology of Xerosis

Xerosis is a result of decreased water content of the stratum corneum, which leads to abnormal desquamation of corneocytes. For the skin to appear and feel normal, the water content of this layer must be above 10% (54). Water is lost through evaporation to the environment under low humidity conditions and must be replenished by water from the lower epidermal and dermal layers (55). The stratum corneum must have the ability to maintain this moisture or the skin will feel rough, scaly, and dry. However, this is indeed a simplistic view, as there are minimal differences between the amount of water present in the stratum corneum of dry and normal skin (56). Xerotic skin is due to more than simply low water content (57). Electron micrographic studies of dry skin demonstrate a stratum corneum that is thicker, fissured, and disorganized. It appears that the scaly appearance of xerotic skin is, in part, due to the failure of corneocytes to desquamate appropriately.

Other disease states, such as psoriasis and atopic dermatitis, also demonstrate abnormal barrier function due to ceramide distribution (58,59). Interestingly, xerosis tends to increase with age, due to a lower inherent water content of the stratum corneum (60). However, this fact does not totally account for the scaliness and roughness of aged skin; probably an abnormal desquamatory process is also present (61).

B. Epidermal Barrier Repair

Once the epidermal barrier has been damaged, signals must be transmitted to the intracellular machinery to initiate repair or reconditioning of the skin. Remoisturization of the skin occurs in four steps: initiation of barrier repair, alteration of surface cutaneous moisture partition coefficient, onset of dermal–epidermal moisture diffusion, and synthesis of intercellular lipids (62). It is generally thought that a stratum corneum containing between 20% and 35% water will exhibit the softness and pliability of normal stratum corneum (63).

Perturbations within the barrier must be sensed before the onset of lamellar body secretion and a cascade of cytokine changes associated with adhesion molecule expression and growth factor production (64). Thus, if skin with barrier perturbations is occluded with a vapor-impermeable wrap, the expected burst in lipid synthesis is blocked. However, occlusion with a vapor-permeable wrap does not prevent barrier recovery (65). Therefore, transepidermal water loss is necessary to initiate synthesis of lipids to allow barrier repair and skin reconditioning to occur (66,67).

C. Mechanisms of Moisturization and Skin Conditioning

Once skin damage has occurred and the barrier damaged, reconditioning can occur only if the loss of moisture is retarded. This is the goal of moisturizers, which function temporarily until skin integrity can be reestablished. There are three physiological mechanisms for rehydrating the stratum corneum: the use of occlusives, humectants, and hydrophilic matrices (68).

1. Occlusives

Occlusives function to condition the skin by impairing evaporation of water to the atmosphere. They are generally oily substances through which water cannot pass. Occlusive substances can be broken down into categories (69):

Hydrocarbon oils and waxes: petrolatum, mineral oil, paraffin, squalene
Silicone oils
Vegetable and animal fats
Fatty acids: lanolin acid, stearic acid
Fatty alcohol: lanolin alcohol, cetyl alcohol
Polyhydric alcohols: propylene glycol
Wax esters: lanolin, beeswax, stearyl stearate
Vegetable waxes: carnauba, candelilla
Phospholipids: lecithin
Sterols: cholesterol

The most occlusive of the above chemicals is petrolatum (70). It appears, however, that total occlusion of the stratum corneum is undesirable. While the transepidermal water loss can be completely halted, once the occlusive is removed, water loss resumes at its preapplication level. Thus, the occlusive moisturizer has not allowed the stratum corneum to repair its barrier function (71). Petrolatum does not appear to function as an impermeable barrier; rather, it permeates throughout the interstices of the stratum corneum, allowing barrier function to be reestablished (72).

2. Humectants

A dehydrated epidermis can also be conditioned through the use of humectants, substances which attract water, mimicking the role of the dermal glycosaminoglycans. Examples of topical humectants include glycerin, honey, sodium lactate, urea, propylene glycol, sorbitol, pyrrolidone carboxylic acid, gelatin, hyaluronic acid, vitamins, and some proteins (73,74).

Topically applied humectants draw water largely from the dermis to the epidermis and rarely from the environment, under conditions where the ambient humidity exceeds 70%. Water that is applied to the skin in the absence of a humectant is rapidly lost to the atmosphere (75). Humectants may also allow the skin to feel smoother by filling holes in the stratum corneum through swelling (76). However, under low humidity conditions, humectants, such as glycerin, will actually draw moisture from the skin and increase transepidermal water loss (77).

3. Hydrophilic Matrices

Hydrophilic matrices are large-molecular-weight substances that form a barrier to cutaneous water evaporation. Hyaluronic acid, a normal component of the dermal glycosaminoglycans, is a physiological hydrophilic matrix, while colloidal oatmeal is a synthetic hydrophilic matrix.

D. Mechanisms of Emolliency and Skin Conditioning

Some moisturizing substances can also function as emollients by temporarily filling the spaces between the desquamating corneocytes (78). Emolliency is important because it allows the skin surface to feel smooth to the touch. Emollients can be divided into several categories: protective emollients, fatting emollients, dry emollients, and astringent emollients (79). Protective emollients are substances such as diisopropyl dilinoleate and isopropyl isostearate that remain on the skin longer than average and allow the skin to feel smooth immediately upon application. Fatting emollients, such as castor oil, propylene glycol, jojoba oil, isostearyl isostearate, and octyl stearate, also leave a long-lasting film on the skin, but they may feel greasy. Dry emollients, such as isopropyl palmitate, decyl oleate, and isostearyl alcohol, do not offer much skin protection but produce a dry feel. Lastly, astringent or drying emollients, such as the dimethicones and cyclomethicones, isopropyl myristate and octyl octanoate, have minimal greasy residue and can reduce the oily feel of other emollients.

The skin is unique in that it produces its own conditioning agent, sebum, and can repair damage due to surfactants and solvents. Hair, on the other hand, receives some sebum from the scalp, but externally applied conditioning agents are necessary at the ends of the hair shafts, especially if the hair is long. Conditioning of the hair is especially important because the damaged hair shaft undergoes no repair processes.

V. HAIR DAMAGE AND CONDITIONING

Hair damage results from both mechanical and chemical trauma that alters any of the physical structures of the hair. Conditioning agents cannot enhance repair, since repair does not occur, but can temporarily increase the cosmetic value and functioning of the hair shaft until removal of the conditioner occurs with cleansing. The cuticle is the main hair structure affected by conditioning agents (80). An intact cuticle is responsible for the strength, shine, smoothness, softness, and manageability of healthy hair. A layer of sebum coating the cuticle also adds to hair shine and manageability.

A. Mechanism of Hair Damage

Healthy, undamaged hair is soft, resilient, and easy to detangle, due to the tightly overlapping scales of the cuticle (81). Cuticular loss, known as weathering, is due to the trauma caused by shampooing, drying, combing, brushing, styling, chemical dyeing, permanent waving, and environmental factors such as exposure to sunlight, air pollution, wind, sea water, and chlorinated swimming pool water (82,83). Conditioning the hair can mitigate this hair damage by improving sheen, decreasing brittleness, decreasing porosity, and increasing strength.

B. Mechanisms of Hair Conditioning

There are several mechanisms by which conditioners can improve the cosmetic value of the weathered hair shaft: increasing shine, decreasing static electricity, improving hair strength, and providing ultraviolet (UV) radiation protection.

1. Shine

Shiny hair is visually equated with healthy hair even though the health of the hair follicle cannot be assessed due to its location deep within the scalp. This shine is due to light reflected by the smooth surface of the individual hair shafts and large-diameter, elliptical hair shafts with a sizable medulla (84). Conditioners, such as those containing polymer film-forming agents, can increase hair shine primarily by increasing adherence of cuticular scale to the hair shaft and filling in the spaces between cuticular defects (85).

2. Static Electricity

Combing or brushing of the hair allows the individual hair shafts to become negatively charged, creating static electricity and preventing the hair from lying smoothly in a given style. Fine hair is more subject to static electricity than coarse hair, due to the larger surface area of the cuticle. Conditioning agents, such as quaternary ammonium compounds, are able to minimize static electricity electrically (86).

3. Strength

Conditioning of the hair can also attempt to slightly increase hair strength by allowing hydrolyzed proteins of molecular weight 1000 to 10,000 to diffuse into the hair shaft through defects in the protective cuticular scale (87). The source of the protein is not as important as the protein particle size (88). However, the protein readily diffuses out of the hair shaft with cleansing, as the nonliving status of the hair shaft precludes permanent incorporation. Proteins can also be used to temporarily reapproximate split hair shaft ends, known as trichoptilosis, resulting from loss of the cortex, required for hair shaft strength, and exposure of the soft keratin of the medulla.

4. Photoprotection

While the hair is made up of nonliving material and cannot develop cancerous changes, its cosmetic value can be diminished through excessive exposure to the sun. Dryness, reduced strength, rough surface texture, loss of color, decreased luster, stiffness, and brittleness of the hair are all precipitated by sun exposure. Hair protein degradation is induced by light wavelengths from 254 to 400 nm (89). Chemically, these changes are thought to be due to ultraviolet light-induced oxidation of the sulfur molecules within the hair shaft (90). Oxidation of the amide carbon of polypeptide chains also occurs, producing carbonyl groups in the hair shaft (91). This process has been studied extensively in wool, where it is known as "photoyellowing" (92,93).

Bleaching, or lightening of the hair color, is common in both brunette and blonde individuals who expose their hair to ultraviolet radiation (94). Brunette hair tends to develop reddish hues due to photooxidation of melanin pigments, while blonde hair develops photoyellowing. The yellow discoloration is due to photodegradation of cystine, tyrosine, and tryptophan residues within the blonde hair shaft (95). This points to a need for the development of photoprotective conditioning products for the hair.

VI. CONCLUSIONS AND FUTURE DEVELOPMENTS

Future developments regarding skin and hair conditioning agents rely on a better understanding of the biology of these structures. Only by elucidating mechanisms for enhanced growth and repair can product development occur. A wealth of information on the mechanisms of skin barrier formation is accumulating, leading to the realization that many disease states (atopic dermatitis, xerotic eczema) may be due to faulty sebum production and/or improper formation of the intercellular lipids. Future dermatological research may find a topical method of replacing missing substances and restoring normal function. A better understanding of defective corneocyte sloughing encountered in mature individuals is leading to the development of topical moisturizers designed to increase desquamation and restore a smooth skin surface.

Most hair damage occurs as a result of grooming habits and chemical exposure for cosmetic purposes. An evaluation of hair structure and biology points to the need for better protective mechanisms from cuticular damage and UV damage to maintain the cosmetic value of the hair. Methods for enhancing product substantivity for hair keratin are necessary to provide long-term protection that is somewhat resistant to cleanser removal.

Through the cooperative efforts of dermatologists and cosmetic chemists, better hair and skin conditioning agents can be developed. The dermatologist needs to elucidate the mechanism through which products can enhance functioning, while the cosmetic chemist needs to identify substances and develop formulations to accomplish the desired end.

REFERENCES

1. Myers RJ, Hamilton JB. Regeneration and rate of growth of hair in man. Ann NY Acad Sci 1951; 53:862.
2. Unna PG. Beitrage zur histologie und entwickiengsgeschichte der menschlichen oberhat und ihrer anhangsgebilde. Arch fur microscopisch Anatomie und Entwickiungsmach 1876; 12:665.
3. Danforth CH. Hair with special reference to hypertrichosis. AMA Arch Dermatol Syphil 1925; 11:494.
4. Pinkus H. Embryology of hair. In: Montagna W, Ellis RA, eds. The Biology of Hair Growth. New York: Academic Press, 1958.
5. Sengel P. Morphogenesis of Skin. Cambridge: Cambridge University Press, 1976.
6. Spearman RIC. Hair follicle development, cyclical changes and hair form. In: Jarrett A, ed. The Hair Follicle. London: Academic Press, 1977:1268.
7. Rook A, Dawber R. Diseases of the Hair and Scalp. Oxford: Blackwell Scientific Publications, 1982:5–6.
8. Holbrook KA. Structure and function of the developing human skin. In: Goldsmith LE, ed. Physiology, Biochemistry and Molecular Biology of the Skin. 2d ed. Vol. 1. New York: Oxford University Press, 1991:63.
9. Breathnach AS. Embryology of human skin. J Invest Dermatol 1971; 57:133.
10. Smack DP, Korge, James WD. Keratin and keratinization. J Am Acad Dermatol 1994; 30:85–102.
11. Briggaman RA. Biochemical composition of the epidermal-dermal junction and other basement membranes. J Invest Dermatol 1982; 78:1.
12. Jarret A, ed. The Physiology and Pathophysiology of the Skin. Vol. III. The Dermis and Dendrocytes. London: Academic Press, 1974.
13. Smith LO, Holbrook KA, Byers PH. Structure of the dermal matrix during development and in the adult. J Invest Dermatol 1982; 79:93S–104S.
14. Halprin KM. Cyclic nucleotides and epidermal cell proliferation. J Invest Dermatol 1979; 73:180–183.
15. Bornstein P, Sage H. Structurally distinct collagen types. Ann Rev Biochem 1980; 49:957–1003.
16. Mercer EH. Keratin and Keratinisation. Oxford: Pergamon, 1961.
17. Barry BW. Dermatological Formulations: Percutaneous Absorption. New York and Basel: Marcel Dekker, 1983.
18. Wertz PW, Downing DT. Glycolipids in mammalian epidermis: structure and function in the water barrier. Science 1982; 4566:126.
19. Elias PM. Epidermal lipids, barrier function, and desquamation. J Invest Dermatol 1983; 80:44S.
20. Menon GK, Feingold KR, Elias PM. Lamellar body secretory response to barrier disruption. J Invest Dermatol 1992; 98:270.
21. Elias PM. Lipids and the epidermal permeability barrier. Arch Dermatol Res 1981; 270:95–117.
22. Holleran WM, Man MO, Wen NG, Gopinathan KM, Elias PM, Feingold KR. Sphingolipids are required for mammalian epidermal barrier function. J Clin Invest 1991; 88:1338–1345.

23. Downing DT. Lipids: their role in epidermal structure and function. Cosmet Toilet 1991 (Dec.); 106:63–69.
24. Grubauer G, Elias PM, Feingold KR. Transepidermal water loss: the signal for recovery of barrier structure and function. J Lipid Res 1989; 30:323–333.
25. Petersen RD. Ceramides key components for skin protection. Cosmet Toilet 1992 (Feb.); 107:45–49.
26. Brod J. Characterization and physiological role of epidermal lipids. Int J Dermatol 1991; 30:84–90.
27. Lampe MA, Williams ML, Elias PM. Human epidermal lipids: characterization and modulation during differentiation. J Lipid Res 1983; 24:131–140.
28. Long SA, Wertz PW, Strauss JS, et al. Human stratum corneum polar lipids and desquamation. Arch Dermatol Res 1985; 277:284–287.
29. Elias PM, Brown BE, Ziboh VA. The permeability barrier in essential fatty acid deficiency: evidence for a direct role for linoleic acid in barrier function. J Invest Demratol 1980; 75:230–233.
30. Brown IA. A scanning electron microscope study of the effects of uniaxial tension on human skin. Br J Dermatol 1973; 89:383–393.
31. Sams WM, Smith JG. Alterations in human fibrous connective tissue with age and chronic sun damage. In: Montagna W, ed. Advances in Biology of Skin. Vol. VI. Ageing. Oxford: Pergamon, 1965:199–210.
32. Dawber R, Van Noote D. Hair and Scalp Disorders. Philadelphia: Lippincott, 1995:4.
33. Giacometti L. The anatomy of the human scalp. In: Montagna W, Dobson RL, eds. Advances in Biology of Skin. Vol IX. Hair Growth. Oxford: Pergamon Press, 1969:97.
34. Van Scott EJ, Ekel TM, Auerbach R. Determinants of rate and kinetics of cell division in scalp hair. J Invest Dermatol 1963; 41:269.
35. Durward A, Rudall KM. The vascularity and patterns of growth of hair follicles. In: Montagna W, Ellis RA, eds. The Biology of Hair Growth. New York: Academic Press, 1958:189.
36. Siatoh M, Uzuka M, Sakamoto M. Human hair cycle. J Invest Dermatol 1970; 54:65.
37. Benfenati A, Brillanti F. Sulla distribuziona della guandole sebacee nella cute del corpo umano. Arch Ital Dermatol Sifilogr Venereol 1939; 15:33–42.
38. Kligman AM. The human hair cycle. J Invest Dermatol 1959; 33:307.
39. Orentreich N. Scalp hair regeneration in man. In: Montagna W, Dobson RL, eds. Advances in Biology of Skin. Vol. IX. Hair Growth. Oxford: Pergamon Press, 1969:99.
40. Witzel M, Braun-Falco O. Uber den haarwurzelstatus am menschlichen capillitium unter physiologischen bedingungen. Archiv fur clinische und experimentelle Dermatologie 1963; 216:221.
41. Dawber R, Van Neste D. Hair and Scalp Disorders. Philadelphia: Lippincott, 1995:15.
42. Robbins CR. Chemical and Physical Behavior of the Hair. New York: Van Nostrand-Reinhold, 1979:7.

43. Odland GF. Structure of the skin. In: Goldsmith IA, ed. Physiology, Biochemistry, and Molecular Biology of the Skin. 2d ed. Oxford: Oxford University Press, 1991:46.
44. Braun-Falco O. The fine structure of the anagen hair follicle of the mouse. In: Montagna W, Dobson RL, eds. Advances in Biology of Skin. Vol. IX. Hair Growth. Oxford: Pergamon Press, 1969:chap 29.
45. Marhle G, Orfanos GE. The spongious keratin and the medullary substance of human scalp hair. Archiv fur Dermatologische forschung 1971; 241:305.
46. Wolfram U, Lindemann MKO. Some observations on the hair cuticle. J Soc Cosmet Chem 1971; 22:839.
47. Swift JA. The histology of keratin fibres. In: Asquith RA, ed. Chemistry of Natural Protein Fibres. London: Wiley, 1977:chap 3.
48. Lindelof B, Forstind B, Hedblad M, et al. Human hair form: morphology revealed by light and scanning electron microscopy and computer-aided three-dimensional reconstruction. Arch Dermatol 1988; 124:1359–1363.
49. Brooks G, Lewis A. Treatment regimes for styled Black hair. Cosmet Toilet 1983 (May); 98:59–68.
50. Johnson BA. Requirements in cosmetics for black skin. Dermatol Clin 1988; 6: 409–492.
51. Mexander P, Hudson PF, Earland C. Wool: Its Chemistry and Physics. 2d ed. London: Chapman & Hall, 1963.
52. Rook A, Dawber R. Diseases of the Hair and Scalp. Oxford: Blackwell Scientific Publications, 1982:36–37.
53. Meredith R, Hearle J. Physical Methods of Investigating Textiles. New York: Interscience, 1959.
54. Boisits EK. The evaluation of moisturizing products. Cosmet Toilet 1986 (May); 101:31–39.
55. Wu MS, Yee DJ, Sullivan ME. Effect of a skin moisturizer on the water distribution in human stratum corneum. J Invest Dermatol 1983; 81:446–448.
56. Wildnauer RH, Bothwell JW, Douglass AB. Stratum corneum biomechanical properties. J Invest Dermatol 1971; 56:72–78.
57. Pierard GE. What does “dry skin” mean? Int J Dermatol 1987; 26:167–168.
58. Motta S, Monti M, Sesana S, Mellesi L, Ghidoni R, Caputo R. Abnormality of water barrier function in psoriasis. Arch Dermatol 1994; 130:452–456.
59. Imokawa G, Abe A, Jin K, Higaki et al. Decreased level of ceramides in stratum corneum of atopic dermatitis: an etiologic factor in atopic dry skin? J Invest Dermatol 1991; 96:523–526.
60. Potts RO, Buras EM, Chrisman DA. Changes with age in the moisture content of human skin. J Invest Dermatol 1984; 82(1):97–100.
61. Wepierre J, Marty JP. Percutaneous absorption and lipids in elderly skin. J Appl Cosmetol 1988; 6:79–92.
62. Jackson EM. Moisturizers: what’s in them? How do they work? Am J Contact Dermatitis 1992; 3(4):162–168.
63. Reiger MM. Skin, water and moisturization. Cosmet Toilet 1989 (Dec.); 104:41–51.
64. Nickoloff BJ, Naidu Y. Perturbation of epidermal barrier function correlates with initiation of cytokine cascade in human skin. J Am Acad Dermatol 1994; 30: 535–546.

65. Elias PM. Epidermal lipids, barrier function, and desquamation. J Invest Dermatol 1983; 80:44s–49s.
66. Jass HE, Elias PM. The living stratum corneum: implications for cosmetic formulation. Cosmet Toilet 1991 (Oct.); 106:47–53.
67. Holleran W, Feingold K, Man MO, Gao W, Lee J, Elias PM. Regulation of epidermal sphingolipid synthesis by permeability barrier function. J Lipid Res 1991; 32:1151–1158.
68. Baker CG. Moisturization: new methods to support time proven ingredients. Cosmet Toilet 1987 (Apr.); 102:99–102.
69. De Groot AC, Weyland JW, Nater JP. Unwanted Effects of Cosmetics and Drugs Used in Dermatology. 3d ed. Amsterdam: Elsevier, 1994:498–500.
70. Friberg SE, Ma Z. Stratum corneum lipids, petrolatum and white oils. Cosmet Toilet 1993 (July); 107:55–59.
71. Grubauer G, Feingold KR, Elias PM. Relationship of epidermal lipogenesis to cutaneous barrier function. J Lipid Res 1987; 28:746–752.
72. Ghadially R, Halkier-Sorensen L, Elias PM. Effects of petrolatum on stratum corneum structure and function. J Am Acad Dermatol 1992; 26:387–396.
73. De Groot AC, Weyland JW, Nater JP. Unwanted Effects of Cosmetics and Drugs Used in Dermatology. 3d ed. Amsterdam: Elsevier, 1994:498–500.
74. Spencer TS. Dry skin and skin moisturizers. Clin Dermatol 1988; 6:24–28.
75. Rieger MM, Deem DE. Skin moisturizers. II. The effects of cosmetic ingredients on human stratum corneum. J Soc Cosmet Chem 1974; 25:253–262.
76. Robbins CR, Fernee KM. Some observations on the swelling of human epidermal membrane. J Soc Cosmet Chem 1983; 37:21–34.
77. Idson B. Dry skin: moisturizing and emolliency. Cosmet Toilet 1992 (July); 107: 69–78.
78. Wehr RF, Krochmal L. Considerations in selecting a moisturizer. Cutis 1987; 39: 512–515.
79. Brand HM, Brand-Garnys EE. Practical applciation of quantitative emolliency. Cosmet Toilet 1992 (July); 107:93–99.
80. Goldemberg RL. Hair conditioners: the rationale for modern formulations. In: Frost P, Horwitz SN, eds. Principles of Cosmetics for the Dermatologist. St. Louis: Mosby, 1982:157–159.
81. Garcia ML, Epps JA, Yare RS, Hunter LD. Normal cuticle-wear patterns in human hair. J Soc Cosmet Chem 1978; 29:155–175.
82. Zviak C, Bouillon C. Hair treatment and hair care products. In: Zviak C, ed. New York: Marcel Dekker, 1986:115–116.
83. Rook A. The clinical importance of "weathering" in human hair. Br J Dermatol 1976; 95:111–112.
84. Robinson VNE. A study of damaged hair. J Soc Cosmet Chem 1976; 27:155–161.
85. Finkelstein P. Hair conditioners. Cutis 1970; 6:543–544.
86. Idson B, Lee W. Update on hair conditioner ingredients. Cosmet Toilet 1983 (Oct.); 98:41–46.
87. Fox C. An introduction to the formulation of shampoos. Cosmet Toilet 1988 (Mar.); 103:25–58.
88. Spoor HJ, Londo SD. Hair processing and conditioning. Cutis 1974; 14:689–694.

89. Arnoud R, Perbet G, Deflandre A, Lang G. ESR study of hair and melanin keratin mixtures: the effects of temperature and light. Int J Cosmet Sci 1984; 6:71–83.
90. Jackowicz J. Hair damage and attempts to its repair. J Soc Cosmet Chem 1987; 38:263–286.
91. Holt LA, Milligan B. The formation of carbonyl groups during irradiation of wool and its relevance to photoyellowing. Textile Res J 1977; 47:620–624.
92. Launer HF. Effect of light upon wool. IV. Bleaching and yellowing by sunlight. Textile Res J 1965; 35:395–400.
93. Inglis AS, Lennox FG. Wool yellowing. IV. Changes in amino acid composition due to irradiation. Textile Res J 1963; 33:431–435.
94. Tolgyesi E. Weathering of the hair. Cosmet Toilet 1983 (Mar.); 98:29–33.
95. Milligan B, Tucker DJ. Studies on wool yellowing. Part III. Sunlight yellowing. Textile Res J 1962; 32:634.

3
The Role of Biological Lipids in Skin Conditioning

Peter M. Elias
University of California, San Francisco, and Veterans Affairs Medical Center, San Francisco, California

I. EVOLVING CONCEPTS OF THE STRATUM CORNEUM BARRIER

A. The Two-Compartment Model of the Stratum Corneum

Because of its loosely organized appearance in tissues subjected to routine fixation, dehydration, and embedding, the stratum corneum was not considered to be important for normal permeability barrier formation until about 35 years ago. However, when epidermis is frozen-sectioned and the cornified envelopes of corneocytes are either swollen at alkaline pH or stained with fluorescent, lipophilic dyes, the stratum corneum appears as a compact structure with geometric, polyhedral squames arranged in vertical columns that interdigitate at their lateral margins (1,2). (Corneocytes are hardened cells viewed as closely packed bundles of keratin filaments.) Moreover, isolated sheets of stratum corneum possess both unusually great tensile strength and very low rates of water permeability (3). Hence, about 30 years ago the homogeneous film or "plastic wrap" concept of the stratum corneum emerged (4). However, since 1975 the stratum corneum has been recognized as comprising a structurally heterogeneous, two-compartment system, with lipid-depleting corneocytes embedded in a lipid-enriched, membraneous extracellular matrix. More recently, the stratum corneum has come to be appreciated as a dynamic and metabolically interactive tissue.

1. Evidence for Intercellular Lipid Sequestration

The stratum corneum is viewed currently as a layer of protein-enriched corneocytes embedded in a lipid-enriched, intercellular matrix (5), the so-called bricks-and-mortar model (6). The evidence for such protein-lipid sequestration is based on freeze-fracture replication, histochemical, biochemical, cell fractionation, cell separation, and physical (4) chemical studies (reviewed in Ref. 6). Freeze-fracture reveals stacks of intercellular bilayers in the intercellular spaces, where transmission electron microscopy previously had revealed only empty spaces. Moreover, histochemical stains also display the membrane domains of the stratum corneum as enriched in neutral lipids, but only when these stains are applied to frozen sections. Furthermore, analysis of isolated peripheral membrane domains showed directly that: (a) the bulk of stratum corneum lipids are in the stratum corneum interstices; (b) the lipid composition of these preparations is virtually identical to that of whole stratum corneum; (c) the freeze-fracture pattern of membrane multilayers, previously described in whole stratum corneum, is duplicated in the membrane preparations. Finally, X-ray diffraction and electron-spin resonance studies also localized all of the bilayer structures, as well as physiological, lipid-based thermal phenomena, to these membrane domains (7).

The two-compartment model also explains the ability of cells in the outer stratum corneum to take up water (i.e., the lipid-enriched lamellar bilayers act as semipermeable membranes). However, the two-compartment, lipid-versus-protein model also requires updating based on recent evidence for microheterogeneity in these domains, e.g., the presence of extracellular proteins, such as desmosomal components, and abundant enzymatic activity within the intercellular spaces.

2. Cellular Basis for Lipid-Protein Sequestration

Since its earliest descriptions (reviewed in Ref. 7), hypotheses have abounded about the function of the epidermal lamellar body. These ellipsoidal organelles, measuring about $^1/_3 \times ^1/_2\,\mu$m, appear initially in the first supra basal cell layer, the stratum spinosum, and continue to accumulate in the stratum granulosum, accounting for about 10% of the volume of the granular cell cytosol. In the outer granular layer, lamellar bodies move to the lateral and apical surfaces, where they are poised to undergo rapid exocytosis (8,9). The lamellar body contains parallel membrane stacks enclosed by a limiting trilaminar membrane. Whereas each lamella appears to be a "disk" in cross sections, with a major electron-dense band separated by electron-lucent material divided centrally by a minor electron-dense band, recent studies have shown instead that lamellar body contents comprise a single membrane structure folded in an accordion-like fashion.

To date, the factors, that regulate lamellar body secretion are not known. Acute perturbations of the barrier result in lamellar body secretion from the outermost granular cell, accompanied by a striking paucity of these organelles in the cytosol (10). By 1–2 hr, abundant nascent lamellar bodies appear in the cytosol. Low rates of secretion apparently occur under basal conditions, while both organellogenesis and secretion are accelerated under stimulated conditions. Whether separate factors regulate the processes of basal versus stimulated secretion, as in other epithelia, is unknown.

Cytochemists provided the first clues about the contents, and therefore the potential functions, of this organelle (11,12). They found lamellar bodies to be enriched in sugars and lipids, thereby generating the initial hypothesis that their contents might be important for epidermal cohesion and waterproofing. Tracer perfusion studies initially demonstrated the role of the lamellar body secretory process in the formation of the barrier. Indeed, the outward egress of water-soluble tracers through the epidermis is blocked at sites where secreted lamellar body contents have been deposited, and no other membrane specializations, such as tight junctions, are present at these locations to account for barrier formation (13).

Biochemical studies further support a role for organelle contents in barrier formation (14–16). Partially purified lamellar body preparations are enriched in glucosylceramides, free sterols, and phospholipids, species which account for almost all of the stratum corneum intercellular lipids. They do not, however, appear to be enriched in cholesterol sulfate, which may reach the intercellular spaces by an alternative mechanism. Moreover, degradation of the plasma membrane during terminal differentiation could result in in-situ delivery of additional lipids to the pool of lipids already available for formation of intercellular bilayers, and/or for the formation of the covalently bound envelope of the stratum corneum (see below).

In addition to lipids, the lamellar body is enriched in certain hydrolytic enzymes, including acid phosphatase, certain proteases, a family of lipases, and a family of glycosidases (reviewed in Ref. 6). As a result of its enzyme content, the lamellar body has been considered a type of lysosome, but evidence for this concept is lacking. Moreover, lamellar bodies lack certain acid hydrolases characteristic of lysosomes, such as arylsulfatases A and B and beta-glucuronidase (17). The same enzymes that are concentrated in lamellar bodies occur in high specific activity in whole stratum corneum, and are further localized specifically to intercellular domains both biochemically and cytochemically (18). As will be discussed below, the enzymes present in lamellar bodies may fulfill dual roles in both barrier formation and desquamation (19–21). The co-localization of "pro-barrier" lipids and various lipases (phospholipase A, sphingomyelinase, steroid sulfatase, acid lipase, and glucosidases) to the same tissue compartment appears to mediate the changes in lipid composition and structure that occur during transit through the stratum corneum interstices (22,23). Whereas the function of lamellar body-derived proteases in the cellular interstices has not been investigated, they could either activate lamellar body-derived hydrolases by conversion of pro-enzymes to active forms of the enzymes, and/or be involved in desmosomal degradation (see below).

Steroid sulfatase, the enzyme responsible for desulfation of cholesterol sulfate, is not enriched in isolated lamellar body preparations, yet somehow, this microsomal enzyme (7) reaches membrane domains in the stratum corneum (24). It is possible that the enzyme is present in lamellar bodies, but that it is lost or destroyed during organelle isolation. However, it also is possible that steroid sulfatase may be transferred from microsomes to the limiting membrane of the lamellar body. Either process could result in either "splicing" of enzyme into the corneocyte periphery or constitutive delivery to the extracellular spaces.

An inadequately studied consequence of lamellar body secretion relates to changes in: (a) the intercellular volume, and (b) the surface area:volume ratio of the stratum corneum and individual corneocytes resulting from coordinated exocytosis of lamellar bodies. Preliminary studies suggest that the intercellular compartment is greatly expanded (5–10% of total volume) (25) in comparison to the volume of the interstices in other epithelia (1–2%). Moreover, the stratum corneum interstices serve as a selective "sink" for exogenous lipophilic agents, which can further expand this compartment. Finally, although not studied to date, the splicing of the limiting membranes of lamellar bodies into the plasma membrane of the granular cell results in an obligatory, massive expansion of the surface area:volume ratio of individual corneocytes, which could explain the remarkable water-holding capacity of corneocytes.

B. Microheterogeneity of the Intercellular Spaces

1. Lamellar Bilayer Generation, Maturation, and Substructure

Lamellar body exocytosis delivers the precursors of these bilayers to the intercellular spaces at the stratum granulosum–stratum corneum interface (26). A transition then can be seen from lamellar body-derived sheets into successively elongated membranes with the same substructure as lamellar body sheets, which unfurl parallel to the plasma membrane (27–29) (Figure 1). End-to-end fusion of lamellar body-derived membrane sheets continues within the first two layers of the stratum corneum, giving rise to broad, uninterrupted membrane sheets, which is followed by compaction of adjacent membrane sheets into lamellar bilayer unit structures. This change in structure correlates with a sequence of changes in lipid composition, i.e., from the polar lipid-enriched mixture of glycosphingolipids, phospholipids, and free sterols present in lamellar bodies to the more nonpolar mixture, enriched in ceramides, free

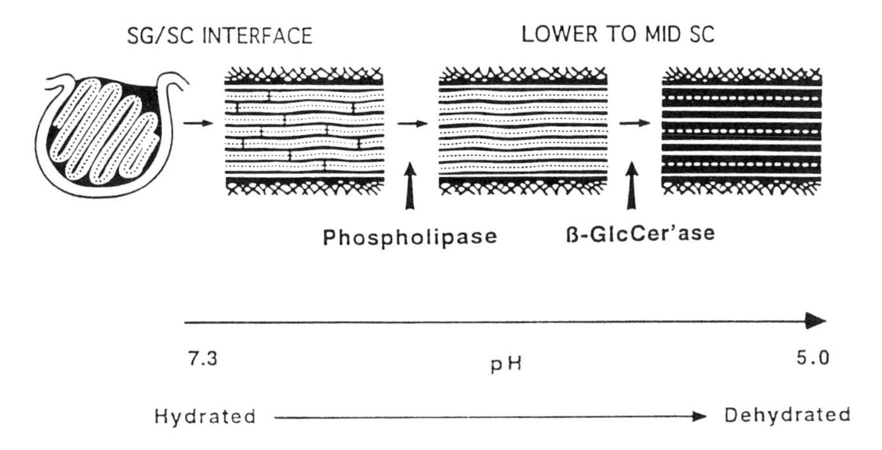

Figure 1 Extracellular processing of polar lipids to nonpolar lipids is required for the sequential membrane modifications that lead to barrier formation. In addition, changes in extracellular pH and hydration may contribute to this sequence.

sterols, and free fatty acids, that is present in the remainder of the stratum corneum (30). An explanation for the structural changes, consistent with the compositional changes and enzyme localization data, is that the initial end-to-end fusion of unfurled lamellar body-derived sheets may be mediated by the initial degradation of phospholipids to free fatty acids by phospholipase A_2, which is present in abundance in lamellar bodies and the stratum corneum interstices (Figure 1). The subsequent transformation of elongated disks into the broad multilamellar membrane system required for barrier function is associated with the further, complete hydrolysis of residual glucosylceramides to ceramides.

Until recently, elucidation of membrane structure in mammalian stratum corneum was impeded by the extensive artifacts produced during processing for light or electron microscopy (reviewed in Ref. 6). Following the application of freeze-fracture replication to the epidermis, the stratum corneum interstices were found to be replete with a multilamellar system of broad membrane bilayers. Further detailed information about intercellular lamellar bilayer structures has resulted from the recent application of ruthenium tetroxide postfixation to the study of stratum corneum membrane structures (31–33). Despite its extreme toxicity to structural proteins, with ruthenium tetroxide postfixation the electron-lucent lamella appears as pairs of continuous leaflets alternating with a single fenestrated lamella. Each electron-dense lamella is separated by an electron-dense structure of comparable width. The entire multilamellar complex lies external to a hydrophobic envelope containing covalently bound ceramides (see below). The lamellar spacing, or repeat diffraction, correlates extremely well with independent measurements of these domains by X-ray diffraction (34).

Because the repeat distance is more than twice the thickness of typical lipid bilayers, each lamellar repeat unit appears to consist of two opposed bilayers. Multiples of these units (up to three) occur frequently in the stratum corneum interstices, and simplifications of the basic unit structure, with deletion of one or more lamellae, occur at the lateral surfaces of corneocytes, i.e., at three cell junctures (35). Dilatations of the electron-dense lamellae, which correspond to sites of desmosomal hydrolysis, also are visualized well with ruthenium staining, and may comprise a "pore pathway" for percutaneous drug and xenobiote movement. Correlation of images obtained with ruthenium tetroxide, biochemical methods, X-ray diffraction methods, and other physical-chemical methods (e.g., ESR and NMR) ultimately should provide an integrated model of the architecture of the stratum corneum intercellular membrane system, as well as important new insights about alterations in membrane structure responsible for altered permeability states and pathological desquamation.

2. *Covalently Bound Envelope*

The membrane complex immediately exterior to the cornified envelope replaces the true plasma membrane during terminal differentiation (36). Although a portion of this trilaminar structure survives exhaustive solvent extraction, it is destroyed by saponification (37). Lipid extracts of saponified fractions yield at least two very long-chain, omega-hydroxy acid-containing ceramides that are believed to be covalently attached to glutamine residues in the cornified envelope (38). Although the covalently bound envelope is enriched in omega-hydroxy acid-containing ceramides, its complete composition is not known, since both the initial solvent extraction of the intercellular lamellae and subsequent saponification could remove or destroy certain constituents of this structure. Since this envelope persists after prior solvent extraction has rendered the stratum corneum porous, it does not itself provide a barrier. However, it may function as a scaffold for the deposition and organization of lamellar body-derived, intercellular bilayers. Finally, the origin of the covalently bound envelope remains unknown. It could originate from the pool of lipids deposited during lamellar body secretion, and/or by in-situ degradation of plasma membrane sphingolipids, such as sphingomyelin.

C. Extracellular Lipid Processing

As noted above, the sequestration of lipids within the intercellular spaces of the stratum corneum results from the secretion of the lipid-enriched contents of lamellar bodies from the outermost granular cell. These organelles also contain selected hydrolytic enzymes which appear to regulate the formation of a component permeability barrier as well as desquamation. Barrier formation requires the transformation of the initially secreted lipids (predominantly glucosylceramides, cholesterol, and phospholipids) into a more nonpolar mixture, enriched in ceramides, cholesterol, and free fatty acids. This process may require concomitant acidification of the extracellular domains (39). Such an acidic milieu may be required for optimal activation of certain of the key hydrolases, beta-glucocerebrosidase, acid lipase, and sphingomyelinase. Whether epidermal extracellular phospholipases also display optimal activity at an acidic pH is unknown. Once activated, these enzymes generate the requisite nonpolar lipid mixture that forms the hydrophobic, intercellular lamellar bilayers (Figure 1).

Proof of the role of intercellular beta-glucocerebrosidase in this process has been provided by the use of both enzyme-specific inhibitors (conduritols) and a transgenic murine model (40,41). In both approaches, lack of enzyme activity leads to a barrier abnormality, which appears to be attributable to accumulation of glucosylceramides (not depletion of ceramides). This biochemical change is accompanied by the persistence throughout the stratum corneum interstices of immature membrane structures. Interestingly, these immature or incompletely processed membrane structures also appear in a subgroup of Gaucher disease (type II), which is characterized by drastically reduced enzyme levels and ichthyosiform skin lesions (42). Such immature, glycosylated membrane structures, although inadequate to meet the demands of terrestrial life, nevertheless appear to suffice in mucosal epithelia (43,44) and in the stratum corneum of marine cetaceans, which both display a high glycosylceramide-to-ceramide ratio (45). Pertinently, endogenous 13-glucosidase levels are reduced in oral mucosa (46). Thus, the persistence of glucosylceramides may indicate less stringent barrier requirements in these locales and/or additional functions of glucosylceramides unique to these tissues.

Extracellular processing of phospholipids also is required for barrier homeostasis. As with beta-glucocerebrosidase, pharmacological inhibitors of phospholipase A_2 both delay barrier recovery after prior disruption (47) and induce a barrier abnormality in intact murine epidermis (48). However, in contrast to beta-glucocerebrosidase, the biochemical abnormality responsible for the barrier defect is product depletion rather than substrate accumulation; i.e., co-applications of palmitic acid (but not linoleic acid) with the phospholipase inhibitors normalize barrier function. Thus, generation of nonessential free fatty acids by phospholipase-mediated degradation of phospholipids is required for barrier homeostasis. In summary, there is indisputable evidence that activity of at least two stratum corneum extracellular enzymes is required for barrier homeostasis. Whether extracellular processing of triglycerides, cholesterol esters, cholesterol sulfate, other acylglycerides, sphingomyelin, and/or glycerophospholipids (e.g., lysolecithin) by their respective hydrolases (i.e., acid lipase, cholesterol ester hydrolase, steroid sulfatase, sphingomyelinase, phospholipase A_1, and lysolecithinase) is (are) required for barrier homeostasis is unknown. Moreover, the role of active or passive acidification mechanisms in triggering the extracellular processing of lipids in a sequential or coincident fashion also remains to be explored.

II. EPIDERMAL LIPIDS

A. Role of Lipids in the Permeability Barrier

The importance of stratum corneum lipids for barrier integrity has been known since the old observation that topical applications of organic solvents produce profound alterations in barrier function. More recently, the importance of bulk stratum corneum lipids for the barrier has been demonstrated by: (a) the inverse relationship between the permeability of the stratum corneum to water and water-soluble molecules at different skin sites (e.g., abdomen versus palms and soles) and the lipid content of that site (49); (b) the observation that organic solvent-induced perturbations in barrier function occur in direct proportion to the quantities of lipid removed (50); (c) the observation that stratum corneum lipid content is defective in pathological states that are accompanied by compromised barrier function, such as essential fatty acid deficiency (51, 52); (d) the observation that replenishment of endogenous stratum corneum lipids, following removal by solvents or detergents, parallels the recovery of barrier function (53); and (e) that topically applied stratum corneum lipids normalize or accelerate barrier recovery when applied to solvent-treated, stripped, or surfactant-treated skin (54–56) (see below).

B. Regional Variations in Human Stratum Corneum Lipid Composition

The lipid composition of human stratum corneum lipids displays striking regional variations that could reflect differences in stratum corneum thickness, turnover, desquamation, and/or permeability. However, the barrier properties of these sites are not explicable by either site-related differences in thickness or the number of cell layers in the stratum corneum. Instead, an inverse relationship exists between the lipid weight percentage and the permeability properties of a particular skin site (57,58). In addition to total lipid content, significant regional differences occur in the compositional profile of stratum corneum lipids over different skin sites. For example, the proportion of sphingolipids and cholesterol is much higher in palmoplantar stratum corneum than on the extensor surfaces of the extremities, abdominal, or facial stratum corneum (59). However, the significance of these differences in lipid distribution is not known, because the absolute quantities of each of these fractions is dependent on the lipid weight percentage of the stratum corneum at each anatomical site. Thus, despite the high proportions of sphingolipids and cholesterol in palmar stratum corneum, when adjusted for the 2% lipid weight of this site, the absolute amounts of sphingolipids and cholesterol in the intercellular spaces are still much lower than in other, more lipid-enriched sites. Moreover, functional interpretations require consideration not only of lipid distribution and weight percentages, but also information about site-related variations in the fatty acid profiles of esterified species, and at present these data are not available.

These regional differences in lipid content do have important clinical implications. First, they correlate with susceptibility to the development of contact dermatitis to lipophilic versus hydrophilic antigens at specific skin sites. Whereas allergy to a fat-soluble antigen, such as poison ivy (urushiol), is more likely to occur in lipid-replete sites, allergy to water-soluble antigens, such as those in foods, flowers, and vegetables, occurs more commonly on lipid-depleted sites; e.g., the palms. Second, subjects with atopic dermatitis, who display a paucity of stratum corneum lipids, are less readily sensitized to lipid-soluble as opposed to hydrophilic antigens, such as nickel. Of course, the T-cell abnormalities in atopics also could contribute to differences in sensitization thresholds. Third, percutaneous drug delivery of lipid-soluble drugs, such as topical steroids and retinoids, occurs more readily on lipid-replete sites, such as the face—hence the relatively higher propensity to develop cutaneous side effects, such as steroid atrophy, at these sites. Conversely, lipid-soluble drugs, such as nitroglycerin, scopolamine, clonidine, fentanyl, and nicotine, are delivered transdermally for systemic therapeutic purposes with relative ease over lipid-replete sites. Finally, the low lipid content of palms (and soles) explains the increased susceptibility of these sites to the development of soap/surfactant and hot water-induced dermatitis; i.e., these sites have a defective, lipid-deficient barrier prior to additional lipid removal, which superimposes a further insult.

The distribution of lipids in nonkeratinized, oral mucosal sites, which generally have a higher water permeability than keratinized regions, is different from that in epidermis (60,61). Nonkeratinized regions, such as the buccal epithelia, contain no acylglucosylceramides and acylceramides and only very small amounts of ceramides. Glycosylceramides replace ceramides in nonkeratinizing epithelia, apparently because of an absence of endogenous beta-glycosidase activity (62). Moreover, both keratinizing and nonkeratinizing regions of porcine oral epithelial contain more phospholipids than epidermis (63,64). Thus, the differences in permeability in epidermis versus mucosal epithelial may be explicable by the replacement of ceramides and free fatty acids by glycosylceramides and phospholipids, respectively.

C. Epidermal Lipid Synthesis

1. *Synthesis Under Basal Conditions*

Cutaneous lipid synthesis has been studied extensively both in vivo and in vitro (65,66). These studies have demonstrated, first, that the skin is a major site of lipid synthesis, accounting for 20–25% of total body synthesis (67,68). Second, the skin generates a broad range of lipid species (69). Third, about 70–75% of cutaneous lipid synthesis localizes further to the dermis (70). The epidermis, which accounts for less than 10% of skin mass, accounts for 25–30% of total cutaneous activity. Thus, the epidermis is an extremely active site of lipid synthesis, with about 60–70% of total lipid synthesis occurring in the basal layer (71). Considerable epidermal lipid synthesis, however, continues in all of the nucleated layers of the epidermis (72–74). It is now clear that systemic factors influence cutaneous lipid synthesis minimally. Dramatic hormonal changes, particularly in thyroid, testosterone, or estrogen status, have been shown to alter epidermal lipid synthesis, but it is not clear that these hormones regulate cutaneous lipid synthesis under day-to-day conditions. In addition, changes in circulating sterols from either diet or drugs do not alter epidermal cholesterol synthesis (75), presumably due to the paucity of low-density lipoprotein (LDL) receptors on epidermal cells (76–78). The autonomy of these layers from circulating influences may have evolutionary significance, because it ensures that the differentiating layers are attuned to their own, special functional requirements, i.e., barrier homeostasis (see below). Despite its relative autonomy from the circulation, the epidermis incorporates some circulating lipids, such as plant sterols, essential fatty acids, polyunsaturated fatty acids (PUFAs), and arachidonic acid (the epidermis lacks the 6-delta-desaturase). Although these lipids are indicators of the capacity of the epidermis to take up exogenous lipids, the quantitative contribution of extracutaneous lipids to the epidermal pool appears to be small in comparison to de novo synthesis (79).

2. *Metabolic Response to Barrier Disruption*

Despite its relative autonomy and high basal rates of lipid synthesis, the epidermis responds with a further lipid biosynthetic burst when the permeability barrier is disrupted by topical treatment with either organic solvents, tape stripping, or detergents (80,81). Regardless of the manner of barrier disruption, a biphasic repair response occurs which leads to about 50% restoration of normal barrier function in about 12 hr in humans, with complete recovery requiring 72–96 hr (82). Quite different metabolic events are associated with the rapid versus slow recovery phase. Immediately after barrier disruption, all of the lamellar bodies in the outermost granular cell are secreted (83), and a burst occurs in both cholesterol and fatty acid synthesis (84,85). In contrast, the late phase of barrier recovery is associated with a delayed burst in ceramide synthesis (86), as well as a stimulation of epidermal DNA synthesis (87). That all of these alterations can be attributed to the barrier abnormality is shown by: (a) the localization of the increase in synthesis to the underlying epidermis—dermal cholesterol and fatty acid synthetic rates remain unaffected after barrier disruption; (b) the extent of the increase in lipid and DNA synthesis is proportional to the degree of the barrier abnormality; (c) the burst in lipid synthesis is prevented when the barrier is artificially restored by application of water vapor-impermeable (but not vapor-permeable) membranes, while the burst in DNA synthesis is partially blocked; and (d) in a sustained model of barrier dysfunction, i.e., rodent essential fatty acid deficiency, lipid synthesis is stimulated, and the increase is normalized when the barrier is restored by either linoleic acid replenishment or occlusion (88). These results indicate that alterations in barrier function stimulate epidermal lipid synthesis, and suggest further that transcutaneous water loss might be a direct or indirect regulatory factor.

Although these results demonstrate that barrier function regulates epidermal lipid synthesis, they do not address the basis for this metabolic response. Three of the most abundant lipid species in the stratum corneum are cholesterol, ceramides, and free fatty acids. The epidermal activities of the rate-limiting enzymes for these species, 3-hydroxy-3-methyl glutaryl coenzyme A (HMG CoA) reductase, serine palmitoyl transferase (SPT), acetyl CoA carboxylase (ACC), and fatty acid synthase (FAS), is unusually high (89–91). Moreover, the activities of all these enzymes increase when the barrier is disrupted in both the acute and EFAD models. Whereas the changes in HMG CoA reductase, ACC, and FAS occur shortly after barrier abrogation, and return to normal quickly, the rise in SPT is more delayed and prolonged. Furthermore, the increase in activity of all these enzymes is blocked when the barrier is restored artificially by occlusion with a vapor-impermeable membrane. Not only the total activity of the HMG CoA reductase, but the activation state (phosphorylation state) of this enzyme, and perhaps ACC as well, are regulated by barrier requirements (SPT and FAS are not known to be phosphorylated). Finally, the extent of the increase in the content and activation state of HMG CoA reductase is proportional to the degree of barrier disruption (92). Since the threshold for changes in the activation state of HMG CoA reductase changes with lesser perturbations in the barrier than those required to increase enzyme content, reversible phosphorylation may allow both rapid responses to barrier requirements and/or fine-tuning after minor insults to the barrier. Finally, the changes in enzyme activity are preceded by changes in the mRNA for at least two of these enzymes, HMG CoA reductase (93) and FAS (94). These data suggest that barrier requirements regulate epidermal lipid synthesis by modulating the content, activation state, and mRNA of the key regulatory enzymes of cholesterol, fatty acid, and sphingolipid synthesis. Thus, acute change in the barrier initiate a sequence of events, including rapid lamellar body secretion and increased lipid synthesis, which lead ultimately to barrier restoration.

Despite these specific increases in selected enzyme proteins and some nonenzyme proteins, such as beta-actin, bulk protein synthesis does not increase significantly after acute barrier perturbations. Moreover, applications of the protein synthesis inhibitors, puromycin and cycloheximide, at doses which inhibit epidermal protein synthesis up to 50%, do not interfere with the kinetics of barrier repair (95). Thus, the stimulation of lipid and DNA synthesis represents a selective, rather than a nonspecific consequence of barrier disruption.

3. Requirements of Specific Lipids Versus Lipid Mixtures for Barrier Function

Whereas metabolic studies clearly show that epidermal cholesterol, fatty acid, and ceramide synthesis are modulated by alterations in barrier function, the demonstration that each of these lipids is required for the barrier requires assessment of function after selective deletion of each of these species. Utilizing pharmacological inhibitors of their rate-limiting enzymes, each of the three key lipids has been shown to be required for barrier homeostasis (96–98). Deletion of any of these species leads to abnormal barrier recovery and/or abnormal barrier homeostasis in intact skin.

Having shown that each lipid is required individually, the next issue is whether they function cooperatively, i.e., whether they must be supplied together in proportions comparable to those present in the stratum corneum. Indeed, when cholesterol, free fatty acids, ceramides, or even acylceramides are applied alone to solvent-perturbed skin, they aggravate rather than improve the barrier. Likewise, any two-component system of the three key stratum corneum lipids is deleterious. In contrast, three-component mixtures of the key lipids, or two-component mixtures of acylceramides and cholesterol, allow normal barrier recovery and can even accelerate barrier recovery, depending on the final proportion of the key lipids. The mechanism for the aggravation and amelioration of barrier function by the physiological lipids is the same: they are quickly absorbed into the nucleated cell layers, and incorporated into nascent lamellar bodies. Whereas incomplete mixtures yield abnormal lamellar body contents, and disorder intercellular malellae, complete mixtures result in normal lamellar bodies and intercellular bilayers (99,100).

III. POTENTIAL SIGNALS OF PERMEABILITY BARRIER HOMEOSTASIS

A. Ionic Modulations

The ability of occlusion to block the lipid and DNA synthetic response to barrier disruption suggests that transepidermal water loss is a regulatory signal for barrier homeostasis (101), yet a perturbed barrier recovers normally when exposed to either isotonic, hypertonic, or hypotonic external solutions. In contrast, if calcium, potassium, and to a lesser extent, magnesium and phosphorus, are present in the bathing solution, barrier recovery is impeded. Moreover, calcium and potassium together appear to be synergistic in inhibiting barrier recovery. Since these inhibitory influences are reversed by blockade with inhibitors of both L-type calcium channels and calmodulin, translocation of extracellular calcium into the cytosol appears to be required (102).

The mechanism for the negative ionic signal seems to relate to the presence of a calcium gradient in the epidermis; with barrier disruption, the inhibitory ions are carried out passively into the stratum corneum (103,104). Moreover, it is depletion of calcium, rather than barrier disruption per se, that regulates lamellar body exocytosis, and perhaps lipid synthesis as well. Whether ionic signals influence metabolic events in deeper layers of the epidermis, where DNA synthesis and most of the lipid biosynthetic occur in response to barrier disruption, is not known.

B. Cytokine Alterations

Whereas the epidermis generates a large number of biological response modifiers (BRM) (105), the cytokine IL-α appears to be one of the few that is present in considerable quantities under basal conditions, where it accumulates in the outer epidermis (106). In response to all forms of acute barrier disruption, a rapid increase occurs in the mRNA and protein content of several cytokines, including IL-α (107–111). Furthermore, the pre-formed pool of IL-α is released from granular and cornified cells, independent of new cytokine formation. Likewise, with sustained barrier disruption, as in essential fatty acid deficiency, both the mRNA and protein content of several of the cytokines increase.

Whether these changes in cytokine expression represent in part a physiological response to barrier disruption versus a nonspecific injury response remains unresolved. IL-α, TNF-α, and several other epidermis-derived BRM are potent mitogens (112), and both IL-α and TNF-α regulate lipid synthesis in extracutaneous tissues. Hence, it is tempting to regard these cytokine responses as homeostatic. However, occlusion with a vapor-impermeable membrane does not block the increase in either cytokine mRNA or protein expression after acute barrier abrogations (113). In contrast, sustained occlusion lowers cytokine mRNA and protein levels not only in essential fatty acid-deficient epidermis, but also in normal epidermis (114).

C. Other Potential Signaling Mechanisms

Ions and cytokines represent only two potential families of regulating signals. Among others to be considered, but not yet studied, are nitric oxide, various

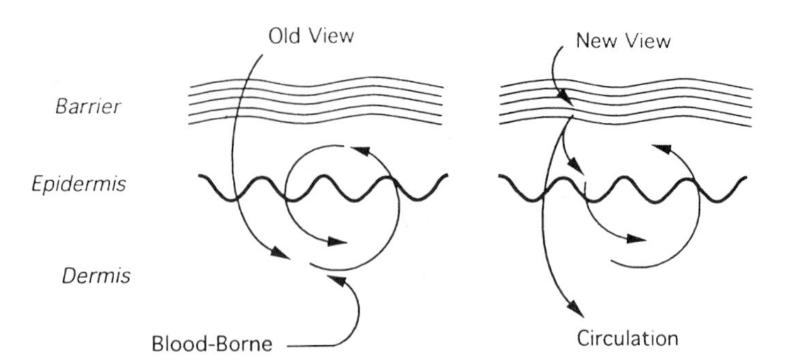

Figure 2 Contrasting outside-in versus inside-out views of triggering of common dermatoses, such as psoriasis and irritant contact dermatitis.

keratinocyte-derived growth factors, neuropeptides, histamine, and eicosanoids. The epidermis synthesizes several potential regulators in these categories. Whether one or more of these will emerge as important regulators of barrier homeostasis remains to be determined.

IV. PATHOPHYSIOLOGICAL CONSEQUENCES OF BARRIER DISRUPTION

Recently, it has been suggested that diffusion into the dermis of one or more of the cytokines or other BRMs generated during barrier abrogation, could initiate or propagate portions of the inflammatory response (115,116). Pertinently, barrier disruption recently was shown to be followed by increased migration of mitotically active Langerhans cells into the epidermis. Moreover, repeated barrier disruption leads to both epidermal hyperplasia and inflammation, changes which, again, are not presented by occlusion. This view provides a relatively new outside→inside concept, as opposed to the dominant inside→outside view, of the pathogenesis of inflammatory skin diseases, such as irritant contact dermatitis, psoriasis, and atopic dermatitis (Figure 2) (117,118). Thus, the leakage, diffusion, or escape of cytokines or other BRMs into the dermis following barrier disruption, could initiate dermal inflammation. These observations have implications beyond the link between barrier function and inflammation: they also provide a possible explanation for an additional group of clinical skin diseases. For example, postinflammatory hyperpigmentation, regardless of specific cause, could be initiated by a barrier-initiated signal cascade (Figure 3). Moreover, if such varied insults as acute sunburn or exfoliative erythiodermas result in sufficient symptoms, fever and malaise can

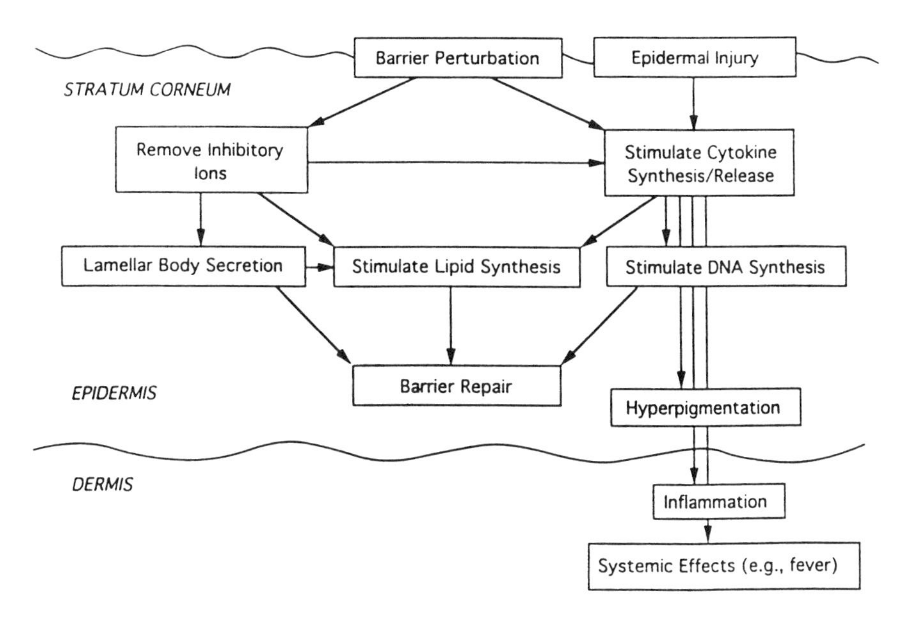

Figure 3 Related mechanisms, such as cytokines, may mediate both homeostatic (repair) and pathological sequences in the skin.

result (Figure 3) (119). Finally, several additional signaling mechanisms can be recruited during this process, further exacerbating the epidermal hyperproliferation and inflammation directly, and, if excessive, further exacerbating the barrier abnormality. This vicious circle is likely to be operative in several important dermatoses, such as psoriasis and a variety of eczemas.

ACKNOWLEDGMENTS

This work was supported by NIH grant AR 19098 and the Medical Research Service, Veterans Administration. Drs. Kenneth R. Feingold, Gopinathan Menon, Erhardt Protsch, Simon Jackson, Nanna Schurer, Seung Lee, Walter Holleran, Man Mao Qiang, Karen Ottey, and many others contributed immeasurably to this research effort. Ms. Sue Allen and Mr. Jason Karpf capably prepared the manuscript.

REFERENCES

1. Feingold KR. The regulation and role of epidermal lipid synthesis. Adv Lipid Res 1991; 24:57–79.

2. Elias PM, Feingold KR. Lipids and the epidermal water barrier: metabolism, regulation, and pathophysiology. Semin Dermatol 1992; 11:176–182.
3. Scheuplein RJ, Blank JF. Permeability of the skin. Physiol Rev 1971; 51:702–747.
4. Elias PM, Feingold KR. Lipids and the epidermal water barrier: metabolism, regulation, and pathophysiology. Semin Dermatol 1992; 11:176–182.
5. Elias PM, Friend DS. The permeability barrier in mammalian epidermis. J Cell Biol 1975; 65:180–191.
6. Elias PM. Epidermal lipids, barrier function, and desquamation. J Invest Dermatol 1983; 80:44–49.
7. Elias PM, Menon GK. Structural and lipid biochemical correlates of the epidermal permeability barrier. Adv Lipid Res 1991; 24:1–26.
8. Odland GP, Holbrook K. The lamellar granules of the epidermis. Curr Probl Dermatol 1987; 9:29–49.
9. Landmann L. The epidermal permeability barrier. Anat Ambryol 1988; 178:1–10.
10. Ibid.
11. Elias PM, Menon GK. Structural and lipid biochemical correlates of the epidermal permeability barrier. Adv Lipid Res 1991; 24:1–26.
12. Odland GP, Holbrook K. The lamellar granules of the epidermis. Curr Probl Dermatol 1987; 9:29–49.
13. Elias PM, Friend DS. The permeability barrier in mammalian epidermis. J Cell Biol 1975; 65:180–191.
14. Menon GK, Feingold KR, Elias PM. The lamellar body secretory response to barrier disruption. J Invest Dermatol 1992; 98:279–289.
15. Wertz PW, Downing DT, Freinkel RK, Traczyk TN. Sphingolipids of the stratum corneum and lamellar granules of fetal rat epidermis. J Invest Dermatol 1984; 83: 193–195.
16. Grayson S, Johnson-Winegar AG, Wintroub BU, Isseroff RR, Epstein EH Jr, Elias PM. Lamellar body-enriched fractions from neonatal mice: preparative techniques and partial characterization. J Invest Dermatol 1985; 85:289–295.
17. Ibid.
18. Elias PM, Menon OK. Structural and lipid biochemical correlates of the epidermal permeability barrier. Adv Lipid Res 1991; 24:1–26.
19. Ibid.
20. Williams ML. Lipids in Normal and Pathological Desquamation. Adv Lipid Res 1991; 24:211–252.
21. Menon GK, Williams ML, Ohadially R, Elias PM. Lamellar bodies as delivery systems of hydrolytic enzymes: implications for normal cohesion and abnormal desquamation. Br J Dermatol 1992; 126:337–345.
22. Elias PM, Menon GK. Structural and lipid biochemical correlates of the epidermal permeability barrier. Adv Lipid Res 1991; 24:1–26.
23. Fartasch M, Bassukas ID, Diepgen TH. Structural relationship between epidermal lipid lamellae, lamellar bodies and desmosomes in human epidermis: an ultrastructural study. Br J Dermatol 1993; 128:1–9.
24. Elias PM, Williams ML, Maloney ME, et al. Stratum corneum lipids in disorders of cornification: steroid sulfatase and cholesterol sulfate in normal desquamation and the pathogenesis of recessive X-linked ichthyosis. J Clin Invest 1984; 74:1414–1421.

25. Menon GK, Price LF, Bommannan B, Elias PM, Feingold KR. Selective obliteration of the epidermal calcium gradient leads to enhanced lamellar body secretion. J Invest Dermatol 1994; 102:789–795.
26. Elias PM, Menon GK. Structural and lipid biochemical correlates of the epidermal permeability barrier. Adv Lipid Res 1991; 24:1–26.
27. Landmann L. The epidermal permeability barrier. Anat Embryol 1988; 178:1–10.
28. Menon GK, Feingold KR, Elias PM. The lamellar body secretory response to barrier disruption. J Invest Dermatol 1992; 98:279–289.
29. Fartasch M, Bassukas ID, Diepgen TH. Structural relationship between epidermal lipid lamellae, lamellar bodies and desmosomes in human epidermis: an ultrastructural study. Br J Dermatol 1993; 128:1–9.
30. Elias PM, Menon GK. Structural and lipid biochemical correlates of the epidermal permeability barrier. Adv Lipid Res 1991; 24:1–26.
31. Fartasch M, Bassukas ID, Diepgen TH. Structural relationship between epidermal lipid lamellae, lamellar bodies and desmosomes in human epidermis: an ultrastructural study. Br J Dermatol 1993; 128:1–9.
32. Madison KC, Swartzendruber DC, Wertz PW, Downing DT. Presence of intact intercellular lamellae in the upper layers of the stratum corneum. J Invest Dermatol 1987; 88:714–718.
33. Hou SYE, Mitra AK, White SH, Menon GK, Ghadially R, Elias PM. Membrane structures in normal and essential fatty acid deficient stratum corneum: characterization by ruthenium tetroxide staining and X-ray diffraction. J Invest Dermatol 1991; 96:215–223.
34. White SH, Mirejovsky D, King GI. Structure of lamellar lipid domains and corneocyte envelopes of murine stratum corneum. An X-ray diffraction study. Biochemistry 1988; 27:3725–3732.
35. Hou SYE, Mitra AK, White SH, Menon GK, Ghadially R, Elias PM. Membrane structures in normal and essential fatty acid deficient stratum corneum: characterization by ruthenium tetroxide staining and X-ray diffraction. J Invest Dermatol 1991; 96:215–223.
36. Elias PM, Menon G. Structural and lipid biochemical correlates of the epidermal permeability barrier. Adv Lipid Res 1991; 24:1–26.
37. Swartzendruber DC, Wertz PW, Kitko DJ, Madison KC, Downing DT. Evidence that the corneocyte has a chemically-bound lipid envelope. J Invest Dermatol 1987; 88:709–713.
38. Chang F, Swartzedrauber DC, Wertz PW, Squier CA. Covalently bound lipids in keratinizing epithelia. Biochim Biophys Acta 1993; 1150:98–102.
39. Chapman SJ, Walsh A. Membrane-coating granules are acidic organelles which possess proton pumps. J Invest Dermatol 1989; 93:466–470.
40. Holleran WM, Takagi Y, Jackson SM, Tran HT, Feingold KR, Elias PM. Processing of epidermal glucosylceramides is required for optimal mammalian cutaneous permeability barrier function. J Clin Invest 1993; 91:1656–1664.
41. Holleran WM, Sidransky E, Menon GK, et al. Consequences of beta-glucocerebrosidase deficiency in epidermis: ultrastructure and permeability barrier alterations in Gaucher disease. J Clin Invest 1994; 93:1756–1764.
42. Ibid.

43. Wertz PW, Cox PS, Squier CA, Downing DT. Lipids of epidermis and keratinized and non-keratinized oral epithelia. Comp Biochem Physiol [B] 1986; 83: 529–531.
44. Wertz PW, Kremer M, Squier SM. Comparison of lipids from epidermal and palatal stratum corneum. J Invest Dermatol 1992; 98:375–378.
45. Elias PM, Menon GK, Grayson S, Brown BE, Rehfeld SJ. Avian sebokeratocytes and marine mammal lipokeratinocytes: structural, lipid biochemical and functional considerations. Am J Anat 1987; 180:161–177.
46. Chang F, Wertz PW, Squier CA. Comparison of glycosidase activities in epidermis, palatal epithelium, and buccal epithelium. Comp Biochem Physiol [B] 1991; 100: 137–139.
47. Mao-Qiang M, Feingold KR, Jain M, Elias PM. Extracellular processing of phospholipids to free fatty acids is required for permeability barrier homeostasis. J Lipid Res. In press.
48. Mao-Qiang M, Jain M, Feingold KR, Elias PM. Secretory phospholipase A_2 activity is required for permeability barrier homeostasis. J Invest Dermatol. In press.
49. Lampe MA, Burlingame AL, Whitney J, et al. Human stratum corneum lipids: characterization and regional variations. J Lipid Res 1983; 24:120–150.
50. Elias PM, Fritsch P, Epstein EH Jr. Staphylococcal scalded skin syndrome: clinical features, pathogenesis, and recent microbiological and biochemical developments. Arch Dermatol 1977; 113:207–219.
51. Grubauer O, Feingold KR, Elias PM. Lipid content and lipid type as determinants of the epidermal permeability barrier. J Lipid Res 1989; 30:89–96.
52. Elias PM, Brown BE. The mammalian cutaneous permeability barrier: defective barrier function in essential fatty acid deficiency correlates with abnormal intercellular lipid deposition. Lab Invest 1978; 39:574–583.
53. Grubauer O, Elias PM, Feingold KR. Transepidermal water loss: the signal for recovery of barrier structure and function. J Lipid Res 1989; 30:323–333.
54. Mao-Qiang M, Feingold KR, Elias PM. Exogenous lipids influence permeability barrier recovery in acetone treated murine skin. Arch Dermatol 1993; 129:728–738.
55. Mao-Qiang M, Brown BE, Wu S, Feingold KR, Elias PM. Exogenous non-physiological vs physiological lipids: divergent mechanisms for correction of permeability barrier dysfunction. Arch Dermatol. In press.
56. Yang L, Elias PM, Feingold KR. Topical stratum corneum lipids accelerate barrier repair after tape stripping, solvent treatment, and some but not all types of detergent treatment. Br J Dermatol. In press.
57. Lampe MA, Burlingame AL, Whitney J, et al. Human stratum corneum lipids: characterization and regional variations. J Lipid Res 1983; 24:120–150.
58. Elias PM, Fritsch P, Epstein EH Jr. Staphylococcal scalded skin syndrome: clinical features, pathogenesis, and recent microbiological and biochemical developments. Arch Dermatol. 1977; 113:207–219.
59. Lampe MA, Burlingame AL, Whitney J, et al. Human stratum corneum lipids: characterization and regional variations. J Lipid Res 1983; 24:120–150.
60. Wertz PW, Cox PS, Squier CA, Downing DT. Lipids of epidermis and keratinized and non-keratinized oral epithelia. Comp Biochem Physiol [B] 1986; 83:529–531.

61. Wertz PW, Kremer M, Squier SM. Comparison of lipids from epidermal and palatal stratum corneum. J Invest Dermatol 1992; 98:375–378.
62. Chang F, Wertz PW, Squier CA. Comparison of glycosidase activities in epidermis, palatal epithelium, and buccal epithelium. Comp Biochem Physiol [B] 1991; 100: 137–139.
63. Wertz PW, Cox PS, Squier CA, Downing DT. Lipids of epidermis and keratinized and non-keratinized oral epithelia. Comp Biochem Physiol [B] 1986; 83:529–531.
64. Wertz PW, Kremer M, Squier SM. Comparison of lipids from epidermal and palatal stratum corneum. J Invest Dermatol 1992; 98:375–378.
65. Yardley HI, Summerly R. Lipid composition and metabolism in normal and diseased epidermis. Pharmacol Ther 1981; 13:357–383.
66. Ziboh VA, Chapkin RS. Metabolism and function of skin lipids. Prog Lipid Res 1988; 27:81–105.
67. Turley SD, Anderson JM, Dietschy JM. Rates of sterol synthesis and uptake in the major organs of the rat in vivo. J Lipid Res 1981; 22:551–569.
68. Feingold KR, Wiley MH, Moser AH, et al. De novo sterologenesis in the intact primate. J Lab Clin Med 1982; 100:405–410.
69. Nicolaides N. Skin lipids. Science 1974; 186:19–26.
70. Feingold KR, Brown BE, Lear SR, Moser AH, Elias PM. Localization of de novo sterologenesis in mammalian skin. J Invest Dermatol 1983; 81:365–369.
71. Monger DJ, Williams ML, Feingold KR, Brown BE, Elias PM. Localization of sites of lipid biosynthesis in mammalian epidermis. J Lipid Res 1988; 29:603–612.
72. Monger DJ, Williams ML, Feingold KR, Brown BE, Elias PM. Localization of sites of lipid biosynthesis in mammalian epidermis. J Lipid Res 1988; 29:603–612.
73. Proksch E, Elias PM, Feingold KR. Localization and regulation of epidermal HMO CoA reductase activity by barrier requirements. Biochem Biophys Acta 1991; 1083:71–79.
74. Holleran WM, Gao WN, Feingold KR, Elias PM. Localization of epidermal sphingolipid synthesis and serine paimitoyl transferase activity. Arch Dermatol Res 1995; 287:254–258.
75. Wu-Pong S, Elias PM, Feingold KR. Influence of altered serum cholesterol levels and fasting on cutaneous cholesterol synthesis. J Invest Dermatol 1994; 102:799–802.
76. Ponec M, Havekes L, Kempenaar J, Vermeer BJ. Cultured human skin fibroblasts and keratinocytes: differences in the regulation of cholesterol synthesis. J Invest Dermatol 1983; 81:125–130.
77. Mommaas-Kienhuis AM, Grayson S, Wijsman MC, Vermeer BJ, Elias PM. LDL receptor expression on keratinocytes in normal and psoriatic epidermis. J Invest Dermatol 1987; 89:513–517.
78. Williams ML, Rutherford SL, Mommaas-Kienhuis AM, Grayson S, Vermeer BJ, Elias PM. Free sterol metabolism and low density lipoprotein receptor expression as differentiation markers in cultured human keratinocytes. J Cell Physiol 1987; 13332:428–440.
79. Feingold KR. The regulation and role of epidermal lipid synthesis. Adv Lipid Res 1991; 24:57–79.
80. Ibid.

81. Elias PM, Feingold KR. Lipids and the epidermal water barrier: metabolism, regulation, and pathophysiology. Semin Dermatol 1992; 11:176–182.
82. Grabauer O, Elias PM, Feingold KR. Transepidermal water loss: the signal for recovery of barrier structure and function. J Lipid Res 1989; 30:323–333.
83. Menon GK, Feingold KR, Elias PM. The lamellar body secretory response to barrier disruption. J Invest Dermatol 1992; 98:279–289.
84. Menon OK, Feingold KR, Moser AH, Brown BE, Elias PM. De novo sterologenesis in the skin. II. Regulation by cutaneous barrier requirements. J Lipid Res 1985; 26:418–427.
85. Grubauer O, Feingold KR, Elias PM. Relationship of epidermal lipogenesis to cutaneous barrier function. J Lipid Res 1987; 28:746–752.
86. Holleran WM, Feingold KR, Mao-Qiang M, Gao WN, Lee JM, Elias PM. Regulation of epidermal sphingolipid synthesis by barrier requirements. J Lipid Res 1991; 32:1151–1158.
87. Proksch E, Feingold KR, Mao-Qiang M, Elias PM. Barrier function regulates epidermal DNA synthesis. J Clin Invest 1991; 87:1668–1673.
88. Feingold KR, Brown BE, Lear SR, Moser AH, Elias PM. Effect of essential fatty acid deficiency on cutaneous sterol synthesis. J Invest Dermatol 1986; 87:588–591.
89. Holleran WM, Feingold KR, Mao-Qiang M, Gao WN, Lee JM, Elias PM. Regulation of epidermal sphingolipid synthesis by barrier requirements. J Lipid Res 1991; 32:1151–1158.
90. Proksch E, Elias PM, Feingold KR. Regulation of 3-hydroxy-3-methyl-glutaryl-coenzyme A reductase activity in murine epidermis: modulation of enzyme content and activation state by barrier requirements. J Clin Invest 1990; 85:874–882.
91. Ottey K, Wood LC, Elias PM, Feingold KR. Cutaneous permeability barrier disruption increases fatty acid synthetic enzyme activity in the epidermis of hairless mice. J Invest Dermatol 1995; 104:401–405.
92. Proksch E, Elias PM, Feingold KR. Regulation of 3-hydroxy-3-methyl-glutaryl-coenzyme A reductase activity in murine epidermis: modulation of enzyme content and activation state by barrier requirements. J Clin Invest 1990; 85:874–882.
93. Jackson SM, Wood LC, Lauer S, et al. Effect of cutaneous permeability barrier disruption on HMG CoA reductase, LDL receptor and apoprotein E mRNA levels in the epidermis of hairless mice. J Lipid Res 1992; 33:1307–1314.
94. Ottey K, Wood LC, Elias PM, Feingold KR. Cutaneous permeability barrier disruption increases fatty acid synthetic enzyme activity in the epidermis of hairless mice. J Invest Dermatol 1995; 104:401–405.
95. Choi S-J, Jackson SM, Elias PM, Feingold KR. The role of protein synthesis in permeability barrier homeostasis. In: Olikawara A, et al., eds. The Biology of the Epidermis: Molecular and Functional Aspects. Amsterdam: Elsevier, 1992:11–19.
96. Holleran WM, Mao-Qiang M, Gao WN, Menon OK, Elias PM, Feingold KR. Sphingolipids are required for mammalian barrier function: II. Inhibition of sphingolipid synthesis delays barrier recovery after acute perturbation. J Clin Invest 1991; 88:1338–1345.
97. Feingold KR, Mao-Qiang M, Menon OK, Cho SS, Brown BE, Elias PM. Cholesterol synthesis is required for cutaneous barrier function in mice. J Clin Invest 1990; 86:1738–1745.

98. Mao-Qiang M, Elias PM, Feingold KR. Fatty acids are required for epidermal permeability barrier function. J Clin Invest 1993; 92:791–798.
99. Mao-Qiang M, Brown BE, Wu S, Feingold KR, Elias PM. Exogenous non-physiological vs physiological lipids: divergent mechanisms for corection of permeability barrier dysfunction. Arch Dermatol. In press.
100. Mao-Qiang M, Feingold KR, Elias PM. Exogenous lipids influence permeability barrier recovery in acetone treated murine skin. Arch Dermatol 1993; 129:728–738.
101. Grubauer O, Elias PM, Feingold KR. Transepidermal water loss: the signal for recovery of barrier structure and function. J Lipid Res 1989; 30:323–333.
102. Lee SH, Elias PM, Proksch E, Menon GK, Mao-Qiang M, Feingold KR. Calcium and potassium are important regulators of barrier homeostasis in murine epidermis. J Clin Invest 1992; 89:530–538.
103. Menon GK, Lee S, Elias PM, Feingold KR. Localization of calcium in murine epidermis following disruption and repair of the permeability barrier. Cell Tissue Res 1992; 270:503–512.
104. Menon GK, Elias PM, Feingold KR. Integrity of the permeability barrier is crucial for maintenance of the epidermal calcium gradient. Br J Dermatol 1994; 130: 139–147.
105. Kupper TS. Immune and inflammatory processes in cutaneous tissues: mechanisms and speculations. J Clin Invest 1990; 86:1783.
106. Hauser C, Saurat J-H, Schmitt JA, Jaunin F, Dayer JH. Interleukin-I is present in normal human epidermis. J Immunol 1986; 136:3317–3322.
107. Wood LC, Jackson SM, Elias PM, Orunfeld C, Feingold KR. Cutaneous barrier perturbation stimulates cytokine production in the epidermis of mice. J Clin Invest 1992; 90:482–487.
108. Tsai JC, Feingold KR, Crumrine D, Wood LC, Grunfeld C, Elias PM. Permeability barrier disruption alters the localization and expression of TNF-alpha protein in the epidermis. Arch Derm Res 1994; 286:242–248.
109. Wood LC, Feingold KR, Sequeira-Martin SM, Elias PM, Orunfeld C. Barrier function coordinately regulates epidermal IL-1 mRNA levels. Exp Dermatol 1994; 3:56–60.
110. Nicholoff BJ, Naider Y. Perturbation of epidermal barrier function correlates with initiation of cytokine cascade in human skin. J Am Acad Dermatol 1994; 30:535–546.
111. Elias PM, Holleran WM, Feingold KR, Menon OK, Ohadially R, Williams ML. Normal mechanisms and pathophysiology of epidermal permeability barrier homeostasis. Curr Opin Dermatol 1993; 231–237.
112. Kupper TS. Immune and inflammatory processes in cutaneous tissues: mechanisms and speculations. J Clin Invest 1990; 86:1783.
113. Wood LC, Feingold KR, Sequeira-Martin SM, Elias PM, Orunfeld C. Barrier function coordinately regulates epidermal IL-1 mRNA levels. Exp Dermatol 1994; 3:56–60.
114. Tsai J-C, Feingold KR, Crumrine D, Wood LC, Grunfeld C, Elias PM. Permeability barrier disruption alters the localization and expression of TNF-alpha protein in the epidermis. Arch Derm Res 1994; 286:242–248.

115. Wood LC, Jackson SM, Elias PM, Orunfeld C, Feingold KR. Cutaneous barrier perturbation stimulates cytokine production in the epidermis of mice. J Clin Invest 1992; 90:482–487.
116. Nickoloff BJ, Naider Y. Perturbation of epidermal barrier function correlates with initiation of cytokine cascade in human skin. J Am Acad Dermatol 1994; 30:535–546.
117. Scheuplein RJ, Blank IH. Permeability of the skin. Physiol Rev 1971; 51:702–747.
118. Ghadially R, Halkier-Sorenson L, Elias PM. Effects of petrolatum on stratum corneum structure and function. J Am Acad Dermatol 1992; 26:387–396.
119. Kupper TS. Immune and inflammatory processes in cutaneous tissues: mechanisms and speculations. J Clin Invest 1990; 86:1783.

4
Petrolatum: Conditioning Through Occlusion

David S. Morrison
Penreco, The Woodlands, Texas

I. INTRODUCTION

Skin conditioning agents can be described as materials which improve the appearance of dry or damaged skin. Many of these products are designed to remain on the skin for a length of time in order to reduce flaking and to act as lubricants. These conditioning agents help maintain the soft, smooth, flexible nature of what is perceived as healthy, young-looking skin.

Occlusive skin conditioning agents perform in such a manner that the evaporation of water from the skin surface (the stratum corneum) into the external environment is substantially blocked. This occlusivity helps to increase the water content of the skin, giving it the desired supple appearance. Typically, occlusive agents are lipids (molecules which are oil-soluble, and consist predominantly of hydrogen and carbon atoms), which, due to their insolubility in water, provide the best barrier to water vapor transport. The mechanism of skin moisturization by these lipids is based on their tendency to remain on the skin's surface over time to provide a long-lasting occlusive film.

Petrolatum (also called petroleum jelly and paraffin jelly) has been known for decades to be an excellent occlusive agent. Consisting of an extremely complex mixture of hydrocarbon molecules, it is obtained from the dewaxing of refined petroleum. This semisolid, unctuous material provides a substantial barrier to moisture which is not easily breached, thus decreasing the loss of water from the skin to the environment (transepidermal water loss, or TEWL). This distinguishing attribute of petrolatum, coupled with its low cost, is the primary reason why this material is a common ingredient in skin care products. Petrolatum requires no special handling or formulation considerations, and has an excellent safety profile. These characteristics are evidenced by petrolatum's continued use for decades as both a cosmetic ingredient and a finished cosmetic product.

When compared to the role of petrolatum in skin care, its position as a hair conditioner in the United States is more limited. Currently, much of its use in hair care products is directed toward the ethnic hair care market. Prior to the 1960s and 1970s, the use of oil-based hair products was more frequent, as these articles were commonly utilized throughout the entire American population. Even today, anhydrous hair preparations are regularly used in other cultures around the world. The primary purpose of using petrolatum in hair dressings, hair grooms, and other hair conditioning products is to hold the hair in place and to add shine, neither of which can be easily (or inexpensively) obtained with any other single ingredient. Additional functions of petrolatum as a hair conditioner are to facilitate styling and to improve the texture of hair that has been damaged by chemical and physical means (such as hair treated with high-pH relaxers). Outside the regular use of petrolatum in ethnic hair products (such as pomades, hair dressings, hair "food," and brilliantines), this ingredient is used in other hair care products, primarily emulsion styling creams and lotions, and anhydrous styling creams and gels. Although petrolatum has a strong tradition in the hair care market, water-based products using synthetic ingredients as styling aids, hair-holding agents, and overall conditioners have garnered a significant share of the consumer's spending on hair care. This has been due mostly to fashion changes and, in more recent years, advancements in polymer technology and in the synthesis of new functional personal care ingredients for hair.

II. PETROLATUM: ORIGIN AND HISTORY

Having been known for thousands of years, crude petroleum (crude oil) has a varied and extensive history (1–5). It is believed to have been found, while drilling for salt, by the Chinese, who used the material in ca. 1700 B.C. for lighting. Asphalt and other heavy, nonvolatile fractions of petroleum (such as bitumen and pitch) were used more often and in earlier times, since these materials do not evaporate with age as do the more volatile components of petroleum used in lighting and as fuels. These heavy materials are thought to have been used as mortar and in other adhesive applications from before ca. 2200 B.C. to at least the second century B.C. by the Assyrians and Babylonians. Not surprisingly, much of the archeological evidence of asphalt use has been discovered in the region of southern Mesopotamia ("the cradle of civilization") between the Tigris and Euphrates Rivers, as well as in ancient Persia.

In addition to being employed for these purposes, asphalt was used around 1000 B.C. by the Egyptians in some of their mummification procedures, as a protective material and to fill body cavities.

It also has been reported that the use of asphalt was proposed by the Assyrians for some undetermined medicinal purposes, and that certain lawbreakers had their heads "anointed" with the molten material! Since few people today would consider these applications as "skin care," the date of the first known cosmetic use of petroleum products is not very clear. It is known, however, that certain other ingredients were used to enhance beauty in Biblical times (such as olive oil for hair preparations, and plant extracts and minerals for coloring purposes) (6–9). In fact, several books of the Bible make reference to the ancient custom of painting the eyelids and using mascara-like materials (such as kohl) to highlight the eyes (10). By emphasizing their eyes in this manner, women made them appear larger and more attractive. Job, who was believed to have lived around 1845 B.C., had a daughter named Keren-Happuch, meaning "container of eye paint," which also attests to the use of cosmetics during that time (11).

The manufacture of petrolatum as a specific substance occurred only after August 1859, when Edwin L. Drake drilled the first modern oil well (69 ft deep) in Titusville, Pennsylvania. Prior to this time, significant quantities of crude oil had been obtained primarily from tar sands and other surface or near-surface oil deposits.

Petrolatum was probably first used as a cosmetic (i.e., skin care) ingredient in the several years preceding Robert A. Chesebrough's patent for petrolatum (12). In this patent, titled "Improvement in Products from Petroleum," Chesebrough states that during the manufacture of petrolatum, the product's color will change over time from pure white to, eventually, a "deep claret" as the filter material ("bone black") becomes more and more saturated. The petrolatum is described as a useful material for several leather treatment processes, and is "a good lubricator, and may be used to great advantage on all kinds of machinery." In addition, Chesebrough remarks that the purest form of Vaseline (which he named this semisolid material) "is also adapted to use as a pomade for the hair, and will be found excellent for that purpose, one of its chief recommendations being that it does not oxidize." He then goes on to mention this substance's utility as a skin care treatment, citing, for example, its use on chapped hands.

The name Chesebrough chose for the petrolatum which he manufactured [Vaseline (13)] apparently was derived from the German word for water (*Wasser*) and the Greek word for oil (14). The basis for this name was a theory mentioned by Chesebrough in an earlier U.S. patent. He believed that Vaseline (i.e., petrolatum), consisting of carbon and hydrogen, was formed by the combination of hydrogen (from the decomposition of water) and carbon (from certain minerals in the earth). Today, it has been well established that petrolatum, and all the thousands, if not millions, of molecular components of crude petroleum, have their origins not from water, but from organic materials which have decomposed naturally under the surface of the earth's crust. A significant body of scientific evidence points to petroleum as having been derived from once-living organisms (plant and animal life), thus indicating that crude petroleum and its components are truly natural materials.

A short 2 years after Chesebrough's patent, a paper presented at the Philadelphia College of Pharmacy reported that petrolatum showed some promise as a pharmaceutical product and was "certainly gaining favor with physicians" (14). This was the beginning of petrolatum's use in pharmaceutical ointments, and the author of an 1875 paper published in the *Proceedings of the American Pharmaceutical Association* reported that this material was "without a superior" and was most often used in the treatment of burns and scalds. Over 100 years later, petrolatum still remains an extremely useful ingredient for similar medical applications (vide infra). In 1880, petrolatum was first listed in the *United States Pharmacopeia*, indicating this material's wide acceptance as a useful pharmaceutical ingredient.

One advantage of petrolatum cited by Chesebrough was its outstanding resistance to oxidation. Prior to Chesebrough's invention, lard (animal fat) and other unsaturated lipids were often used as ointments and salves. These materials, not being preserved with antioxidants, quickly turned rancid, often leaving the user with a discolored, smelly product. At the February 2, 1876, meeting of the Pharmaceutical Society, one speaker stated that petrolatum "may be kept indefinitely without becoming rancid, and this, together with its indifference to chemicals and its readiness to take up any perfume, is sufficient to recommend it for pharmaceutical and toilet purposes in place of the fats generally used" (15). Thus, physicians and other medical practitioners readily accepted petrolatum as a great improvement over their currently used materials. This natural immunity to oxidation is yet another characteristic of petrolatum which helps it retain its current popularity.

In the late 1800s and early 1900s, most of the developments regarding petrolatum concerned its manufacture. Improvements were made in the dewaxing of oils to yield the solid hydrocarbon residues for petrolatum production, and in the filtration of petrolatum. More recent advances in oil refining and manufacturing have simplified the production of petrolatum, improved its yield, and allowed more consistent products to be made. Even with current improvements, today's petrolatum has retained the natural characteristics of the original which Robert Chesebrough saw as being so valuable to hair and skin care. These timeless properties are what have kept this material as popular as it is today.

Table 1 Some Major Noncosmetic Applications of Petrolatum

Agricultural chemicals	Munitions
Aluminum mills and containers	Paint coatings
Animal feed	Paper applications
Automotive	Pharmaceuticals
Bait	Plastics industry
Candles	Polishes
Chemical processing	Printing ink
Concrete	Rubber industry
Electrical	Rust preventatives
Food processing	Sealants
Industrial lubricants	Solders
Leather processing	Textiles
Modeling clay	Veterinary applications

III. NONCOSMETIC USES OF PETROLATUM

In his patent on petrolatum, Chesebrough lists this material's use as a lubricant. While petrolatum is still used as a lubricant in certain industries, such as in aluminum mills and when manufacturing aluminum cans, it is used in a multitude of other products, processes, and industries. Table 1 lists many of the noncosmetic applications of petrolatum.

In the agriculture industry, petrolatum is used as a fungicide carrier and in fruit and vegetable coatings. These coating properties also allow it to be used in rust-preventative compounds for food processing machinery and in automotive undercoatings.

Its lubrication properties are useful in the automotive industry, especially during reassembly of transmission parts. In the candle industry, petrolatum is used as a lubricant and also for aiding in the dissolution of fragrances, as well as for minimizing wax shrinkage when clear jar candles are cooled. Petrolatum's characteristics enable it to be used in textile manufacture, where it is employed as a thread lubricant and in fiber finishes. Electrical applications benefit from petrolatum, where it is used as a lubricant, insulation, and rust preventative in electrical junction boxes. Petrolatum also can be used as a food-grade grease, as a rust preventative in food processing, and as a food release agent. Its use in industrial applications is primarily in applications where a remaining tacky film is desirable, such as in penetrating oils and in gasket lubricants and seals. Petrolatum's lubrication properties and its waterproofing abilities also are useful in the manufacture of munitions.

Since minerals and other additives in animal feed are usually in powder form and have a tendency to dust and separate, petrolatum is used as a grain dedusting agent and as a binder for pellets, cubes, and blocks in the processing of animal feed. It is also used in the manufacture of fishing lures and for wood impregnation to aid in preservation. Marine base paints often contain petrolatum, as do water sealants, lacquers, and primers. Petrolatum has been used for decades as a concrete curing compound. It helps retain moisture in the concrete during the curing process, which gives the final product its maximum compressive strength. Petrolatum is useful in paper manufacture, specifically for the manufacture of butcher paper, carbon paper, and mimeograph forms.

As cited by Chesebrough, petrolatum is useful in leather tanning (12). Here, petrolatum is used as an emollient to soften the hide and make it more pliable. Other leather applications include shoe polishes, leather conditioners for finished leather goods, and in waterproofing compounds for boots. Besides softening leather, petrolatum is used as a softener for modeling clay, while reducing its drying time. Petrolatum is used as a plasticizer, as a mold release agent in plastic and rubber processing, and as an additive in printing ink solvents that reduce tack. It also is used in solder flux.

In the pharmaceutical industries, petrolatum is useful in dental adhesives and in medicated ointments (both over-the-counter and prescription). It is used as a release agent for tablets and as a hospital lubricant for certain applications. Petrolatum is used as a general protective salve in veterinary applications, such as a lubricant for horses' leg shackles, and as cow udder ointment.

The unique nature of petrolatum enables it to meet the requirements of the above uses while still remaining a workhorse of the cosmetic industry. What is in this material that makes it so useful for many different applications?

IV. THE COMPOSITION OF PETROLATUM

Due to the chemical complexity of the constituents of crude oil and the vast number of different molecules in the refined fractions of petroleum, many definitions have been presented to describe the composition of petrolatum. This unctuous, semisolid material can be described as a complex, homogeneous but colloidal mixture of solid and high-boiling liquid hydrocarbons, obtained from petroleum, which is transparent when spread in a thin layer. Petrolatums typically have melting points of between 35 and 60°C (depending on the amount of solid hydrocarbon present), with molecular masses ranging from 450 to 1000. At these masses, classification of the components by molecular type becomes extremely difficult due to the almost endless possibilities of substitution and isomers of the various molecules. In fact, even the simplified nomenclature of petroleum components becomes blurred due to overlap between different species.

The components in petrolatum can be generally classified as naphthenics, paraffinics, and isoparaffinics. The naphthenics are saturated ring structures (not aromatics), the paraffinics are straight-chain hydrocarbons, and the isoparaffinics are branched-chain hydrocarbons. (Examples of these types of molecules are shown in Figure 1.) The difficulty in the discrimination of such molecules becomes evident when a ring structure contains several chains, both straight and branched. How is this molecule classified? Naphthenic because of the ring structure, paraffinic because of the straight chain, or isoparaffinic due to the branched chain? In practice, when the naphthenic and paraffinic constituents of a petroleum fraction are determined, one structure or another will typically predominate, thus allowing a determination to be made. This is done despite the inevitable fact that some molecules will always be in the "gray area" between categories. Fortunately, understanding petrolatum and its key properties does not require the determination of its constituents to extreme molecular detail.

Generally, the most practical knowledge of the composition of petrolatum is based on the amounts of solid and liquid components in the material. The solid components are obviously mineral waxes (e.g., paraffin and microcrystalline wax), while the liquid component is a heavy mineral oil. [It should be noted, however, that even this border between the solid wax hydrocarbons and liquid mineral hydrocarbons "is neither definite nor scientific" (16). One can easily identify many saturated hydrocarbon molecules which melt at or near ambient temperatures.] The paraffin waxes are commonly recognized as paraffinic components, due to their brittleness. This lack of ductility arises from the ease by which the paraffinic molecules can align themselves and crystallize, due to the overall lack of significant branching. On the other hand, microcrystalline waxes are isoparaffinic, will not crystallize easily due to molecular branching, and so are not as brittle as the paraffin waxes.

Despite understanding the oil and wax composition of petrolatum, it is known that "synthetic" petrolatum cannot be created by simply combining paraffin wax and mineral oil. When this is attempted, the blend does not remain uniform and separates. Thus, it was believed at one time that a third substance was present which kept the wax-and-oil mixture stable. This material was named "protosubstance" by F. W. Breth, who claimed the discovery of this component in an unpublished study in 1925 (17).

The "protosubstance" was obtained by repeatedly extracting petrolatum with acetone, but no adequate explanation for how protosubstance works was ever found. In addition, no amount of the extracted protosubstance could convert a mixture of mineral oil and wax into petrolatum. Today, while there still may be some believers in the protosubstance theory, the general consensus is that no substantial scientific evidence has been published which conclusively proves the existence of protosubstance and the requirement that it be present for petrolatum to exist. Therefore, claims to the presence of protosubstance are usually met with skepticism.

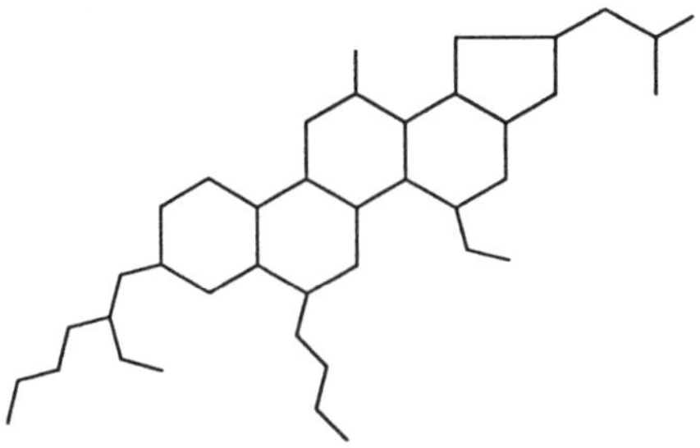

Naphthenic Hydrocarbon

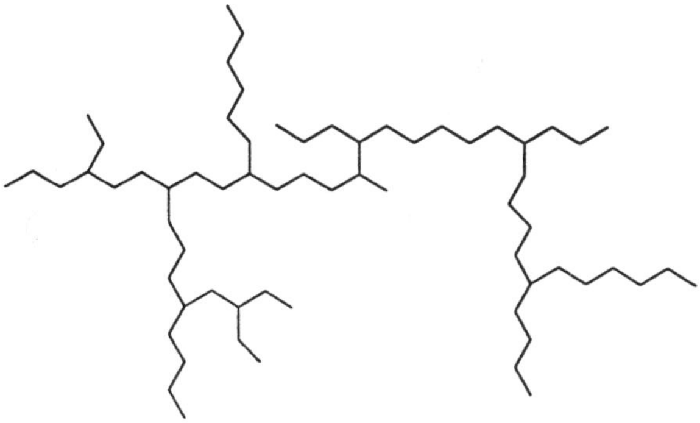

Isoparaffinic Hydrocarbon

n-$C_{36}H_{74}$

Paraffinic Hydrocarbon

Figure 1 Examples of hydrocarbon types found in petrolatum.

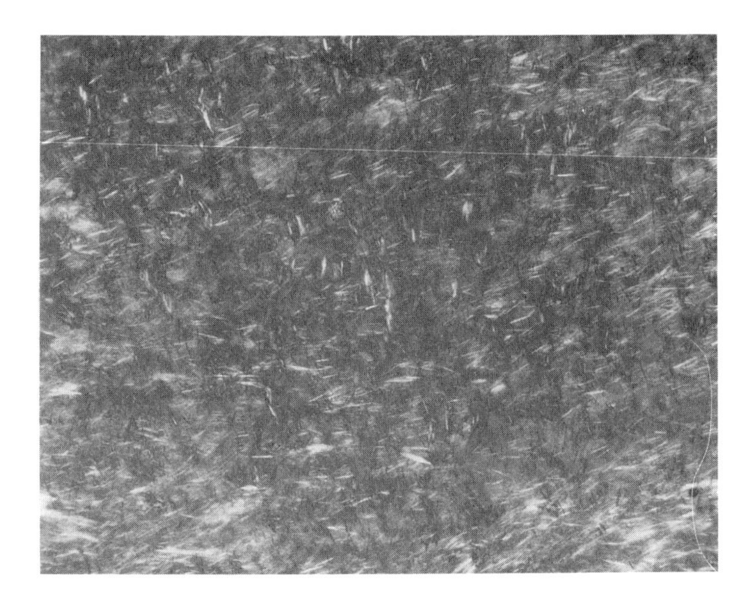

Figure 2 Microphotograph of petrolatum (100×). Note the wax crystals.

Figure 3 Microphotograph of petrolatum (500×). Wax crystals are clearly seen.

Although mineral oil and paraffin wax cannot be mixed to create petrolatum, various types and grades of petrolatum can be prepared which contain differing amounts of wax and oil. This allows numerous blends to be prepared, each designed to meet a specific need. Obviously, petrolatums with high levels of wax have high melting points and hard consistencies, while those with more oil are softer and more fluidlike. The waxes which are present in petrolatum are clearly evident when samples are viewed under a microscope. Figure 2 shows a microphotograph (100×) of a petrolatum, while Figure 3 is the same material photographed at 500×. In both photos, the wax crystals can be seen quite easily.

In order for petrolatum to be labeled "Petrolatum, U.S.P." or "White Petrolatum, U.S.P.," it must conform to the requirements described in the *United States Pharmacopoeia*. One of the tests it must meet is a consistency test. Since adding too much oil or too much wax to a petrolatum may cause the final product to fail this U.S.P. requirement, care must be taken when blending petrolatums to ensure that the product's consistency does not fall outside the required range if the final material is to be U.S.P. petrolatum. Note that the U.S.P. also has other requirements for petrolatums, but these are easily found in the *United States Pharmacopoeia* and will not be discussed here, as they are outside the scope of this chapter.

V. REFINING AND PRODUCTION

As stated previously, petrolatum is a purified, semisolid mixture of hydrocarbons taken from petroleum (crude oil). Since petroleum is obtained from the earth, and since the materials it contains are simply refined (i.e., separated from impurities) and not synthesized, petrolatum can be considered a natural material in the truest sense of the word (18).

The exact method of production of petrolatum varies depending on several factors. These factors include the type of crude oil used (the chemical compositions of crude oils from different sources vary considerably, as do the boiling range and other physical properties which affect the conditions used to refine the petroleum) and the types of petroleum products desired at the end of the refining process. However, a general process for the manufacture of petrolatum, from crude oil to the final product, is described below.

Figures 4, 5, and 6 show the general scheme of a petroleum refinery process, with focus on the production of petrolatum. First, a crude oil is subjected to atmospheric distillation, which removes gases and lighter refined products (i.e., fuels, such as gasoline, kerosene, and diesel fuel) from the bulk of the crude oil product at atmospheric pressure. The remaining oil is sent to a vacuum distillation unit so that the heavier fractions (e.g., lubricating oil fractions) can be removed without the extreme temperatures which would be required

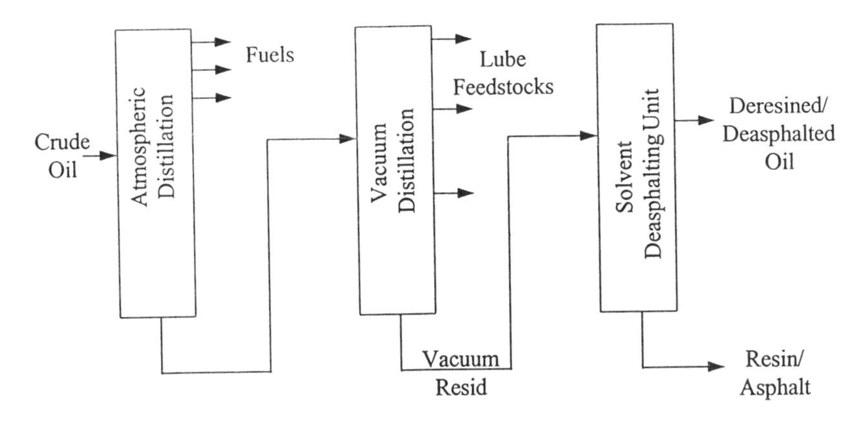

Figure 4 Petroleum refinery process No. 1.

to perform this operation in an atmospheric distillation tower. Once these lube oil fractions are removed, the remaining material, often called "vacuum resid," is then taken to a solvent deasphalting/deresining unit. In this process, a solvent is used to extract the heavy oil from resins and asphalts which would be detrimental components in the finished petroleum products.

The deresined or deasphalted oil is purified further by one of two methods (Figure 5). It can be hydrotreated, which is simply a hydrogenation step to convert unsaturated molecules to saturated ones and to remove heteroatoms such as sulfur and nitrogen compounds. Another method is to extract these same impurities by mixing the oil with a suitable solvent. The more polar materials

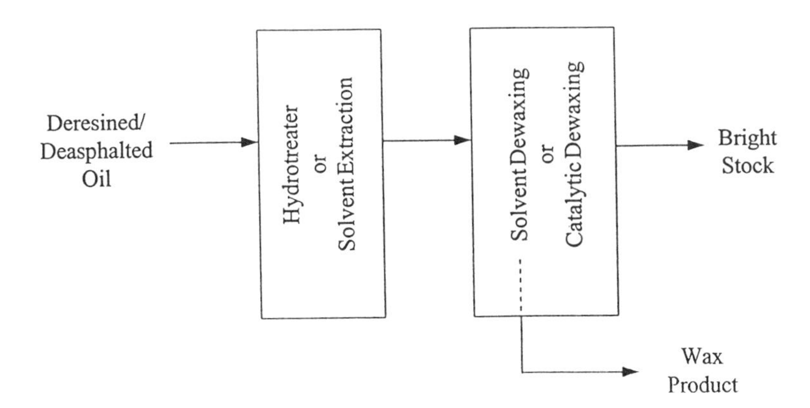

Figure 5 Petroleum refinery process No. 2.

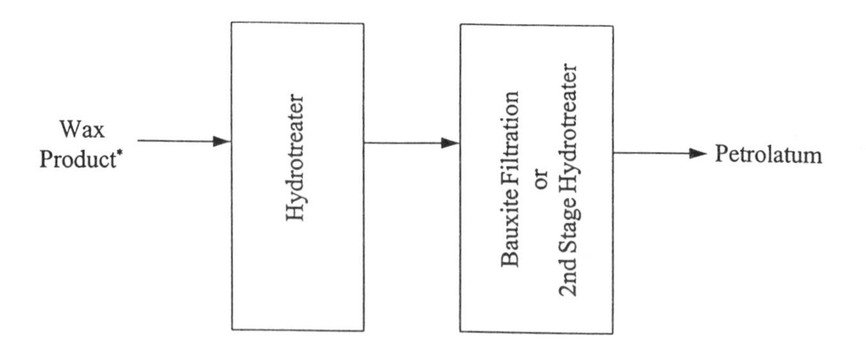

*Blended with other components as needed to make various grades and types of petrolatum.

Figure 6 Petroleum refinery process No. 3.

(unsaturated and heteroatom compounds) will preferentially be dissolved in the solvent, thus leaving behind an oil which is essentially free from unwanted components. Following this process, the oil is treated in a dewaxing step.

The dewaxing of oil can be done in one of two ways. When the goal of the refinery is to minimize production of waxes and wax-containing materials, a catalytic dewaxing process is used. In this case, the oil is treated with a metal catalyst which causes the waxes to break apart into smaller molecules. Depending on the size of these newly created molecules, they can be captured as petroleum gases, fuels, or lubricating oils. This is an efficient method to maximize the output of these materials.

When petrolatum or waxes are to be produced, this step is a solvent dewaxing, wherein the oil is heated and mixed with a hot solvent. The waxy material from the oil will be dissolved in the solvent, which then is cooled and/or centrifuged to remove the waxes. The heavy oil which remains is essentially free of waxes which could limit the oil's use as a lubricant. This oil is called bright stock and often is used as a component in high-viscosity oil products. The wax product yielded by this process is then further refined as shown in Figure 6.

Depending on the desired properties of the final petrolatum material, the wax product obtained from the solvent dewaxing step can be blended with other components such as oils and other waxes. This wax product or wax product blend is hydrotreated once again to further remove unsaturation and heteroatom compounds, followed by either bauxite filtration or a second hydrotreatment step to remove the ever-decreasing amounts of impurities. This will produce a petrolatum which is pure enough to meet U.S.P. standards and is safe enough for direct use on human lips.

Different "variations" of petrolatum can be obtained (higher or lower melting points, different consistencies, etc.) by changing the formula of the wax product which is to be purified, as shown in Figure 6. Another method of preparing different petrolatums involves simply blending the final petrolatum product with wax and/or oil. Since the petrolatum is already present as the base material, additional wax or oil can be mixed without the aforementioned problems of separation which are encountered when attempts are made to prepare petrolatum from just oil and wax.

Of course, in every step along this lengthy production process, the refinery operator can make adjustments which will vary the types and amounts of products formed. Thus, the process is quite flexible and allows the maximum utilization of crude oil while simultaneously providing maximum yield of the desired petroleum products. Such an efficient process enables the refinery to make multiple salable products with essentially no waste. Given the cost of crude oil, it is not surprising that the petroleum refinery recycles the processing material as much as possible and utilizes every drop of the crude oil.

VI. APPLICATIONS OF PETROLATUM

A. Skin Conditioning as a Cosmetic Ingredient

Two key characteristics of many skin conditioning agents are that they eliminate dry skin and maintain moist, healthy skin. Petrolatum is one ingredient which has performed these roles superbly for decades. In fact, the first public document which described the benefits of petrolatum on dry skin was probably Chesebrough's 1872 patent (12), in which he cited its excellence for use on chapped hands. The current literature overflows with evidence that petrolatum acts on the skin to make it more moist, supple, and both visually and physiologically appealing. Even newspapers and magazines tout the benefits of spreading petrolatum onto dry skin, especially during the winter season when the need for skin moisturizers is great due to a combination of low humidity, indoor heat, and reduced fluid intake.

While some studies indicate possible pharmacological mechanisms involving petrolatum and the skin, this material's excellent occlusivity is the primary reason why it is used as a skin conditioner. Blank showed in 1952 that water in the stratum corneum is a primary factor for supple, smooth, healthy-looking skin (19); therefore, ingredients which are effective at keeping moisture in skin (such as petrolatum) will undoubtedly be excellent skin conditioners.

1. General Information

This conviction is widely held throughout the cosmetic industry. Petrolatum is known as an excellent moisturizer, and this "moisturization" (i.e., reduction of transepidermal water loss, TEWL) of the skin by petrolatum arises from its occlusion. Thus, it is frequently found in many skin care formulations, especially since liquid occlusive agents (e.g., lipophilic oils) are less effective moisturizers than petrolatum (20). Hydration of the stratum corneum, such as that provided by petrolatum, may reach all strata of the skin under certain circumstances, which clearly relates to a favorable skin physiology (21).

Numerous authors have cited petrolatum's conditioning of the skin via occlusivity. Steenbergen reported that petrolatum is the most efficient substance for retaining moisture in the skin and allowing it to hydrate to the point where the dry skin condition is overcome (22). A recent column by Fishman (23) stated that, for treating dry skin, petrolatum is a "classic emollient" and is "extremely efficient" at doing its job. White petrolatum was noted as being the best occlusive moisturizer in an article by Fisher and co-workers (24). Additionally, white petrolatum was recommended as the ideal moisturizer for people with abnormal skin, defined as that segment of the population with cosmetic dermatitis, since petrolatum is such a good moisturizer and it needs no preservative. An excellent paper on dry skin also has been published by Idson (25). In this fairly comprehensive article, the author reviews dry skin, the structure and function of skin, emolliency, and moisturization. Even though oils which contain fatty acid glycerides contribute to overall skin flexibility, Idson writes, their occlusion (and thus moisturization) falls far short of that provided by petrolatum. The efficiency and superiority of petrolatum as a skin conditioning agent was once again referred to, as this material was called "the most efficient occludent and emollient for protecting dry skin and allowing it to hydrate again." A chart in this paper placed petrolatum at the top of a list of emollient preparations, based on the ability to "protect" dry skin.

It should be noted that the ability of petrolatum to protect and moisturize dry skin also applies to lips. This occlusive substance is probably the most commonly used ingredient in lip balms, and in some instances, petrolatum is essentially the only ingredient. Chapped lips (i.e., "dry skin") are ideal sites for petrolatum application due to both its safety as well as its skin conditioning and moisturization.

Petrolatum also was mentioned in an interview in a 1990 issue of *Dermatology Times* regarding nonprescription products for improving the appearance of skin, hair, and nails (26). This article stated that traditional moisturizers are products which typically contain occlusive ingredients, such as petrolatum, which reduce water vapor loss from the skin.

2. *In-Vivo Studies*

These and other reports of skin-hydrating effects from the occlusivity of petrolatum have been proven in clinical trials using human subjects. These in-vivo studies conceivably give the best verification of the skin conditioning ability of petrolatum, since they are as close to "real-world" situations as are typically available.

In 1978, Kligman published an article describing a regression method for evaluating moisturizers (27). This method, which has gained wide acceptance (28), was used to assess the efficacy of a number of moisturizers (both formulated products and ingredients). Not surprisingly, petrolatum was found to be the best moisturizer, which was defined as "a topically applied substance or product that overcomes the signs and symptoms of dry skin." The data Kligman obtained from the evaluation of ingredients and finished products enabled him to conclude the following: "When it comes to efficacy, petrolatum is the unrivaled moisturizer. No material in our experience exceeds it in relieving ordinary xerosis." Some of the materials used in this study were petrolatum, mineral oil, olive oil, a commercial hand cream, and a commercial face cream. These were applied to dry skin once daily for 3 weeks. Lanolin also was studied and was applied twice daily for 3 weeks. The results, shown in Figure 7, indicate that petrolatum is a superior moisturizer both at the end of treatment (3 weeks) and at 4 weeks (1 week beyond the end of treatment). The petrolatum gave the greatest decrease in xerosis grade (the greatest improvement in dry skin conditions) from pretreatment to the end of treatment, as well as to 1 week after treatment had stopped. Thus, petrolatum was shown to

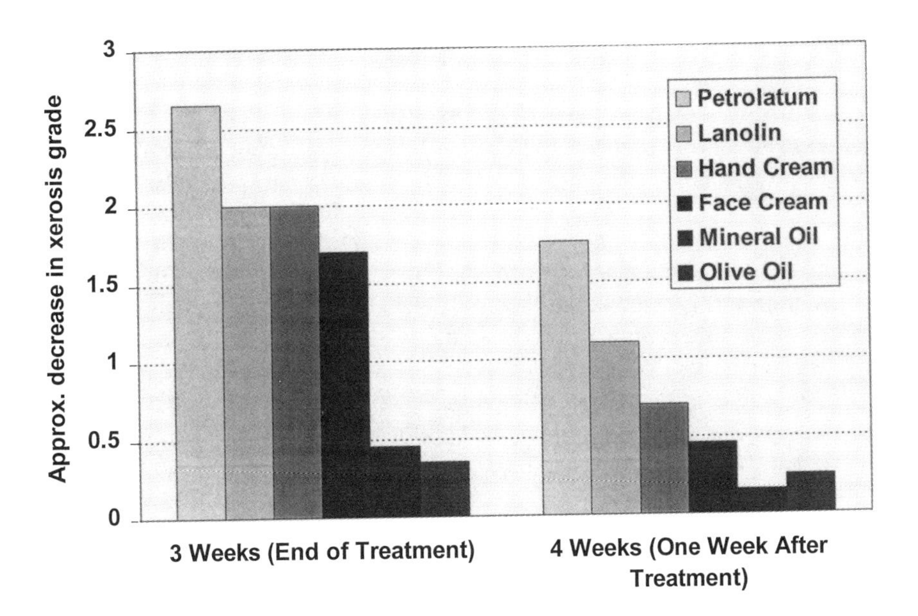

Figure 7 Approximate decrease in xerosis grade (from pretreatment grade) of various materials. A larger value means a greater improvement in xerosis (27).

exhibit excellent relief for dry skin both during and after application of the material.

Kligman also stated that the mode of action of petrolatum is not just occlusion of moisture from the skin, since wrapping the skin with an impermeable plastic film does not benefit dry skin. He theorized that petrolatum may have some pharmacological effects which contribute to its role as a moisturizer, but the mode of such action has not been determined.

If such pharmacological effects do occur, they are likely confined to the skin surface and stratum corneum. Elias and co-workers have shown that hydrocarbons such as petrolatum, when applied topically, do not penetrate to the deeper layers of the skin (29). This holds true for both intact and acetone-damaged skin. When skin is damaged by acetone, it was found that petrolatum (applied topically) permeates all levels of the stratum corneum, implying that, in addition to moisturizing the skin, it helps repair the damaged tissue (30). The petrolatum actually increased the rate of healing the damaged skin, whereas a vapor-impermeable film impairs this process. Like Kligman, these authors concluded that petrolatum is not acting only as an inert, occlusive barrier to skin moisture.

In 1992, Loden published a study on the increase in skin hydration after lipid-containing materials were applied (31). The materials tested were petrolatum, an oil-in-water cream containing an unusually high concentration of lipids (66%), and an oil-in-water cream containing 27% lipids. The occluding properties of these materials were determined by application to skin, followed by determination of TEWL 40 min later. This author found that petrolatum reduced water loss from the skin by nearly 50%, whereas the other materials reduced water loss by only 16% (Figure 8). While this result is expected based on the well-known occlusivity of petrolatum, one of the most interesting aspects of this study was the evaluation of TEWL following *removal* of the products 40 min after application. In this case, petrolatum gave a very high TEWL relative to the oil-in-water creams, indicating that the skin was hydrated by petrolatum to a greater extent than by the creams. Not only was the occlusion of petrolatum determined, the direct hydration of the skin by this occlusion was verified as well.

In another study on the effectiveness of cosmetic products and their ability to alleviate dry skin conditions, petrolatum was evaluated as an emulsified ingredient. This emulsion was compared to other emulsions which contained either urea or alpha-hydroxy acids (AHAs) (32). The panelists had their skin analyzed over a 4-week period by "expert assessors," who judged the severity of skin dryness visually. It was found that an emulsion containing 15% petrolatum was equivalent to a 10% urea emulsion in reducing dryness when measured weekly over 4 weeks. A second 4-week study compared the 15% petrolatum emulsion with an emulsion containing 6% AHAs, and again, the two

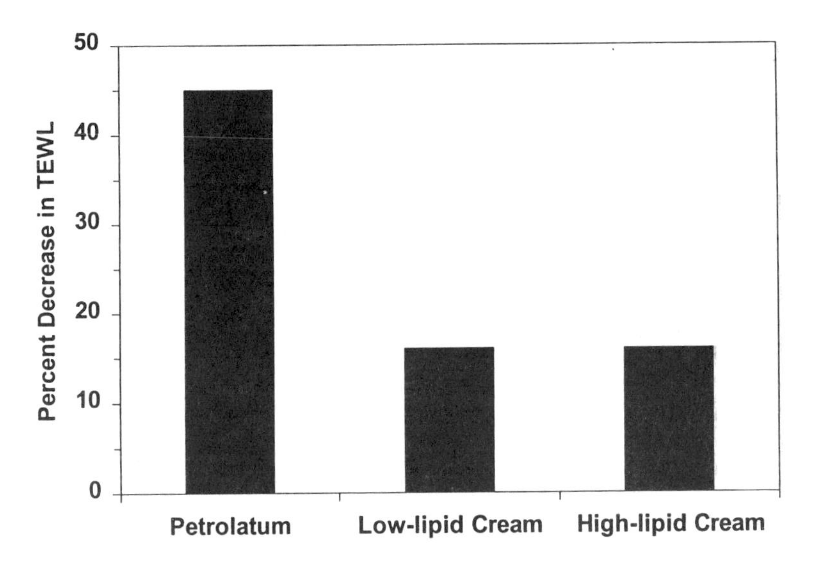

Figure 8 The effects of three emollients on TEWL (31).

products were essentially identical in performance as determined by a reduction in the visual characteristics of skin dryness. These results suggest that routes other than occlusion may be used to reduce the appearance of dry skin, but it is evident that the decades of proven safety and the economy of petrolatum would invite formulators to select it over the urea and AHAs when developing an emulsion product designed primarily to eliminate or reduce dry skin.

The occlusivity of "oil films" was the subject of a 1979 publication by Tsutsumi and co-workers, who compared three physical forms of hydrocarbons and their reductions in TEWL when applied to human skin (33). The hydrocarbons they studied were liquid (mineral oil), solid (paraffin wax, m.p. 48°C), and semisolid (petrolatum). These authors found that for all three materials, increasing the amount of hydrocarbon applied to the skin increased the occlusivity as measured by a Servo Med evaporimeter, but the occlusivity eventually leveled off. While this finding is not unexpected, comparing all three materials at the same concentration was especially telling. Figure 9 shows the results of this study when 1 mg of hydrocarbon was used per square centimeter of skin (applied as a solution in a volatile solvent) and TEWL (occlusivity) was measured at 15 and 60 min after application. This figure indicates that petrolatum clearly outperformed the other materials. Although the solid paraffin hydrocarbon showed very good occlusivity when applied uniformly, its tendency to crack (and thus lose its barrier properties) and overall poor esthetics do not lend it to common cosmetic use as a primary emollient/moisturizer/conditioning agent.

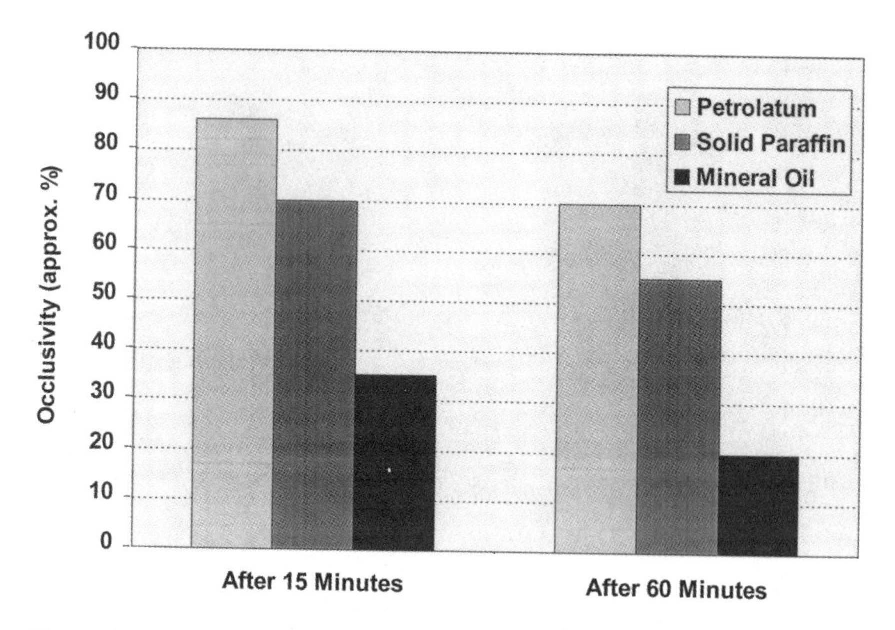

Figure 9 Barrier properties of three forms of hydrocarbon (33).

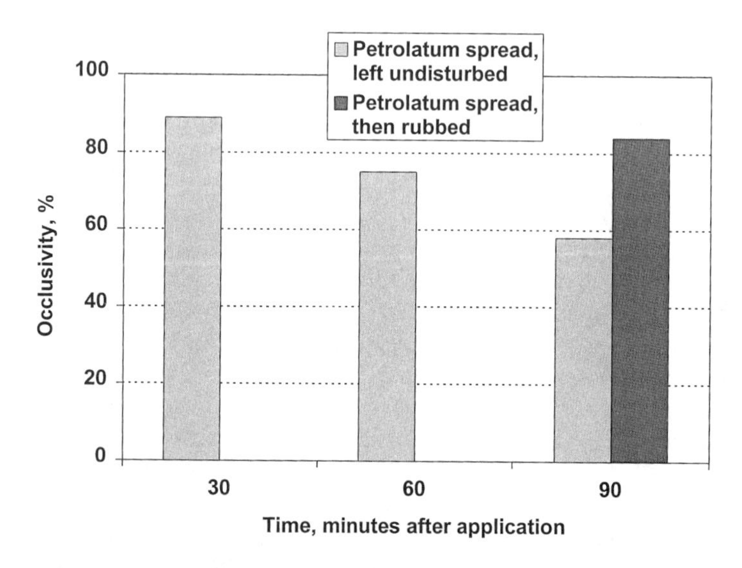

Figure 10 How rubbing petrolatum 90 min after application affects its occlusivity (33).

In a noteworthy sidelight to this study, the authors also reported on the effect of rubbing petrolatum once it had been applied to the skin and left undisturbed for a period of time. A sample of petrolatum was placed on the skin by simply spreading it into a film. The occlusivity of this film was measured at 30, 60, and 90 min after application. The occlusivity dropped from 89% (at 30 min) to 58% (at 90 min). However, when this film was rubbed at 90 min after application, the occlusivity *increased* to 84% (Figure 10)! What is occurring here? The authors suggest that the film of petrolatum may become discontinuous over time. Another possibility is that the water vapor begins to create molecular scale "channels." In either case, an increase in water vapor permeation is taking place. Whatever the reason for this increase in vapor permeation, rubbing the film restores occlusivity by creating a more solid, continuous film, thus reducing the vapor transport across the petrolatum.

A paper by Rietschel published in *The Journal of Investigative Dermatology* in 1978 also reported on the in-vivo evaluation of skin moisturizers (34). This author determined that occlusion (such as with petrolatum) is an effective method for treating dry skin, with or without hydration of the skin prior to application of the occluding material. The suppression of moisture loss from the skin was plainly seen in Rietschel's studies and is shown in Figure 11, taken from Rietschel's paper.

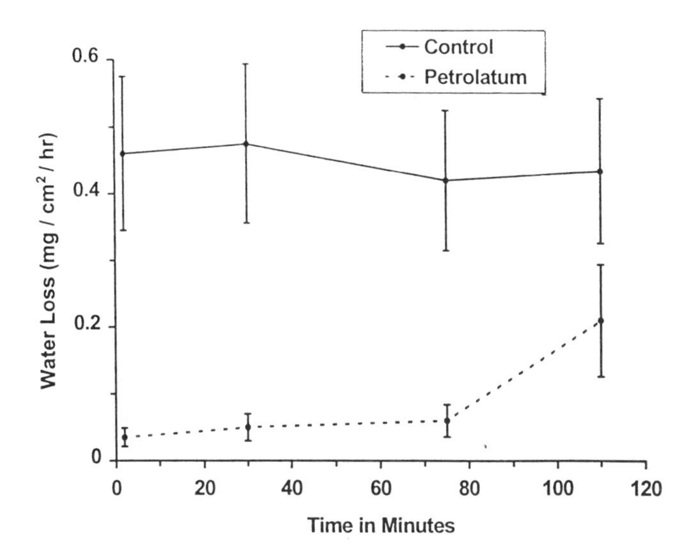

Figure 11 The suppression of detectable moisture by petrolatum (0.02 mL petrolatum applied to 6.25 cm^2 of normal skin) attributed to occlusion. (Reprinted from Ref. 34 with permission.)

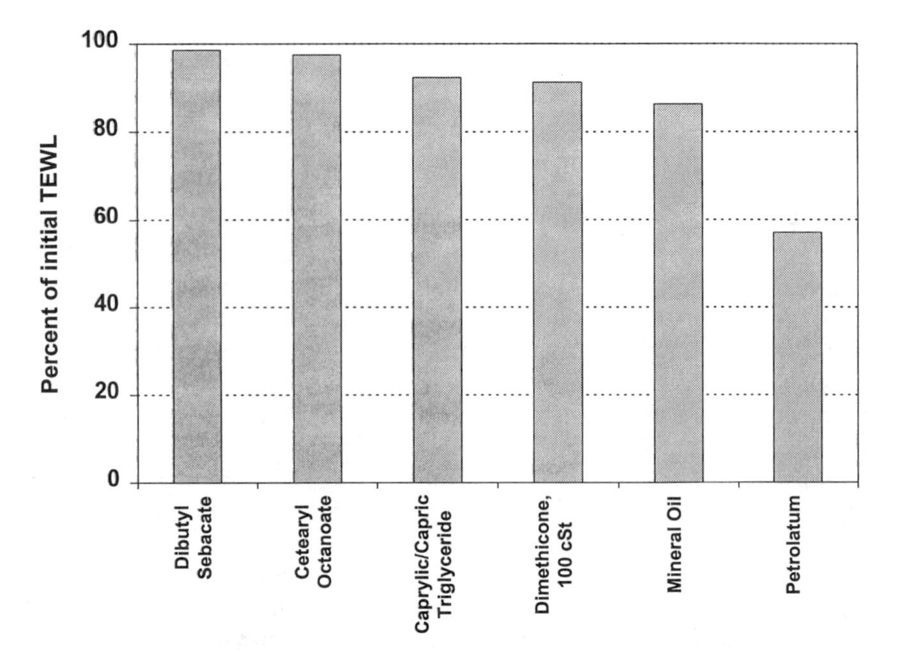

Figure 12 In-vivo moisturizing abilities of various lipid emollients as measured by TEWL (36). Petrolatum gives a greater reduction in TEWL than the other emollients.

This occlusivity of petrolatum has been evaluated in other studies. For example, hairless rats have been used in an evaluation of the hydrating effects of cosmetic preparations (including petrolatum) as measured by cutaneous impedance (35). An initial increase in impedance is seen with petrolatum, which corresponds to this material's resistivity. This resistivity indicates that the petrolatum is highly impermeable to water, and the intensity and duration of this impedance increase was determined to be proportional to the amount of petrolatum used.

Frömder and Lippold have published a study of water vapor transmission and occlusivity using both in-vivo and in-vitro techniques to evaluate lipids often used as ointment ingredients (36). White petrolatum was found to be the best lipophilic excipient at reducing TEWL in vivo (Figure 12) and at preventing water vapor transmission in vitro when compared to mineral oil, dimethicone, and several emollient esters (Figure 13).

3. *In-Vitro Studies*

This often-cited reduction of water vapor permeation by petrolatum has been detected by others using in-vitro studies. One particular article has praised the

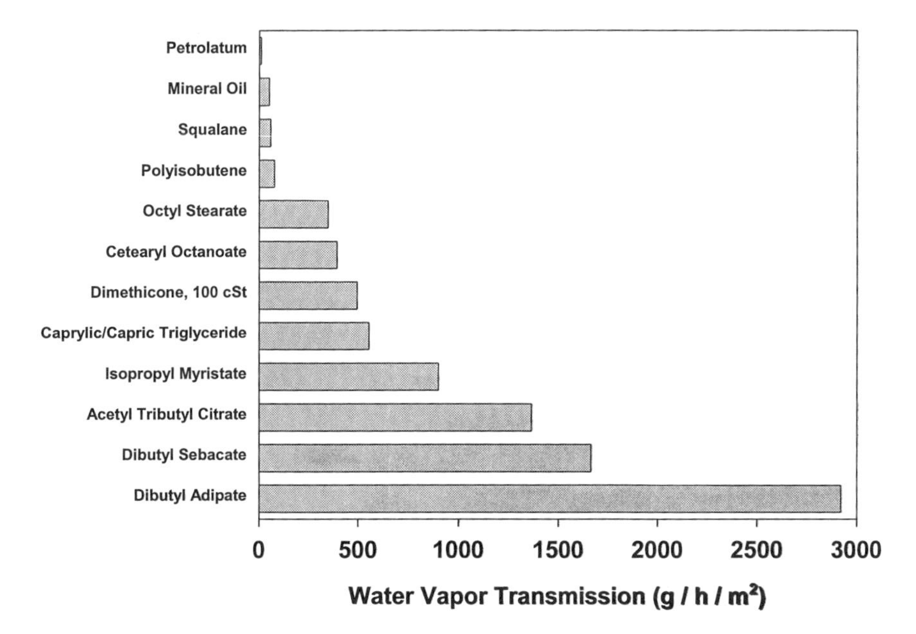

Figure 13 A study of in-vitro water vapor permeation (36). Notice that the nonpolar hydrocarbons (especially petrolatum) are significantly better at preventing water vapor transmission than dimethicone and the more polar esters.

moisturizing properties of mineral oil and petrolatum based on such studies. Tranner and Berube, in 1978, reported the results from a test which measured the reduction of water loss by various materials across a nylon film used as a skin substitute (37). The results from this screening test (performed at 20% relative humidity) are shown in Figure 14 and verify the superior ability of petrolatum to prevent skin water loss by acting as an occlusive vapor barrier.

In a study on the effects of petrolatum, white mineral oil, and gelled white oils on stratum corneum lipids, Friberg and Ma showed that petrolatum is retained in the outer layer of skin (38). Using a liquid crystal model of the stratum corneum, these authors indicated that the petrolatum reaches all levels of the stratum corneum, but it is not incorporated in the layered lipid structure of the skin. The strength of petrolatum as a skin moisturizer by acting as a barrier to vapor permeation also was noted.

Another in-vitro test method was developed by Obata and Tagami (39) and has been used to assess topically applied skin moisturizers. This method uses human stratum corneum combined with moistened filter paper. This unique apparatus gave a water gradient across the stratum corneum comparable to that found in vivo. Using high-frequency conductance measurements, it was

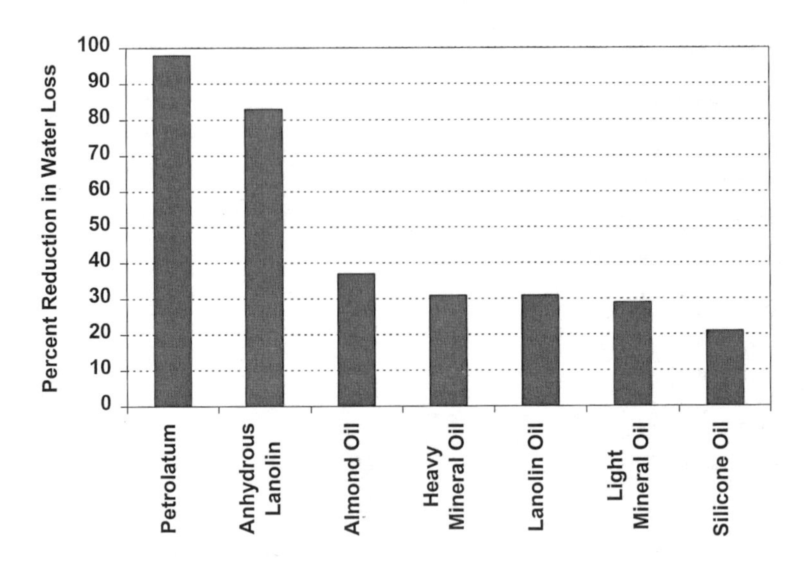

Figure 14 Measurements of water vapor loss across a nylon film treated with various lipids (37). Petrolatum provides the best barrier to water vapor, so it reduces water vapor loss by a greater percentage (>95%) than the other lipids.

determined that an oil-in-water emulsion and a 10% urea cream both immediately increased hydration of the stratum corneum, followed by a decrease in hydration over time. This immediate increase in hydration was ascribed to delivery of water to the stratum corneum by the high water content of these two emulsions. Conversely, a large increase in hydration was not seen by petrolatum, but the hydration of the skin model slowly increased for up to 1 hr. At 1 and 2 hr after application of the test material, petrolatum outperformed each of the emulsions in skin hydration as determined by this test method. While it does not give an immediate boost to skin moisture, petrolatum hydrates the skin within minutes to a greater extent and is much "longer-lasting" due to its occlusivity. From this information alone, an obvious scenario presents itself to the skin care formulator: obtain the "best of both worlds" by incorporating petrolatum in an oil-in-water emulsion.

B. Skin Conditioning as a Medical Ingredient

Occlusion and the subsequent moisturization of skin are properties furnished by petrolatum which give it wide acceptance as a skin conditioner. These benefits of petrolatum also have encouraged physicians and others in the medical field to look to this material as an ingredient for use in topical therapeutic products. For example, petrolatum has been recommended as a moisturizer for diabetics with dry feet (40). The unique occlusive properties of petrolatum also have enabled it to be used frequently as a healing aid, primarily as an ingredient in wound care dressings. Previously, it was noted that petrolatum helps repair damaged skin (30). These authors are not alone in their perception that petrolatum is good for traumatized skin.

The fact that petrolatum bridges the gap between simple skin moisturization and the treatment of wounded skin was nicely summarized by Kligman and Kligman (41). They stated that this material is "a work-horse moisturizer in a variety of settings, for instance after chemical peels and dermabrasions, superficial burns, skin grafts, atopic dry skin and other xerotic rashes." The established use of petrolatum in skin care, including medical applications, was reiterated very clearly by Kligman not long thereafter (42). He noted that petrolatum has many uses in the protection of skin against chemical and physical trauma, such as in "the treatment of cuts, abrasions and burns. It is also a moisturizer par excellence."

1. Uses on Burns and Wounds

The recommendation to use petrolatum on skin burns goes back many years. It is very likely that this application of petrolatum was initiated not long after Chesebrough's patent, when the medical and pharmaceutical communities discovered the advantages petrolatum had over the commonly used (at that time) lard-based ointments. In the 1940s, petrolatum was still being used in the treatment of skin burns. A textbook of laboratory experiments in organic chemistry recommends using borated petrolatum dressings to treat physical burns, as well as for chemical burns such as bromine and organic materials (e.g., phenol) (43).

Petrolatum has been found to be a favorite burn dressing ingredient for many medical professionals. However, due to the painful, slow-to-heal nature of severely burned skin and its tendency to stick to many burn dressings, medical researchers often look for new or improved materials to be used for burn dressings. In studies like these, the experimental dressings are compared to a standard, which often is a petrolatum-impregnated gauze (44–46). Similarly, petrolatum-based materials have been used as comparisons in the development of dressings for skin graft sites (47,48). One alternative burn dressing which has been developed is a cellulose product which is impregnated with a petrolatum-containing emulsion, and is claimed to be nonadhering (49). Petrolatum also has been claimed as a base for a medicinal salve which can be incorporated into a gauze dressing, then applied directly onto traumatized skin (50), and as a base for a topical antibiotic ointment (51). Both products are said to be useful for treating epidermal trauma, including burns and wounds.

Other wound care applications of petrolatum include the use of this material as a dressing in major cancer surgery of the head and neck (52), and as an ointment base in the removal of skin grafts from the scalp (53). The ointment promotes speedy hemostasis of split-thickness skin donor wounds, thus allowing additional skin grafts to be easily taken from nearby areas, without obstruction or delay caused by bleeding from the earlier donor sites.

2. Uses to Protect Skin

Both "petrolatum" and "white petrolatum" have been recognized by the U.S. Food and Drug Administration (FDA) as over-the-counter (OTC) skin-protectant drug products suitable for topical administration (54). When present at 30–100% of the final product, these materials may be considered OTC skin-protectant active ingredients. The FDA describes a skin protectant as "a drug which protects injured or exposed skin or mucous membrane surface from harmful or annoying stimuli."

This skin protection feature has been exploited in many ways. For example, the moisturizing ability of petrolatum has been used to great benefit in protecting the skin of premature infants (55). This protection comes primarily in the form of enhancing the epidermal barrier, since this barrier does not become functionally mature until between 32 and 34 weeks' estimated gestational age. The topical application of a preservative-free anhydrous ointment (ingredients: petrolatum, mineral oil, mineral wax, and wool wax alcohol) was found to improve the health of premature infants significantly. The patients who received this treatment had decreases in three areas studied: TEWL, the severity of dermatitis, and bacterial colonization of axillary skin. In addition, the authors found that these infants also had fewer microorganism-positive cultures of blood and cerebrospinal fluid. It was postulated that the impaired epidermal barrier of premature infants is an easy route by which bacterial infection can occur, and the application of the ointment either serves as a protective skin barrier or causes the skin to improve its natural barrier. By reducing dermatitis, the ointment also prevents skin cracking which could permit easier bacterial invasion. Thus, the application of preservative-free topical ointment was shown to be beneficial and was recommended therapy for premature infants.

In other areas of skin protection, petrolatum is not used to prevent water vapor from exiting the skin into the environment; rather, its primary function as a skin protectant is to stop environmental materials from entering the skin. The skin conditioning result of a protectant and lubricant is that the skin remains healthy and is protected from harmful external influences such as friction, chemicals (e.g., detergents), and even dry air.

The use of petrolatum as a skin protectant in a medical setting is often as a barrier for incontinent patients. Petrolatum acts as a barrier on the skin from toxins and excessive moisture in these situations, which is especially important since human stool contains caustic enzymes and pathogens (56). Petrolatum has been claimed as a base for an ointmentlike material which is useful as a skin protectant against other biochemical hazards, especially on the hands of medical professionals under surgical gloves (57).

The protection afforded by petrolatum against nonbiogenic chemical hazards also has been documented. A 1990 article reported on the evaluation of materials which were tested for chemical penetration by FT-IR spectroscopy (58). The length of time for ethyl disulfide to penetrate through a barrier thickness of 0.45 mm was determined, and it was found that petrolatum resisted penetration for 22 min, while a polyunsaturated fat allowed ethyl disulfide through the barrier after only 5 min. The effective skin protection characteristics of petrolatum also have been shown in studies involving exposure to sodium lauryl sulfate (59) and industrial sealants (60).

One of the most surprising reports about petrolatum's skin protection was published by Kligman and Kligman in 1992 (41). This article referred not to petrolatum's barrier properties against physical or chemical hazards, but to its protection against UVB-induced skin damage. These authors found that when petrolatum was applied to mouse skin prior to UVB irradiation, a 95% reduction in tumor yield was found. In addition, the incidence of tumors decreased by 21% when petrolatum was applied after irradiation. It was concluded that the number of sunburn cells were reduced by the petrolatum, and that this material also protects against damage to the skin.

3. Skin Lubricant Uses

The greaselike consistency of petrolatum and its crude oil origins are obvious indications that this material can perform as a skin lubricant. This property, while certainly meaningful in general skin care, becomes extremely important in the medical field. The prevention of ulcerated skin such as blisters and bed sores is often an ongoing battle in hospitals all over the world.

This greasiness of petrolatum has often been cited as a beneficial quality in skin lubrication. It has been noted that the greasier (or heavier) the product, the better it acts as a lubricant (20). Additionally, these authors stated that petrolatum was the recommended lubricant for bedridden patients. It also can be used as a rectal thermometer lubricant for children (61,62), and petrolatum has been claimed (likely for its lubrication and protection) as part of a low-irritation shave cream formula (63).

Lubrication of the skin is very important in many athletic activities, and petrolatum has been frequently recommended (64,65). Runners (66,67) and backpackers often use petrolatum as a blister preventative on their feet, while the lubricating and skin protecting properties of petrolatum are utilized by open-water swimmers to prevent chafing (68).

C. Hair Conditioning with Petrolatum

As mentioned previously, the primary use of hair conditioning with petrolatum is in the ethnic hair care market, as the main ingredient in anhydrous pomades, solid and semisolid brilliantines, opaque anhydrous hair dressings, and solid pressing oils (69–73). Since pomades are based almost exclusively on petrolatum, they can sometimes be difficult to remove from the hair. Thus, emulsifiers are sometimes added to the pomade to facilitate this process. Some cosmetic manufacturers have even been producing "rinsable pomades," which allow the user to remove the pomade with a minimum of shampooing. Anhydrous pomades typically are used without any other hair treatment, after washing the hair or after chemically treating the hair (74). Solid and semisolid brilliantines are used in a manner similar to pomades.

Pressing oils are added to the hair immediately before the hair is straightened, typically via a hot comb which stretches out the oiled hair. Although this is an effective method for straightening hair, it does not last very long. (Relaxers are generally preferred for straightening hair for longer time periods.)

The mechanism of how petrolatum conditions in these types of products is straightforward: physically holding the hair in place, providing lubrication for ease of styling and combability, enhancing gloss (oil-based products are naturally shiny), and creating a barrier to protect the hair from breakage, external environmental damage, and to keep it from drying out. Petrolatum also can prevent static "flyaway," by creating a film which insulates the hair from the charges on the surrounding hairs (73). Conditioning with these products has been known for decades, and they are still popular today. Most of the more recent advances in hair conditioning with petrolatum have come in the form of emulsion products. Within the ethnic market, petrolatum is also a very common ingredient in hair relaxer formulations, acting as a scalp protectant. Since petrolatum also aids combing and adds shine to the hair in relaxer formulations, it can be considered a hair conditioner to some extent in these products (75).

The use of petrolatum as a pomade is well known, but it is still occasionally trumpeted in the newspapers as "the secret" for holding ethnic hair in place and adding shine (76). Others vary the ingredients to produce slightly different derivatives which are also used to condition the hair and scalp (77,78). Scientists, however, have also developed many hair care product formulations containing petrolatum to benefit from this material's properties while minimizing its greasiness. For example, petrolatum can be used to retain moisture and enhance the texture of the hair when used in an emulsified form (79), which is often more esthetically pleasing than an anhydrous preparation.

These conditioning properties of petrolatum have been revealed elsewhere. A rinse-off conditioner formula which contains petrolatum was shown to give the hair a more smooth and moist texture than an identical formula without the petrolatum (80). One particular hair setting cream which claims to provide gloss, antistatic qualities, and good combability has been developed, and surprisingly incorporates petrolatum at 10% while a silicone fluid is present at only 5% (81). A unique "rinse-on" conditioner which incorporates petrolatum as a particularly preferred ingredient also has been developed (82). This composition exists as an oil-in-water emulsion, but inverts to water-in-oil when it is rubbed onto the hair. Conditioning, moisturization, and protection are imparted to the hair from this nongreasy product.

One of the greatest improvements which has been made in hair relaxer formulations is the reduction of irritation from the highly alkaline relaxing agent. Since a strong base is typically needed for these products to perform well, reducing irritation by using a milder active ingredient will give only limited success in relaxing the hair. Even in the case of breakthrough formulas, scalp protectants are still needed to minimize irritation. Examples of improved hair relaxer formulas have been disclosed (83–87), but the unique nature of petrolatum ensures that it is present in both the classic and the newer products. Thus, it is often a preferred ingredient in relaxer formulas.

VII. FORMULATING WITH PETROLATUM

A. Typical Procedures

Just as the mechanism of hair conditioning with petrolatum is straightforward, so is the procedure of formulating with this material. Since petrolatum is naturally resistant to oxidation (due to its saturated hydrocarbon composition), no special precautions are necessary when using it to prepare skin and hair care cosmetics. The material is usually melted to obtain homogeneity with other oily ingredients and can withstand high temperatures, but long-term storage at elevated temperatures or exposure to excessively high temperatures is not recommended and may eventually cause product decomposition or some smoking (at very high temperatures).

In anhydrous cosmetics such as lip balms, lipsticks, pomades, and topical ointments, the usual formulation procedure for incorporating petrolatum is simple:

> Mix the petrolatum with the other lipophilic ingredients, and heat with stirring at the desired temperature until the mixture is melted and homogeneous. Proceed with further formulation (e.g., adding color or active ingredients) or package the anhydrous product at the desired temperature.

For emulsion products, formulating with petrolatum is just as easy.

> Mix the petrolatum with the other oil phase ingredients and heat to the desired temperature with stirring. Once this phase is uniform and the other, lipophobic

Table 2 Selected Personal Care Products That May Contain Petrolatum as a Skin or Hair Conditioner

After-sun products	Lip balms and creams
AHA creams and lotions	Lip glosses
Antiperspirants and deodorants	Lipsticks
Baby creams and lotions	Makeup foundations and powders
Brilliantines	Makeup removers
Cleansing creams	Mascaras
Cold creams	Massage products
Concealers	Moisturizers
Dermatological products (e.g., wound care)	Night creams
Diaper rash ointments	OTC topical pharmaceuticals
Dry skin treatments	Pomades
Eye makeup products	Protective creams and lotions
Face creams and lotions	Scalp creams and lotions
Hair conditioners	Self-tanners
Hair dressings	Shaving and aftershave products
Hair relaxers and straightening products	Sunscreen and sunblock products
Hair styling aids	Suntan products
Hand and body creams and lotions	Topical ointments
Insect repellents	Waterless hand cleaners

> ingredients have been suitably prepared, combine the phases in the desired order with sufficient mixing to yield a product of desired consistency.

In other words, the petrolatum is treated just like any other oil-phase ingredient when formulating emulsions, be they water-in-oil, oil-in-water, microemulsions, or multiple emulsions.

B. Examples of Formulated Products

Sources of personal care formulations which incorporate petrolatum abound throughout the literature and within the cosmetic industry. Examples of petrolatum-containing formulations can be found in many of the references listed at the end of this chapter, in supplier formulations, and in trade magazines and journals (88). While it is not comprehensive, Table 2 lists many types of personal care products in which petrolatum finds application as a hair or skin conditioner.

The patent literature is a useful source for uncovering personal care products of unique function or formulation. Because of the wide acceptance and utility of petrolatum as a skin (and, to a lesser extent, hair) conditioner, patented products which contain oily materials will almost always include petrolatum as a claimed oil-phase ingredient. Some recent examples of patents which use petrolatum include anhydrous skin care products (89), emulsion skin and hair care products (90–103), hydroxy acid products (104,105), and skin-lightening cosmetics (106–110). Petrolatum also has been claimed as an ingredient in deodorants (111), self-tanners (112), lotioned tissue paper (113, 114), powder cosmetics (115), and in a sebum secretion promoter (116), even though its use in these products may be perceived as rather unconventional.

VIII. SAFETY

From its infancy, petrolatum has been safely used millions of times on human hair, skin, and lips. Its safety should be without question, based solely on over 100 years' use in real-life situations. However, this material, while known industry-wide (and probably worldwide) as a moisturizer without comparison, is frequently required to justify its presence in cosmetics when the issue of comedogenicity arises.

A. Comedogenicity and Acnegenicity

Comedogenicity, or the clogging of skin pores, is often associated (wrongly so) with petrolatum because of its physical characteristics. The heaviness and greasiness of petrolatum appear to indicate that this material, or a cosmetic ingredient which contains it, will be comedogenic or even acnegenic (causes acne). After all, if a consumer has oily skin and acne, why should he or she just put more oil on it? This argument may seem valid to the uninitiated; however, skin oil (sebum) does not have the same chemical composition as petrolatum.

In fact, at cosmetic industry conferences and seminars, many misinformed people (scientists and nonscientists alike) are often heard saying things like, "Well, everyone knows that petrolatum clogs the pores." The evidence clearly shows otherwise. Some crude versions of petrolatum which were manufactured during its early years may have contained impurities and caused some problems, but the petrolatum from current production methods is *completely noncomedogenic*. Petrolatum samples once reported to be comedogenic were later found to be "false positives," and these substances were not comedogenic at all, either in the broadly used rabbit ear assay or in humans (117).

In addition, studies done by Fulton (118) and Lanzet (119) showed that petrolatum gave no comedogenic response whatsoever in the rabbit ear assay. Interestingly, Fulton found that many of the ingredients which cosmetic formulators often use as nonoily substitutes for petrolatum (e.g., when preparing "oil-free" products) are strongly comedogenic. Since dilution effects on the comedogenic potential of ingredients cannot be easily determined, it is necessary for the formulator to test the complete, fully formulated product for comedogenicity. Nevertheless, it is reassuring to know that petrolatum will not cause comedone formation, so it can be considered a safe ingredient to use. Petrolatum has also been cited as a moisturizer for patients with acne and acne-prone skin (117,118).

The issue of comedogenicity was considered at a 1988 American Academy of Dermatology Invitational Symposium on Comedogenicity (120). At this conference, the attendees stated that "neither the consumer nor the physician can assess whether the formulation will be acnegenic by simple inspection of the product or by examining the list of ingredients. Furthermore, the product's physical characteristics, such as oiliness or viscosity, are not in themselves predictors of an acnegenic response." Thus, it is clear that the physical properties of petrolatum do not predict acnegenicity or comedogenicity (117,120).

B. Allergenicity

It also is known that petrolatum does not cause allergic reactions. In fact, this neutrality is why petrolatum is commonly used as a carrier for the topical application of hydrophobic materials in human patch tests for sensitization (121, 122), in studies on transdermal absorption (123), and in the treatment of certain diseases (124). A study also has been performed to evaluate the relationship between materials applied to the skin of infants and eventual atopic dermatitis (125). These authors found no evidence that the use of emollients (petrolatum was used most frequently) on children caused the eventual development of atopic dermatitis (AD). They concluded that petrolatum "can safely be used in the skin care of AD susceptible individuals."

C. Irritation

Since petrolatum is frequently used as an ointment base and as a vehicle for topical patch applications, it may be expected that petrolatum is nonirritating to skin. This is the case, and such nonirritation has been proven by decades of human use. Therefore, the irritation potential of petrolatum has understandably been rarely studied over the years. In 1979, Motoyoshi determined that petrolatum is a nonirritant on both miniature swine skin (an acceptable substrate for determining skin irritation) and human skin (126). Petrolatum has also been compared to other materials in an evaluation of their dermatitic properties when used as cutting waxes (127). When tested fresh and after reconstitution, petrolatum produced no significant dermatological lesions, thus attesting to its nonirritancy.

It is clear that petrolatum is a safe material for all skin and hair care use, and holds no potential for comedogenicity, allergenicity, or irritation. In addition, studies have shown that petrolatum and other topical hydrocarbons remain in the stratum corneum and do not penetrate to the deeper layers of skin (30,123). The human use of this material on skin and hair for over a century is more evidence that might ever be necessary to determine the safety of a product, and it shows that petrolatum is indeed an exceedingly safe material for skin and hair care applications, either neat or as a part of a formulated personal care product.

IX. CONCLUSIONS AND THE FUTURE OF PETROLATUM IN COSMETICS

Petrolatum has been shown to be a safe and effective ingredient for hair and skin conditioning. Its ability to moisturize is unrivaled, and it has been used continuously in hair products for decades, despite frequent changes in fashion and hair styles. Petrolatum is a superb wound ointment and an excellent skin protectant. Therefore, it is doubtful that this ingredient will fall out of favor in the foreseeable future. Based on its conditioning properties, its natural origins, its safety, its ease of formulation, and its low cost relative to other materials, this substance should be a constant in the ever-changing field of personal care ingredients. Although petrolatum has changed little since it was first reported by Chesebrough, it has never been duplicated synthetically. The unique physical properties and conditioning abilities of petrolatum will bolster its use in cosmetic products for years to come. Indeed, the future is bright for petrolatum, a true workhorse of the personal care industry.

REFERENCES

1. Speight JG. The Chemistry and Technology of Petroleum. New York: Marcel Dekker, 1980:1–4.
2. Uhl WC. Introduction. In: Bland WF, Davidson RL, eds. Petroleum Processing Handbook. New York: McGraw-Hill, 1967:1–1.
3. Kalichevsky VA, Peters EH. The Raw Material, Crude Petroleum and Natural Gas. In: Guthrie VB, ed. Petroleum Products Handbook. New York: McGraw-Hill, 1960:1–5.
4. Speight J. Petroleum (Refinery Processes). In: Kroschwitz JI, exec. ed., Howe-Grant M, ed. Kirk-Othmer Encyclopedia of Chemical Technology. Vol. 18. 4th ed. New York: Wiley, 1996:434.
5. Jakubke H-D, Jeschkeit H, eds. Concise Encyclopedia Chemistry. Berlin: Walter De Gruyter, 1993.
6. Sage LL. Cosmetics—past, present, future. In: DeNavarre MG, ed. The Chemistry and Manufacture of Cosmetics. Vol. III. 2d ed. Wheaton IL: Allured, 1988:1–6.
7. Rieger MM. Cosmetics. In: Kroschwitz JI, exec. ed., Howe-Grant M, ed. Kirk-Othmer Encyclopedia of Chemical Technology. Vol. 7. 4th ed. New York: Wiley, 1996:573.
8. Busch P. The historical development of cosmetic science. In: Umbach W, ed. Cosmetics and Toiletries: Development, Production and Use. Chichester, UK: Ellis Horwood, 1991:1–7.

9. Wigoder G, general ed. Illustrated Dictionary and Concordance of the Bible. Jerusalem: G.G. The Jerusalem Publishing House, 1986.
10. See: 2 Kings 9:30, Jeremiah 4:30, Ezekiel 23:40, and Proverbs 6:25.
11. Job 42:14.
12. Chesebrough RA. U.S. Patent 127,568 (June 4, 1872).
13. Throughout this chapter, the generic name "petrolatum" will be used in instances where previous authors have called this material by other names, such as Vaseline, Cosmoline, and Petroline.
14. Schindler H. Petrolatum for drugs and cosmetics. Drug Cosmet Ind 1961; 89(1): 36–37, 76, 78–80, 82.
15. Crew BJ. A Practical Treatise on Petroleum. London, 1887: Henry Carey Baird & Co.
16. Sachanen AN. The Chemical Constituents of Petroleum. New York: Reinhold, 1945:284.
17. Meyer E. White Mineral Oil and Petrolatum and Their Related Products. New York: Chemical Publishing Company, 1968:16.
18. A striking analogy between the refining of petroleum and the refining of gold ore has been made. See: Morrison DS, Schmidt J, Paulli R. The scope of mineral oil in personal care products and its role in cosmetic formulation. J Appl Cosmetol 1996; 14:111–118.
19. Blank IH. Factors which influence the water content of the stratum corneum. J Invest Dermatol 1952; 18:433–440.
20. Lazer AP, Lazer P. Dry skin, water, and lubrication. Dermatol Clin 1991; 9(1): 45–51.
21. Rieger M. Skin care: new concepts vs. established practices. Cosmet Toilet 1991; 106(11):55–58, 60–62, 64, 66, 68.
22. Steenbergen C. Petroleum derivatives as moisturizers. Am Cosmet Perf 1972; 87(3):69–70, 72.
23. Fishman HM. Treating dry skin. Household Pers Prod Ind 1994; 31(2):30.
24. Fisher AA, Pincus SH, Storrs FJ, Richman E. When to suspect cosmetic dermatitis. Patient Care 1988; 22(11):29.
25. Idson B. Dry skin: moisturizing and emolliency. Cosmet Toilet 1992; 107(7):69–72, 74–76, 78.
26. Levine N. Advise patients on proper cosmetic use. Dermatol Times 1990 (Oct.):1.
27. Kligman AM. Regression method for assessing the efficacy of moisturizers. Cosmet Toilet 1978; 93(4):27–35.
28. Grove GL. Noninvasive methods for assessing moisturizers. In: Waggoner WC, ed. Clinical Safety and Efficacy Testing of Cosmetics. New York: Marcel Dekker, 1990:121–148.
29. Brown BE, Diembeck W, Hoppe U, Elias PM. Fate of topical hydrocarbons in the skin. J Soc Cosmet Chem 1995; 46:1–9.
30. Ghadially R, Halkier-Sorenson L, Elias PM. Effects of petrolatum on stratum corneum structure and function. J Am Acad Dermatol 1992; 26(3):387–396.
31. Loden M. The increase in skin hydration after application of emollients with different amounts of lipids. Acta Dermatol Venereol (Stockh) 1992; 72:327–330.

32. Prall JK, Theiler RF, Bowser PA, Walsh M. The effectiveness of cosmetic products in alleviating a range of skin dryness conditions as determined by clinical and instrumental techniques. Int J Cosmet Sci 1986; 8:159–174.
33. Tsutsumi H, Utsugi T, Hayashi S. Study on the occlusivity of oil films. J Soc Cosmet Chem 1979; 30:345–356.
34. Rietschel RL. A method to evaluate skin moisturizers in vivo. J Invest Dermatol 1978; 70(3):152–155.
35. Wepierre J. Study of the hydrating effect of cosmetic preparations by measuring cutaneous impedance in the hairless rat. Soap Perfum Cosmet 1977; 50(12): 506–509.
36. Frömder A, Lippold BC. Water vapour transmission and occlusivity *in vivo* of lipophilic excipients used in ointments. Int J Cosmet Sci 1993; 15:113–124.
37. Tranner F, Berube G. Mineral oil and petrolatum: reliable moisturizers. Cosmet Toilet 1978; 93(3):81–82.
38. Friberg SE, Ma Z. Stratum corneum lipids, petrolatum, and white oils. Cosmet Toilet 1993; 108(7):55–59.
39. Obata M, Tagami H. A rapid *in vitro* test to assess skin moisturizers. J Soc Cosmet Chem 1990; 41:235–241.
40. U.S. Department of Health and Human Services. Diabetic neuropathy: the nerve damage of diabetes. Pamphlet, September 1991.
41. Kligman LH, Kligman AM. Petrolatum and other hydrophobic emollients reduce UVB-induced damage. J Dermatol Treatment 1992; 3:3–7.
42. Kligman AM. Why cosmeceuticals? Cosmet Toilet 1993; 108(8):37–38.
43. Adams R, Johnson JR. Elementary Laboratory Experiments in Organic Chemistry. 3d ed. New York: Macmillan, 1940:415.
44. Guilbaud J. European comparative clinical study of Inerpan: a new wound dressing in treatment of partial skin thickness burns. Burns 1992; 18(5):419–422.
45. Gao ZR, Hao ZQ, Li Y, Im MJ, Spence RJ. Porcine dermal collagen as a wound dressing for skin donor sites and deep partial skin thickness burns. Burns 1992; 18(6):492–496.
46. Eloy R, Cornillac AM. Wound healing of burns in rats treated with a new amino acid copolymer membrane. Burns 1992; 18(5):405–411.
47. Sawada Y, Sone K. Beneficial effects of silicone cream on grafted skin. Br J Plast Surg 1992; 45(2):105–108.
48. Sawada Y, Sone K. Benefits of silicone cream occlusive dressing for treatment of meshed skin grafts. Burns 1992; 18(3):233–236.
49. Markintel 1993 (Mar. 1):1–15.
50. Whitham J. Medicinal salve composition. U.S. Patent 5,270,042 (Dec. 14, 1993).
51. Shinault WK. Topical ointment for the treatment of epidermal trauma. U.S. Patent 5,407,670 (Apr. 18, 1995).
52. Phan M, Van der Auwera P, Andry G, et al. Wound dressing in major head and neck cancer surgery: a prospective randomized study of gauze dressing vs sterile vaseline ointment. Eur J Surg Oncol 1993; 19(1):10–16.
53. Sawada Y, Ara M, Yotsuyanagi T. A thrombin ointment that achieves rapid haemostasis of split thickness donor wounds, particularly on the scalp. Burns 1991; 17(3): 225–227.

54. 48 Fed Reg 6820 (Feb. 15, 1983).
55. Nopper AJ, Horii KA, Sookdeo-Drost S, Wang TH, Mancini AJ, Lane AT. Topical ointment therapy benefits premature infants. J Pediatr 1996; 128(5):660–669.
56. Maklebust J. Pressure ulcer update. RN 1991; 54(12):56–64.
57. Lemole GM. Protective gel composition. U.S. Patent 5,019,604 (May 28, 1991).
58. Braue EH Jr, Pannella MG. Topical protectant evaluation of FT-IR spectroscopy. Appl Spectrosc 1990; 44(6):1061–1063.
59. Liu JC, Huang MJ, Sun Y, Chien YW. The effect of barrier creams on the electrical conductivity of excised skin during exposure to detergents. J Soc Cosmet Chem 1987; 38:63–75.
60. Shulakov NA, Novikov VE, Loseva VA, et al. Vaseline protection of the skin from the effects of the sealent Uniherm-6. Gig Tr Prof Zabol 1990; 12:43–44.
61. Nestruk C, Sangiorgio M. Colds, kids & you. Prevention 1992; 44(1):58–67.
62. Kennedy B. Babes in the woods. Rodale's Guide to Family Camping, 1995 (Spring); 1(1):16, 18, 20–24, 128.
63. Curtis AW. Less irritating shaving material. U.S. Patent 5,252,331 (Oct. 12, 1993).
64. Fitness 1995 (May):101.
65. Ramsey ML. Managing friction blisters of the feet. Physician and Sportsmedicine 1992; 20(1):116–121.
66. Pate D. Success secrets for runners. Health & Fitness Magazine, 1994 (Dec.): 34–35.
67. Kuscsik N. That first marathon. Women's Sports and Fitness 1989; 11(8):20–25.
68. Zempel C. Take to the waves! Women's Sports and Fitness 1989; 11(4):66.
69. Goode ST. Hair pomades. Cosmet Toilet 1979; 94(4):71–74.
70. Syed AN. Ethnic hair care: history, trends and formulation. Cosmet Toilet 1993; 108(9):99–102, 104–107.
71. Lochhead RY, Hemker WJ, Castañeda JY. Hair care gels. Cosmet Toilet 1987; 102(10):89–100.
72. Mottram FJ. Hair treatments. In: Butler H, ed. Poucher's Perfumes, Cosmetics and Soaps. Vol. 3. 9th ed. London: Chapman & Hall, 1993:130–169.
73. Jellinek JS. Formulation and Function of Cosmetics. New York: Wiley-Interscience, 1970:457–471.
74. Böllert V, Siemers H. Hair Care Products. In: Umbach W, ed. Cosmetics and Toiletries: Development, Production and Use. Chichester, UK: Ellis Horwood, 1991:190–201.
75. Obukowho P, Birman M. Hair curl relaxers. Cosmet Toilet 1995; 110(10):65–69.
76. Nelson J. I love Vaseline; it comes in tubs and tins, protects babies' bottoms, tames teenage hairdos, softens women's lips and is absolutely essential to life as we know it. Washington Post Magazine 1987 (Jan. 11):w27.
77. McCarthur CM. Hair dressing cosmetic. U.S. Patent 3,932,611 (Jan. 13, 1976).
78. Vernon DM. Hair treatment composition and method. U.S. Patent 4,999,187 (Mar. 12, 1991).
79. Tada T, Myamoto N. Hair conditioners containing vaseline. JP 03 264,516 (Nov. 25, 1991). Chem Abstr 1992; 116:158567p.
80. Fujikawa M, Suzuki N. Hair conditioners containing methylpolysiloxanes, protein derivatives, and vaseline. JP 03 264,515 (Nov. 25, 1991). Chem Abstr 1992; 116: 136011a.

81. Hasegawa M, Yoda E, Yogoshi M, Koresawa T. Hair-setting preparations containing branched esters and silicone oil. JP 07 25,733 (Jan. 27, 1995). Chem Abstr 1995; 122:196547p.
82. Wagman J, Sajic B. Skin and hair conditioner compositions and conditioning method. U.S. Patent 4,551,330 (Nov. 5, 1985).
83. Patel MM. Conditioning and straightening hair relaxer. U.S. Patent 5,476,650 (Dec. 19, 1995).
84. Cowsar DR, Adair TR. Hair relaxer compositions containing strong base and alkaline earth metal hydroxides. WO 95 03,031 (Feb. 2, 1995). Chem Abstr 1995; 122:196539n.
85. Hawkins GR, Simpson CB Jr, Klein GJ. Hair relaxer composition and associated methods. U.S. Patent 5,304,370 (Apr. 19, 1994).
86. Akhtar M. Hair relaxer cream. U.S. Patent 5,171,565 (Dec. 15, 1992).
87. McKaba W, Simpson CB. Quaternary ammonium hydroxide hair relaxer composition. U.S. Patent 4,530,830 (July 23, 1985).
88. Morrison DS. Petrolatum: a useful classic. Cosmet Toilet 1996; 111(1):59–66, 69.
89. Swenson RH. Skin care formulation. U.S. Patent 5,494,657 (Feb. 27, 1996).
90. Cho SH, Frew LJ, Chandar P, Madison SA. Synthetic ceramides and their use in cosmetic compositions. U.S. Patent 5,476,671 (Dec. 19, 1995).
91. Rose W, Zimmerman AC. Petroleum jelly cream. U.S. Patent 5,407,678 (Apr. 18, 1995).
92. Rawlings AV, Watkinson A. Skin care method and composition. U.S. Patent 5,554,366 (Sept. 10, 1996).
93. Amin MK, Butter SS. Topical composition for skin treatment containing an isoquinoline and a vasodilator. GB 2,290,470 (Jan. 3, 1996). Chem Abstr 1996; 124: 155717s.
94. Kikuchi H, Tsubone K. Moisturizing emulsion-type skin cosmetics containing ammonioethyl phosphates and surfactants. JP 08 113,527 (May 7, 1996). Chem Abstr 1996; 125:67275m.
95. Sato K, Mizutani M, Yamane T. Method for manufacturing cosmetic emulsions. JP 08 99,835 (Apr. 16, 1996). Chem Abstr 1996; 125:67210m.
96. Jokura Y, Uesaka T, Honma S, Kato Y, Ishida K. Moisturizing cosmetic containing ceramide and dicarboxylic acid. DE 19,539,016 (Apr. 25, 1996). Chem Abstr 1996; 124:352348y.
97. Sakamoto O, Yanagida T, Yonezawa K. Water-in-oil skin preparations of α-tocopheryl retinoate. JP 08 99,834 (Apr. 16, 1996). Chem Abstr 1996; 125: 41495h.
98. Hikima T. Skin cosmetics containing *Pyracantha fortuneana* fruit extracts and other ingredients. JP 08 133,959 (May 28, 1996). Chem Abstr 1996; 125:123293e.
99. Nakamura T, Kaneki H, Ito K. Water/oil-type cosmetic emulsions. JP 08 126,834 (May 21, 1996). Chem Abstr 1996; 125:123257w.
100. Staeb F, Landenzoerfer G. Cosmetic or dermatologic compositions containing cinnamic acid derivatives and flavonoid glycosides. EP 716,847 (June 19, 1996). Chem Abstr 1996; 125:95577e.
101. Doughty DG, Gatto JA, Nawaz Z, Rolls RGA. Skin care oil-in-water dispersions. WO 96 16,545 (June 6, 1996). Chem Abstr 1996; 125:95591e.

102. Doughty DG, Gatto JA, Weisgerber DJ, Schwartz JR. Topical skin care compositions containing non-occlusive liquid polyol carboxylic acid esters as skin conditioning agents. WO 96 16,637 (June 6, 1996). Chem Abstr 1996; 125: 95592f.
103. Murase T, Hase T, Takema Y, Ogawa A, Oosu H, Tokimitsu I. Cosmetics containing stilbenes for prevention of wrinkles. JP 08 175,960 (July 9, 1996). Chem Abstr 1996; 125:177025n.
104. Smith WP. Low irritant skin-cosmetic composition for daily topical use, its application and manufacture. U.S. Patent 5,520,918 (May 28, 1996).
105. Maurin E, Sera D, Guth G. Cosmetic compositions containing an enzyme and a hydroxyacid precursor. FR 2,725,898 (Apr. 26, 1996). Chem Abstr 1996; 125: 67189m.
106. Yoshioka M, Iwamoto A, Masaki H. Skin-lightening cosmetics containing plant extracts as tyrosinase inhibitors. JP 08 104,646 (Apr. 23, 1996). Chem Abstr 1996; 125:67215s.
107. Itokawa H, Morita H, Takeya K, et al. Skin-lightening cosmetics containing cyclopeptide Pseudostellarins extracted from *Pseudostellaria heterophylla* roots as tyrosinase inhibitors and melanin formation inhibitors. JP 07 324,095 (Dec. 12, 1995). Chem Abstr 1996; 124:241760g.
108. Yagi E, Komazaki H, Shibata Y, Naganuma M, Fukuda M. Skin-lightening cosmetics containing melanin formation inhibitors from *Quararibea amazonia*. JP 08 12,556 (Jan. 16, 1996). Chem Abstr 1996; 124:211546z.
109. Ochiai M, Tada A, Yokoyama Y, Nozawa S. Reduced ionones as melanin formation inhibitors and topical preparations containing them. JP 08 73,334 (Mar. 19, 1996). Chem Abstr 1996; 125:67246c.
110. Tada A, Yokoyama Y, Ochiai M, Nozawa S. Melanin formation inhibitors comprising reduced ionones and skin preparations containing them. JP 08 73,335 (Mar. 19, 1996). Chem Abstr 1996; 125:41476c.
111. McCuaig D. Long-life deodorant composition containing zinc oxide and starch. CA 2,130,967 (Feb. 27, 1996). Chem Abstr 1996; 124:298463s.
112. Philippe M, Tuloup R, De Salvert A, Sera D, Fodor P. Cosmetic compositions containing a precursor of dihydroxyacetone. EP 709,081 (May 1, 1996). Chem Abstr 1996; 125:41447u.
113. Klofta TJ, Warner AV. Lotioned tissue paper and impregnation composition for its manufacture. WO 95 35,412 (Dec. 28, 1995). Chem Abstr 1996; 124:179297f.
114. Klofta TJ, Warner AV. Lotioned tissue paper and impregnation composition for its manufacture. WO 95 35,411 (Dec. 28, 1995). Chem Abstr 1996; 124:179298g.
115. Nakajima H. Water-repellent powders overcoated with solid oily substances and powdery cosmetics containing them. JP 08 59,431 (Mar. 5, 1996). Chem Abstr 1996; 124:352342s.
116. Hanamura A, Sata A. Sebum secretion promoters containing bryonolic acid. JP 07 109,214 (Apr. 25, 1995). Chem Abstr 1995; 123:40736j.
117. Kligman AM. Petrolatum is not comedogenic in rabbits or humans: a critical reappraisal of the rabbit ear assay and the concept of "acne cosmetica." J Soc Cosmet Chem 1996; 47(1):41–48.

118. Fulton JE Jr, Pay SR, Fulton JE III. Comedogenicity of current therapeutic products, cosmetics, and ingredients in the rabbit ear. J Am Acad Dermatol 1984; 10:96–105.
119. Lanzet M. Comedogenic effects of cosmetic raw materials. Cosmet Toilet 1986; 101(2):63–64, 66–68, 70, 72.
120. American Academy of Dermatology Invitational Symposium on Comedogenicity. J Am Acad Dermatol 1989; 20:272–277.
121. Brandrup F, Menne T, Agren MS, Stromberg HE, Holst R, Frisen M. A randomized trial of two occlusive dressings in the treatment of leg ulcers. Acta Dermatol Venereol (Stockh) 1990; 70(3):231–235.
122. Old problems from new sensitizers. Dermatol Times 1991 (Mar.):34.
123. Brown BE, Diembeck W, Hoppe U, Elias PM. Fate of topical hydrocarbons in the skin. J Soc Cosmet Chem 1995; 46(1):1–9.
124. Ceilley RI, Goldberg GN, Prose NS. A guide to pediatric rashes. Patient Care 1989; 23(15):150–160.
125. Macharia WM, Anabwani GM, Owili DM. Effects of skin contactants on evolution of atopic dermatitis in children: a case control study. Trop Doct 1991; 21:104–106.
126. Motoyoshi K, Toyoshima Y, Sata M, Yoshimura M. Comparative studies on the irritancy of oils and synthetic perfumes to the skin of rabbit, guinea pig, rat, miniature swine and man. Cosmet Toilet 1979; 94(8):41–43, 45–48.
127. Ireson JD, Leslie GB, Osborne H, Read M. A laboratory investigation of a selection of lubricant waxes as dermatitic agents. Pharmacol Res Commun 1972; 4(4): 353–356.

5

Humectants in Personal Care Formulation: A Practical Guide

Bruce W. Gesslein
Ajinomoto U.S.A., Inc., Teaneck, New Jersey

I. INTRODUCTION

Compounds that can improve the surface of skin or hair are called conditioning agents. The mechanisms by which conditioning compounds do this vary depending on the type of compound and surface to which they are applied to. One of the primary methods by which conditioning ingredients function is by regulating the amount of moisture in the skin or hair. Since moisture is a key factor in the condition of both hair and skin, ingredients which affect moisture levels are potentially conditioning agents.

Humectants are conditioning agents that regulate water levels on the skin and hair in a distinct way. Due to their chemical nature, they are able to attract and bind water to themselves. This property is known as hygroscopicity. By utilizing this useful property, formulators have been able to incorporate humectants effectively in conditioning products for skin and hair. While there are both inorganic and organic materials which have this property, only organic ones have generally been used in cosmetic products.

Humectants have been said to have the ability to rehydrate the skin when delivered from a cream or lotion. In shampoos and conditioners they are said to soften the hair and in fact to swell the hair shaft. They may also help plasticize the films of certain polymers, and are often used as co-solvents during the solubilization of fragrances or oils into products. Some humectants have the reported ability to enhance the efficacy of preservative systems.

The primary humectant used in personal care products is glycerin (or glycerol). It is a trihydric alcohol that is chiefly derived from fats and oils and as a by-product from soap manufacture. This semisweet, clear liquid has been shown to have conditioning benefits when applied to skin. Additionally, it also provides many other benefits to formulations via its water-regulating ability. Glycerin's versatility coupled with its relatively low cost and low toxicity has made it a favorite among formulators for decades. Other important humectants include sorbitol, propylene glycol, and other polyhydric alcohols. Some metal-organic types have been gaining popularity due to their greater hygroscopic characteristics. These include types like sodium PCA and sodium lactate.

In this chapter we will discuss the properties of humectants commonly used in today's practice. The basic theory of how and why humectants are used will be given, based on the experiments and experience of past investigators. Newer and exciting work will be referenced, and some original work will be outlined. Formulations and test results will be given to demonstrate the matter covered.

II. HISTORICAL DEVELOPMENT

Humectants have been known and used in skin treatment products for thousands of years. One of the earliest applications was a hand cream made up of 50% humectant and 50% water (1). In addition to being the most widely used humectant in personal care formulation, glycerin is also the oldest. In 1779, the Swedish chemist K. W. Scheele (1742–1786) discovered glycerin by accident when he was investigating a soap plaster by heating a mixture of olive oil and lead monoxide. The material he called "sweet principle of fat" we now call glycerin. He later determined that the same reaction which produced glycerin and soap occurred when other metals and glycerides were heated together. Scheele published his findings in 1783 in the *Transactions of the Royal Academy of Sweden*.

In 1811 the French investigator Chevreul (1786–1889) named Sheele's sweet principle of fat glycerin, from the Greek *glykys* (meaning sweet). In 1823 he obtained the first patent pertaining to the production of fatty acids from fats and oils. This patent also contained a method for the recovery of glycerin that was released during the process. Pelouze, another French scientist, determined the empirical formula for glycerin in 1836, and the structure was published 47 years later (2). Over time scientists realized that glycerin was present in all animal and vegetable fats and oils. Improved methods of isolating and producing it on a commercial scale were developed. Soap manufacturers led the way in this endeavor with large-scale glycerin production beginning during the early 1900s. Methods for producing synthetic glycerin were developed as a result of high demand brought on by the war during the 1930s and 1940s.

Since glycerin had desirable humectant properties, was relatively inexpensive, and was readily available, it became incorporated in a wide variety of personal care products. With the advent of World War II, however, glycerin was needed for military use, and other humectants were pressed into service (3). Polyols such as sorbitol and propylene glycol were most used, though they were not perfect glycerin replacements. These materials remain today some of the most widely utilized humectants after glycerin.

III. NONCOSMETIC USES OF HUMECTANTS

The versatility of humectants is evidenced by the numerous ways in which they find application. Since they have the ability to reduce the freeze point of solutions, they are frequently used as antifreezes in cooling systems. They are also used in the manufacture of products such as printing rolls, glues, leather, alcohol, nitroglycerol (dynamite), and tobacco. Humectants ensure a smooth flow of ink in writing instruments and prevent crusting on the pen point. In candy manufacturing, sorbitol is used to extend shelf life by inhibiting the solidification of sugar. They are also used as food sweeteners, texturizers, and as a softener in peanut butter. Finally, humectants have been used to increase the absorption of vitamins in pharmaceutical preparations (4).

IV. HUMECTANCY

The key characteristic of a humectant is hygroscopicity. Hygroscopicity is the ability of a material to hold (or bind) moisture to itself. A useful humectant will retain moisture over a wide range of humidity conditions and for an extensive time period. There are two ways of expressing the hygroscopicity of a humectant: (1) equilibrium hygroscopicity and (2) dynamic hygroscopicity. Both these descriptors are important in choosing a humectant for personal care applications.

A. Equilibrium Hygroscopicity

Equilibrium hygroscopicity is the measure of water held by a humectant when there is equilibrium between the material and the relative humidity of the air. An effective humectant should exhibit a high equilibrium hygroscopicity. Equilibrium hygroscopicity is expressed as the weight percent of water held by the humectant divided by the relative humidity:

$$H_e = \frac{\%(\mathrm{w/w})\ \mathrm{water}}{\mathrm{RH}}$$

To measure the equilibrium hygroscopicity of a humectant, one exposes humectant solutions of known concentration to various controlled relative humidities. The weight of the solution is taken periodically. A curve is then plotted contrasting the weight of water held by the humectant versus the relative humidity. As an example, it is found that at 50% relative humidity the H_e for glycerin is ~25%, and for propylene glycol it is ~20%.

B. Dynamic Hygroscopicity

Dynamic hygroscopicity is an expression of the rate at which a humectant gains or loses water when approaching equilibrium. An ideal humectant should exhibit low dynamic hygroscopicity, which indicates that it is able to retain moisture for a long period of time. Since there are no direct methods, no absolute measures, to determine dynamic hygroscopicity, it is necessary to compare one humectant relative to another. Glycerin is often used as the known humectant. In this work it has been found that sorbitol exhibits the lowest dynamic hygroscopicity of the more commonly used humectants.

V. THE IDEAL HUMECTANT FOR COSMETIC PRODUCTS

While many different materials exhibit humectancy, not all are appropriate for use as a conditioning ingredient. There are several desirable properties that should be exhibited by a humectant. Griffin et al. delineated the desirable properties that an ideal humectant should possess (5). A summary of these important properties follows.

1. A humectant should be able to absorb a great deal of moisture from the air at a broad range of relative humidities.
2. Its moisture content should change little when exposed to large changes in relative humidity.
3. It should be nontoxic.
4. A humectant should exhibit good color properties.
5. It should be of low viscosity to facilitate incorporation into systems.
6. A humectant should be nonreactive with commonly used materials, and it should be noncorrosive to commonly used packaging.
7. It should be readily available and relatively low in cost.

The above list describes an ideal humectant; however, none of the materials used today exhibits all of these properties. Glycerin is perhaps the most nearly "ideal" of all the available humectants.

VI. FORMULATING WITH HUMECTANTS

Humectants are used in formulations for more than just their effects on skin and hair. They have been shown to have an important effect on a product's freezing point and to affect the solubility of certain compounds. Also, they have been shown to increase product viscosity, retard the evaporation of water, and act as a preservative and an emulsion stabilizer.

One of the aspects of humectants that is important, but not always appreciated, is their ability to couple mutually incompatible materials together. The humectant can act as a co-solubilizer for many materials in creating a clear solution. The classic example of this is the coupling ability of glycerin with sodium stearate in the formation of "clear" bar soaps. In fragrance solubilization, materials such as propylene glycol, or dipropylene glycol are often used as a "co-solvent" in conjunction with solubilizing materials such as polysorbate-20.

Polyols are known to depress the freezing point of water solutions. The *CRC Handbook of Chemistry and Physics* includes a number of tables delineating the freezing-point depression of polyols in water solutions by concentration (6). Humectants are added to water-based systems, especially surfactant cleansers, to maintain clarity at low temperatures during shipping and to prevent bottle cracking. The clarity of a surfactant cleansing system is maintained chiefly through the polyol-humectant's ability to act as a co-solvent and antifreeze for fatty materials, making it more difficult for these materials to come out of solution. As the ambient temperature decreases, the longer-chain-length fatty materials freeze out of solution first, followed by the shorter-chain-length materials. At these reduced temperatures, their solubility in the surfactant solution is significantly reduced. The addition of a polyol humectant helps to enhance the solubility of these fats in the system.

Bottle cracking is a phenomenon that may occur when water-based systems in sealed containers with insufficient expansion space are exposed to temperatures low enough for the formulation of ice. As ice crystals form, an expansion of the water contained in the product occurs that may be great enough to stress the packaging to the point of breakage. The polyol-humectants act as antifreeze, much the same as the antifreeze in a car's radiator, by reducing the freezing point of water to a point where ice does not form at the temperatures expected during shipping and storage.

VII. TYPES OF HUMECTANTS

As suggested, humectants are conditioning agents which have the ability to attract and bind water. Because of this they are able to regulate the exchange of moisture between a surface, such as hair or skin, and the air. By increasing the level of moisture in the skin, its condition is improved. Hair also exhibits improved condition when it is moisturized; however, the effectiveness of humectants in hair is less than that of skin.

Humectants can generally be classified as either inorganic, metal-organic, or organic. With the exception of the inorganic class, all these materials find extensive use in personal care formulations. Humectants have a long history of use in personal care formulation. It has been stated that some of the earliest hand creams were a mixture of 50% humectant and 50% water. Nearly every personal care product contains a hygroscopic material for one purpose or another. In some cases the material is intended to protect the product from drying out and shrinking, or to prevent rolling up on application. In other cases the humectant is intended to have a real effect on the skin or the hair.

A. Inorganic Humectants

Inorganic humectants are typically the salts of inorganic acids. Calcium chloride is an example of this class. There is currently little use for this class of humectant in personal care formulation, because they have problems related to their corrosive effects and incompatibility with other raw materials (7). Additionally, since these materials are salts, their incorporation in personal care formulations, at amounts necessary to deliver substantial humectancy, can detract from the other desired properties of the formulation. For example, emulsion products such as creams and lotions have a tendency to destabilize due to the salting-out effect. The effect on surfactant-based cleansing products is well established; they cause the system to go beyond the peak of the salt curve, lowering viscosity and often depressing foam. These materials also tend to increase the potential for eye sting and irritation, and in powdered products, may desiccate the skin to an extent.

B. Metal-Organic Humectants

The metal-organic class of humectants is much more useful. These materials, which contain a mixture of a metal ion and an organic portion, are usually the salts of strong bases and weak organic acids. Sodium lactate and sodium PCA (pyrrolidone carboxylic acid) are typical examples and are perhaps the most widely used of this class of humectants.

C. Organic Humectants

The organic humectant class is the broadest and the most widely used in personal care formulation. These materials are typically polyhydric alcohols and their esters and ethers. Glycerin, propylene glycol, and sorbitol are the most common examples. The simplest unit is ethylene glycol, and a series can be built by the addition of ethylene oxide to the basic unit. This leads to a series composed of ethylene glycol, propylene glycol, glycerin, sorbitol, and polyethylene glycol. The series cannot be extended too far, however, since the ether linkages tend to reduce hygroscopicity. The “strength” of a humectant is principally dependent on the ratio of hydroxy groups to carbons. As this ratio increases, there is an increase in the hygroscopicity.

1. Glycerin

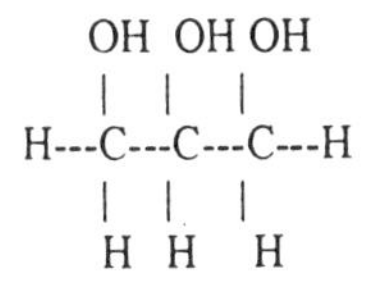

Chemically, glycerin is a trihydric alcohol denoted as 1,2,3-propanetriol. It is a clear, colorless, viscous material which is miscible in water and alcohol. It is a sweet-tasting, naturally occurring compound which is a component of thousands of materials including animal fats and vegetable oils.

Today, glycerin is manufactured chiefly from fats and oils that have been hydrolyzed, transesterified, or saponified. The glycerin that is recovered is in the crude state and undergoes further "purification" through distillation or by an ion-exchange process. Glycerin is also synthesized from propylene. Fats and oils that are tallow- and vegetable-derived (from coconut oil or palm kernel oil) are both used with about equal availability. Trends toward more use of natural products have caused more manufacturers to produce vegetable-derived glycerin. While there are many grade or purities of glycerin, the chief ones used in personal care formulation are > 95% purity and usually 99%. These amounts are based on the pure glycerol content of the material (8).

As an aside, the spelling "glycerin" is commonly used in the United States and is interchangeable with the spelling "glycerine." In Europe, this material is referred to as "glycerol" because that is the active material in glycerin.

Since glycerin has three hydroxyl groups, a number of analogs may be formed. Like most other alcohols, glycerin can form esters, amines, and aldehydes. It has two primary hydroxyl groups and one secondary hydroxyl group. The primary hydroxyls tend to react before the secondary hydroxyl, and the first primary hydroxyl before the second primary hydroxyl. In any reaction, however, the second primary hydroxyl and the secondary hydroxyl will react to some extent before all the reactive groups have been used (9). This accounts for the distribution seen in materials such as glyceryl monostearate, in which there typically is about 70% mono-, 20% di-, and 10% triester. While glycerin is stable to atmospheric oxidation under normal conditions, it can be oxidized by other oxidants.

Chief among the advantages of glycerin in personal care products is that it exhibits good equilibrium hygroscopicity and is virtually nontoxic except at very high concentrations, where a dehydration effect on the skin can be seen. At low temperatures, glycerin tends to supercool rather than crystallize. It does not discolor nor produce off odors with aging. It is a good dispersing agent for pigments in foundation makeups, and lowers the freezing point of lotions and shampoos effectively, which helps maintain clarity of the systems and helps prevent bottle cracking due to freezing.

The principal drawback of glycerin is that it can produce stickiness in creams or lotions. This can be overcome by adjusting its use level or by using it in combination with another humectant such as sorbitol.

2. Propylene Glycol

```
     H    H   H
     |    |   |
H----C----C----C----H
     |    |   |
     H   OH  OH
```

Propylene glycol is the second most widely used humectant in personal care formulations. It is a three-carbon molecule with two hydroxyl groups. It is a clear, colorless, viscous liquid and, like glycerin, is miscible in water. The safety of propylene glycol is well known, and ingestion is harmless because in the body its oxidation results in metabolically useful pyruvic and acetic acids (10).

Propylene glycol is obtained from propylene by the cracking of propane. The propylene is converted to the chlorhydrin with chlorine water. The chlorhydrin is then converted to the glycol with a sodium carbonate solution. Propylene glycol can also be obtained from the heating of glycerin in the presence of sodium hydroxide (11).

The chief advantages of propylene glycol in personal care formulations is that it is less expensive than glycerin and has a better skin feel when used in creams and lotions. It is also a somewhat better coupling agent or co-solvent than is glycerin for fragrance oils. Propylene glycol effectively reduces the "balling up" of oil-in-water creams, giving a better rub-out. Additionally, it has some antimicrobial activity and will enhance the activity of antimicrobial agents. Some disadvantages of propylene glycol are that it is not "naturally" derived and it tends to de-foam surfactant systems. It also greatly reduces the viscosity of personal care preparations.

3. Sorbitol

```
     CH2OH
      |
   H—C—OH
      |
  HO—C—H
      |
   H—C—OH
      |
   H—C—OH
      |
     CH2OH
```

Sorbitol is a hexahydric alcohol which is typically supplied as a 70%-by-weight syrup, although the crystalline form is also available from analytical-reagent suppliers. It was first discovered as a component of the ripe berries of the mountain ash. Commercially, it is obtained by the high-pressure hydrogenation of glucose or by electroreduction.

Sorbitol exhibits the best dynamic hygroscopicity of the humectants, although it has somewhat lower equilibrium hygroscopicity than other materials. The skin feel of sorbitol is soft and nongreasy, with very little tackiness. This makes the use of sorbitol as an emollient possible, especially in systems where there is a high-viscosity and heavier body. It is often used as a cost-effective substitute for glycerin in creams and lotions. Sorbitol is not used as an "antifreeze" in personal care formulations, as glycerin and propylene glycol are.

An interesting property of sorbitol is that it has the ability to chelate iron, copper, and aluminum ions. It is most effective at alkaline pH and in highly acidic media. Sorbitol has little chelation ability at neutral pH. This may be important in cleansing systems where there is concern about high iron content in water.

4. 1,3-Butylene Glycol

$$HOCH_2CH_2\underset{\displaystyle OH}{\underset{|}{C}}HCH_3$$

Like many of the humectants described, butylene glycol is a clear, viscous liquid. It contains two hydroxyl groups and has a slightly sweet taste. Production of butylene glycol is achieved by the catalytic hydrogenation of acetaldehyde. This material is further distilled and purified to produce a food-grade quality.

1,3-Butylene glycol is a material that has become more important in recent years, primarily because of its reported hypoallergenicity. It has a hygroscopicity about that of propylene glycol and acts as an effective antifreeze in liquid systems.

Butylene glycol is a more effective essential oil and fragrance solubilizer than either glycerin or propylene glycol. This is especially true for the citrises and mint oils, for which 1,3-butylene glycol acts as a very effective co-solvent. It also helps retard the loss of fragrance through plastic packaging. As an antimicrobial adjunct, butylene glycol acts to enhance the distribution of preservatives across the water/oil interface better than either propylene glycol or glycerin.

5. Polyethylene Glycols

$$H(OCH_2CH_2)_nOH$$

Where n = 2 through 90,000

A whole range of polyethylene glycols is available to the personal care formulator, from a PEG number average molecular of about 200 to well over 6000.

It is only the shorter polyethylene glycols of about PEG numbers 200 through 2000 that exhibit any real hygroscopicity. As the PEG number increases, the water solubility and humectancy decrease. As an example, PEG-200 is completely water-soluble and has a hygroscopicity of about 70% of that of glycerin, while a PEG-1000 is only about 70% soluble in water and has a hygroscopicity of the order of 35% that of glycerin. At a PEG number of 4000 the material is only about 60% soluble in water and is nonhygroscopic.

The physical forms of the polyethylene glycols range from a liquid through a hard flake, again dependent on the PEG number. The higher the PEG number, the more solid is the material. While the higher polyethylene glycols are not humectants, they may be used as very effective thickeners for water-based systems, especially systems that have high surfactant loading. When they are used in this manner, it is necessary to balance the viscosity desired with the hand feel of the system since the very high polyethylene glycols tend to leave a "greasy" feel on the skin. Often the higher polyethylene glycols are used in conjunction with glycerin or sorbitol to improve humectancy and hand feel.

Polyethylene glycols of up to about 2000 PEG numbers are often used in creams and lotions to prevent rolling or balling of the cream upon application. These shorter PEG numbers have a smooth, lubriciousness feel on the skin. In some brushless shave creams they are used to replace all or part of the normal emollient system and improve wash-off. Polyethylene glycols are toxicologically inert, as exemplified by their use in a USP PEG ointment. This is a 1:1 mixture of PEG-400 and PEG-4000 and is used as a water-washable ointment.

6. Sodium Pyrrolidone Carboxylate (Na-PCA)

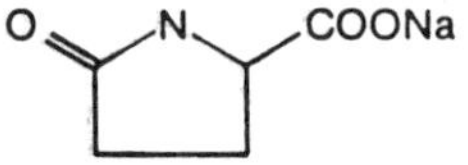

O N COONa

Sodium PCA is the salt of pyrrolidonecarboxylic acid and is found naturally in human skin. It is typically available as a 50% active aqueous solution. When placed on the skin, it has virtually no skin feel and, due to its salt nature, is obviously entirely soluble in water. In the so-called natural moisturizing factor (NMF), a theoretical collection of compounds that are thought to be present in the skin and responsible for ubiquitous moisturization, sodium PCA comprises about 10–15% by weight (12).

Sodium PCA is a highly effective humectant that is applicable for use in all normal creams, lotions, and surfactant-based cleansing products. It has excellent humectancy, showing a moisture-binding capacity 1.5 times greater than glycerin, twice as great as propylene glycol and six times that of sorbitol (13). Its dynamic hygroscopicity is somewhat less than sorbitol and about equal to that of glycerin. These characteristics make it one of the best humectants to use for skin-softening effects. Unlike other humectants discussed, it has no reported effect on preservative systems and no effect as a co-solvent for solubilizing essential oils. One of the prohibiting factors to widespread use of this material is its relatively high cost compared to other humectants.

7. Acetamide MEA

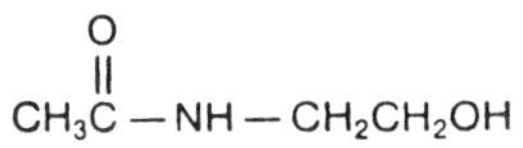

$$CH_3\overset{\overset{\displaystyle O}{\|}}{C}-NH-CH_2CH_2OH$$

Acetamide MEA is the aliphatic amide of acetic acid and monoethanolamine. It has very good toxicological properties, with a primary eye irritation index, a skin irritation index of zero, and an LD_{50} of >24 g/kg. As presented to the personal care formulator, it is a clear liquid with a yellow cast that has a slight acetic acid odor. It is typically available as a 70% active aqueous solution.

Acetamide MEA offers a number of useful properties to the personal care formulator. Incorporation of acetamide MEA in a shampoo has been shown to reduce eye irritation (14). It normally has little effect on the viscosity in these systems while offering good hygroscopicity. Acetamide MEA is a good dye/pigment-dispersing agent that allows for the easy incorporation of these materials into makeups and affords their even deposition upon application.

8. Miscellaneous "Humectants"

Many materials have been claimed to be humectants based on their water-absorbing characteristics when evaluated empirically. Among the many are the collagens, both tropocollagen and the hydrolysates, the keratins, glucose ethers and esters, and various mixtures of materials of botanical nature. In 1980 Deshpande, Ward, Kennon, and Cutie published work done in evaluating these humectants against the known classical materials such as glycerin and sodium lactate (15). In these studies, materials were evaluated in vitro at several humidity conditions ranging from a relative humidity of 20% to one of 90%. At all humidity conditions, the proteins and derivative exhibited poor results and in fact, at relative humidities of 79% or below, they had negative results. The glucose ethers and esters gave good results, as did the lactates and lactylates. It must be noted that at 20% relative humidity, no humectant was found to be effective in this study.

In in-vivo human studies conducted in the Henkel COSPHA USA AWT Laboratories, some interesting results were found (16). Using a dermal phasometry technique, several wheat proteins and their hydrolysates were contrasted with collagen and collagen hydrolysates. In these studies it was found that a 1.5% active wheat protein increased the moisturization of the skin by 20%. Collagen produced similar results. The hydrolysates of both wheat and collagen had little effect on moisturization of the skin.

Table 1 Classic Skin Care Formulations

	Percent in formulations				
	1	2	3	4	5
Mineral oil	15.0	10.0	—	—	—
Isopropyl palmitate	5.0	—	3.0	—	—
Cetearyl alcohol	2.0	1.0	1.5	3.0	2.0
Dicapryl ether	—	3.0	—	5.0	3.0
Coco glycerides	—	2.0	10.0	10.0	12.0
Stearic acid	2.0	1.5	—	—	—
Glyceryl monostearate	0.5	0.5	0.5	—	0.5
Ceteareth-20	—	0.5	2.0	1.5	2.0
Deionized water	qs	qs	qs	qs	qs
Glycerin	5.0	—	3.0	—	—
Propylene glycol	—	3.0	—	—	—

VIII. APPLICATION OF HUMECTANTS

A. Skin Creams and Lotions Using Humectants

Humectants are used in creams and lotions at levels of 1–5% typically to prevent surface dehydration of the product when exposed to air. In some creams, glycerin is used at a 10% loading to enhance the moisturization/remoisturization of the skin. Glycerin was traditionally used as an "active" ingredient in hand and body creams because it tended to reduce roughness, which is due predominantly to the dehydration of the uppermost levels of the stratum corneum. Some classic skin care formulations are given in Table 1.

In experiments conducted by Griffin, Behrens, and Cross and confirmed by Bryce and Sugden (17), it was determined that the amount of humectant to reach equilibrium at 70–75% relative humidity was about a 65% humectant solution. In the same work it was determined that at 40–60% relative humidity, equilibrium would be reached using a 85% humectant solution. Obviously, these are impossible amounts for normal personal care formulation. Bryce and Sturgen reported not only maxima but also minima. In this work it was found that as little as 1% humectant produced some effect. According to this, it seems that the normal use levels of 1–5% humectant in a cream or lotion is a reasonable compromise between maximum hygroscopicity and other desirable properties of the formulation.

Other investigators have found that certain combinations of humectants function more effectively in skin creams than when the humectants are used alone (18). For example, it has been claimed that when a mixture of glycerin,

Table 2 Classic Cleansing Formulations

	Percent in formulations	
	1	2
Hair care		
Ammonium lauryl sulfate	28.0	14.0
Sodium lauryl sulfate	—	14.0
Cocamide DEA	5.0	3.0
Cocamidopropyl betaine	7.5	5.0
Glycerin	5.0	3.0
Deionized water	qs	qs
Citric acid	pH 6.5	pH 6.5
Body cleansing		
Ammonium lauryl sulfate	20.0	35.0
Sodium lauryl sulfate	20.0	—
Cocamidopropyl betaine	5.0	6.0
Cocamide DEA	2.5	3.0
Glycerin	3.0	5.0
Sorbitol	2.0	—
Deionized water	qs	qs
Citric acid	pH 6.0	pH 6.0

sodium lactate, urea, and collagen are incorporated into a formula, they significantly improve the suppleness of skin.

The humectants normally used are of the organic class (glycerin, propylene glycol, sorbitol, 1,3-butylene glycol), and to a lesser extent the metal-organic class (sodium PCA, sodium lactate). Attention must be paid to the use level, since it has been reported that these humectants can actually remove moisture from the skin at very high levels and cause burning.

The addition of humectants to stearic acid-based creams with high stearic acid contents can reduce the problem of flaking. Sorbitol is probably the best in this regard, followed by propylene glycol and glycerin. Glycerin can, however, leave a stickiness to the skin after the emulsion has been applied and allowed to dry or "absorb." Sorbitol normally produces a soft, velvety feel, while propylene glycol is somewhere between glycerin and sorbitol.

B. Hair Applications for Humectants

Humectants are typically added to shampoos, body washes, and bath and shower gels at 2–5% by weight to prevent a surface skin from forming on the product. They also help to reduce the cloud point of the product, ensuring clarity in cold temperatures, and help to prevent package cracking. It must be noted that while there are beneficial effects from humectants in these applications, care must be used with regard to the amounts used. Humectants tend to reduce the foaming of surfactant systems. Since most consumers relate foaming with cleansing efficacy, any material that "de-foams" must be used judiciously. Some classic cleansing formulations are listed in Table 2.

The idea has been propagated that the addition of humectants to a shampoo helps to moisturize the hair shaft. Aside from large, very hygroscopic polymers, it is debatable whether any humectant remains on the hair shaft after shampooing and rinsing. The situation is different, however, with regard to leave-on conditioners, where the humectant will form a very thin film that helps to bind water to the hair shaft. Since the conditioner is of the leave-on type, the humectant film stays in place until it is washed or rinsed out of the hair. Newell describes how a humectant-containing hair treatment can help to restore moisture to severely damaged hair (19).

IX. FUTURE

Humectants will continue to have a place in personal care formulation for the reasons described above. In the near future, more "natural" humectants will be used over the synthetics. There is also a trend toward the use of polymeric humectants that are not based on ethylene oxide. These materials will offer the lasting power associated with polymers and still be very mild to the skin and eyes. Humectants are continually finding more applications in personal care products. This makes it important for the formulator to stay current with the latest technology.

REFERENCES

1. Balsam MS. Cosmetics Science and Technology. New York: Wiley-Interscience, 1972:198.
2. Jungermann E, Sonntag N, eds. Glycerine: A Key Cosmetic Ingredient. New York: Marcel Dekker, 1991:9.
3. Balsam MS, Sagarin E. Cosmetics Science and Technology. New York: Wiley-Interscience, 1972:198.
4. Budavari S, ed. The Merck Index. Rahway, NJ: Merck, 1989:1375–1376.
5. Humectants. In: Harry's Cosmeticology. New York: Chemical Publishing, 1982: 641.
6. Weast R, ed. CRC Handbook of Chemistry and Physics. Boca Raton, FL: CRC Press, 1981.
7. Humectants. In: Harry's Cosmeticology. New York: Chemical Publishing, 1982: 642.
8. Henkel Corp. Glycerine: An Overview. Ambler, PA: Henkel.

9. Pfiser & Pfiser. Advanced Organic Chemistry.
10. Budavari S, ed. The Merck Index. Rahway, NJ: Merck, 1989:1247.
11. Ibid.
12. Ajidew. N-50. Ajinomoto Corp.
13. Takahashi M, et al. A New Method to Evaluate the Softening Effect of Cosmetic Ingredients on the Skin. J Soc Cosmet Chem 19??; 35:171nnn181.
14. Shercomid AME-70 Technical Bulletin. Clifton, NJ: Scher Chemicals, July 1977.
15. 1980 Deshpande, Ward, Kennon, and Cutie published
16. Henkel Cospha USA AWT Laboratories.
17. Bryce and Sugden. J Pharm 1959.
18. Jungermann E, Sonntag N, eds. Glycerine: A Key Cosmetic Ingredient. New York: Marcel Dekker, 1991:346.
19. Newell GP. Method of restoring normal moisture level of hair with severe moisture deficiency. U.S. Patent 4,220,166 (1980).

6
Emollient Esters and Oils

John Carson and Kevin F. Gallagher
Croda, Inc., Edison, New Jersey

I. INTRODUCTION

This chapter will provide an introduction and survey of emollient esters and oils as they are encountered in personal care formulations. The purpose of providing this information is to create a conceptual "road map" which can be used by formulators to distinguish among the performance properties of a large class of ingredients, by relating their performance properties to the chemical and physical characteristics of the ingredients themselves.

According to the Oxford English Dictionary the definition of the adjective emollient is: "that which has the power of softening or relaxing the living animal textures." It is derived from the Latin verb *mollire*, meaning "to soften."

While a dictionary definition is helpful, it may be more useful to see how various industry authors have defined an emollient. Strianse (1) defined an emollient as "an agent which, when applied to a dry or inflexible corneum, will affect a softening of that tissue by inducing rehydration." The implication here is that esters and oils would induce rehydration by reducing the loss of water from the stratum corneum (typically measured by the transepidermal water loss, or TEWL). More recently, Idson has attempted to clarify the difference between a moisturizer and an emollient. His argument bears repeating: "There are only two cosmetic ways to treat a dry skin condition: either by attempting to hydrate it with externally applied water-miscible agents or by lubricating and occluding the skin with water insoluble materials. By usage, the former have come to be called moisturizers and the latter emollients or conditioners."

Since this is a broad class of ingredients which also includes triglyceride oils, the use of these compounds dates back to the earliest use of topically applied natural materials to effect some changes in the condition of skin or hair. As is the case with many other types of formulation components, the history of this class of materials has developed from the initial use of naturally occurring ingredients through the development of synthetic compounds that possess improved performance characteristics. In fact, the well-known cold cream formula of Galen (3) from the second century employed almond oil as the emollient.

Vegetable oils were the predominant emollients up until the late 19th century, when mineral oil and petrolatum became commercially available as products of petroleum oil distillation. Other emollients, like lanolin and hydrogenated vegetable oils, became commercially available in the early part of the 20th century. The development and availability of synthetic esters occurred during the 1930s and 1940s. Many of these esters were spinoffs from the development of synthetic lubricants (4) and so paralleled the intense military activity of this period.

II. NATURALLY OCCURRING ESTERS

Although these materials were the first used, they are still in wide use today because they perform so well. One of the driving forces in their continued popularity is the fact that these materials most closely mimic the lipid components which occur naturally in skin and hair, and are largely responsible for its condition.

A. Triglycerides

Triglycerides compose the largest group of oils and fats found in both vegetable and animal sources. Triglycerides are compounds in which three fatty acid radicals are united by oxygen (in an ester linkage) to glycerine. A sample chemical structure is shown in Figure 1. Since triglycerides occur naturally, we do not typically describe their biochemical synthesis. We do, however, need to be familiar with the breakdown reaction by hydrolysis, since these products include fatty acids and glycerine, and fatty acids are important components for the manufacture of synthetic esters, which will be described in later sections. The alkaline hydrolysis of triglycerides is typically carried out in the presence of water and is known as "fat splitting." A simplified description is provided in Figure 2.

The most important differentiating characteristic among the triglycerides is their fatty acid composition or "distribution." It is the component fatty acids which will contribute the particular properties which we associate with

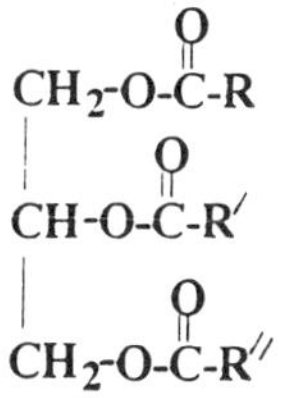

$$
\begin{array}{l}
CH_2\text{-}O\text{-}\overset{O}{\overset{\|}{C}}\text{-}R \\
| \\
CH\text{-}O\text{-}\overset{O}{\overset{\|}{C}}\text{-}R' \\
| \\
CH_2\text{-}O\text{-}\overset{O}{\overset{\|}{C}}\text{-}R''
\end{array}
$$

Figure 1 Triglyceride structure. R, R′, and R″ represent hydrocarbon chains in which the number of carbon atoms in the chain typically ranges from 7 to 21 and most commonly from 11 to 17.

triglycerides (Table 1). For instance, Tristearin is the triglyceride composed entirely of stearic acid ($CH_3(CH_2)_{16}COOH$). Its meling point is 69 to 70°C, and it is solid at room temperature.

Triolein is the triglyceride composed entirely of oleic acid ($CH_3(CH_2)_7CH{=}CH(CH_2)_7COOH$). It has a solidification point of 4°C, and it is liquid at room

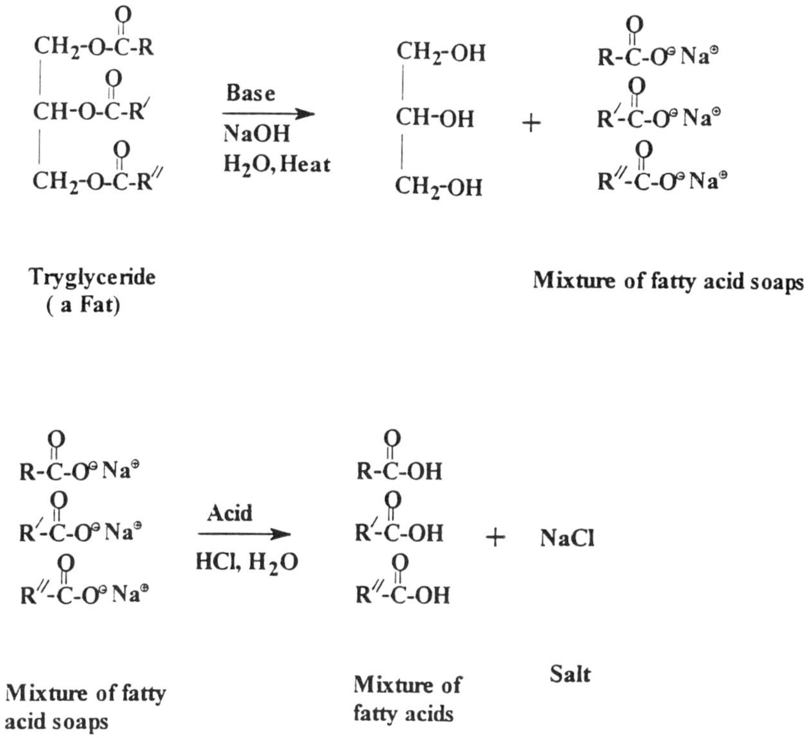

$$
\begin{array}{l}
CH_2\text{-}O\text{-}\overset{O}{\overset{\|}{C}}\text{-}R \\
| \\
CH\text{-}O\text{-}\overset{O}{\overset{\|}{C}}\text{-}R' \\
| \\
CH_2\text{-}O\text{-}\overset{O}{\overset{\|}{C}}\text{-}R''
\end{array}
\xrightarrow[\text{NaOH}\ \ H_2O\text{, Heat}]{\text{Base}}
\begin{array}{l}
CH_2\text{-}OH \\
| \\
CH\text{-}OH \\
| \\
CH_2\text{-}OH
\end{array}
+
\begin{array}{l}
R\text{-}\overset{O}{\overset{\|}{C}}\text{-}O^{\ominus}Na^{\oplus} \\
R'\text{-}\overset{O}{\overset{\|}{C}}\text{-}O^{\ominus}Na^{\oplus} \\
R''\text{-}\overset{O}{\overset{\|}{C}}\text{-}O^{\ominus}Na^{\oplus}
\end{array}
$$

Tryglyceride
(a Fat)

Mixture of fatty acid soaps

$$
\begin{array}{l}
R\text{-}\overset{O}{\overset{\|}{C}}\text{-}O^{\ominus}Na^{\oplus} \\
R'\text{-}\overset{O}{\overset{\|}{C}}\text{-}O^{\ominus}Na^{\oplus} \\
R''\text{-}\overset{O}{\overset{\|}{C}}\text{-}O^{\ominus}Na^{\oplus}
\end{array}
\xrightarrow[HCl,\ H_2O]{\text{Acid}}
\begin{array}{l}
R\text{-}\overset{O}{\overset{\|}{C}}\text{-}OH \\
R'\text{-}\overset{O}{\overset{\|}{C}}\text{-}OH \\
R''\text{-}\overset{O}{\overset{\|}{C}}\text{-}OH
\end{array}
+ \ NaCl
$$

Mixture of fatty acid soaps

Mixture of fatty acids

Salt

Figure 2 Alkaline hydrolysis (saponification) of triglycerides.

Table 1 Melting Points of Some Fatty Acids and Triglycerides

Name	Carbons	MP acid (°C)	MP triglyceride (°C)
Saturated			
Caprylic	8	16.7	8.3
Capric	10	31.6	31.5
Lauric	12	44.2	46.4
Myristic	14	54.4	57.0
Palmitic	16	62.9	63.5
Stearic	18	69.6	73.1
Arachidic	20	75.4	na[a]
Behenic	22	80.0	na
Unsaturated			
Oleic	18:1 cis	16.3	5.5
Elaidic	18:1 trans	43.7	42.0
Linoleic	18:2 cis	–6.5	–13.1
Linolenic	18:3 cis	–12.8	–24.2
Erucic	22:1 cis	33.4	30

[a]Not available.

temperature. Neither Tristearin or Triolein occurs naturally, but one can easily see how the physical form of a triglyceride could be deduced from its fatty acid distribution. The authoritative source for information about triglycerides and their fatty acid distributions is *Bailey's Oils, Fats & Waxes* (5).

1. Vegetable Triglycerides

Vegetable triglycerides, such as olive oil, were probably the original emollients. In general, vegetable triglycerides are called oils, whereas animal triglycerides are more commonly referred to as fats. This nomenclature alone provides a key to their respective physical forms and fatty acid distributions.

Vegetable oils are more likely to be fluid at room temperature. This fluidity is conferred by the composition and nature of their component fatty acids. Vegetable oils typically contain a higher proportion of unsaturated fatty acid triglyceride esters than animal fats. These unsaturated fatty acid esters contain at least one carbon-to-carbon double bond, so they are not "saturated" with hydrogen. The presence of this double bond creates a kink in the fatty chain. This kink, or bend, will make it more difficult for the molecule to form an ordered structure, which is necessary for solidification. A simple comparison between a saturated fatty acid, stearic acid, and an almost identical (but for two hydrogen atoms) unsaturated fatty acid, oleic acid, is shown in Figure 3.

Both the nature of the fatty acid distribution and the degree of fatty acid unsaturation are important concepts which will recur in our attempts to

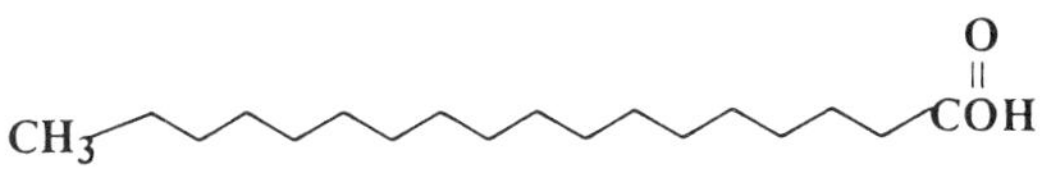

Stearic acid

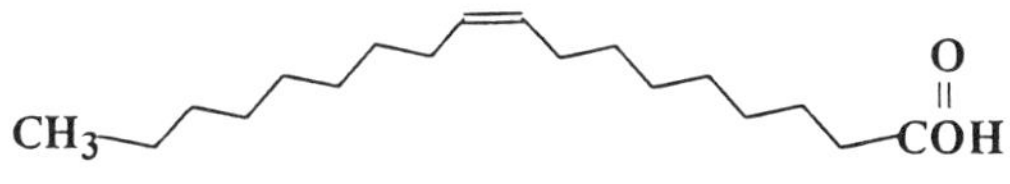

Oleic acid

Figure 3 Structures of stearic and oleic acid.

understand the relationship between the structure and the function of esters, including the synthetic esters we will encounter later.

Since vegetable oils are more likely to be fluid at room temperature, owing to their unsaturated fatty acid ester distribution, they have played a more important role as emollients. Clearly, a fluid material is more likely to be able to "soften" than a solid. Vegetable oils, such as olive oil, have long been used to soften the skin. This use predates the use of emulsions. In fact, one could view the development of the emulsion as the invention of a more aesthetically pleasing and efficacious way of applying vegetable oils to the skin!

Unfortunately, vegetable oils are not the perfect emollients. This is why their use was supplanted first by mineral oils, and more recently by synthetic esters. The advantage of fluidity and hence softening properties that is conveyed by the presence of unsaturated fatty acids in the triglyceride is balanced by the disadvantages conferred by these same unsaturates. The chief disadvantage is the lack of stability of the unsaturates, due to the susceptibility of the carbon to carbon double bond to chemical attack, especially by oxidation.

This oxidation of such double bonds is commonly referred to as rancidity, and it produces the unpleasant odor and taste that accompany these chemical changes. Clearly, this means that where vegetable oils are used, it is important to use antioxidants to avoid the unpleasant effects of rancidity.

Another disadvantage of the use of vegetable oils is their aesthetic ("feel") properties. While vegetable oils are certainly capable of softening the skin, they do so while contributing an oily feel to the final formulation, such as an emulsion. This is due in part to the nature and size of the triglyceride molecule. It contains three fatty chains, and this large proportion of fatty component no doubt contributes to the oily feel of the material.

The relatively large size of the triglyceride molecule is also a disadvantage from the standpoint of emulsification, since it is easier to emulsify a smaller molecule. In general, emulsification is somewhat easier when the emulsifier is larger than the compound to be emulsified.

Before leaving the subject of vegetable oils, it is important to include some brief mention of trace components. These trace components are nontriglyceride in nature and are often described as unsaponifiables, since they will not saponify in the presence of caustic to yield a fatty acid soap. Most often these unsaponifiables consist of sterols, or other hydroxyl containing compounds. Some of these polar compounds can have beneficial properties, such as the sterol fraction on avocado oil (5).

The secondary products of oxidation can also be present in trace quantities, depending upon the components of the oil, its oxidation, and the stability and storage history of the triglyceride. These secondary oxidation products include aldehydes and ketones, which can act as potential irritants, or further catalyze oxidative degradation. Recent advances in chromatographic purifications (6) have made possible the production of highly purified triglycerides that are virtually free of such components. The information in Table 2 provide the comparative fatty acid distribution for a number of well-known and less well known triglyceride oils.

Table 2 Fatty Acid Composition of Several Plant-Sourced Triglyceride Oils

Name	Composition wt% of fatty acids						
	Lauric ($C_{12:0}$)	Myristic ($C_{14:0}$)	Palmitic ($C_{16:0}$)	Stearic ($C_{18:0}$)	Oleic ($C_{18:1}$)	Linoleic ($C_{18:2}$)	Linolenic ($C_{18:3}$)
Almond	—	—	7	—	66	27	—
Apricot kernel	—	—	5	—	62	33	—
Avocado	—	11	3	—	69	15	—
Coconut[a]	48	16	10	2	7	—	—
Evening primrose	—	—	6	2	9	68	15
Olive	—	—	12	2	75	10	—
Peanut	—	—	8	4	62	20	—
Saflower	—	—	5	4	13	76	—
Sesame	—	—	8	5	40	47	—
Soybean	—	—	11	4	25	54	6
Sunflower	—	—	4	2	29	60	—
Wheatgerm	—	—	13	3	14	58	8

[a]Contains 7% caprylic acid ($C_{8:0}$) and 8% capric acid ($C_{10:0}$).

2. Animal Triglycerides

The same concepts that are important for the vegetable triglycerides are also of importance in understanding animal and marine triglycerides. These concepts are mostly related to the fatty acid distribution of the oils or fats and can be summarized as follows:

1. Hydrocarbon chain length of component fatty acids
2. Degree of unsaturation of the component fatty acids
3. Presence of trace or minor components

It is these properties that will determine the physical form of the oil or fat, as well as its feel on the skin and stability.

Animal oils or fats originate from a wide variety of animals and parts of animals. Most animal fats come from the subcutaneous fat layer. Examples include beef tallow and lard. Due to a higher content of saturated fatty acid-containing triglycerides, these materials are typically soft solids rather than fluid oils. An exception would be mink oil, where the subcutaneous fat is "winterized" (essentially chilled and filtered) to remove the more saturated components. This process provides an "oil" of animal origin containing a higher percentage of triglycerides containing palmitic oleic acid esters.

Animal oils can be from sources other than subcutaneous fat. Egg oil and milk (butter) fat are good examples.

Table 3 provides some composition information on a variety of animal fats and oils. It may be useful to remember that although animal fats are not typically used directly as emollients, they can be used as a source of fatty acids for the production of synthetic esters.

Table 3 Fatty Acid Distribution of Some Animal- and Marine-Sourced Triglyceride Oils

Name	Composition wt% of fatty acids							
	Myristic ($C_{14:0}$)	Palmitic ($C_{16:0}$)	Palmitoleic ($C_{16:1}$)	Stearic ($C_{18:0}$)	Oleic ($C_{18:1}$)	Linoleic ($C_{18:2}$)	EPA ($C_{20:5}$)	DHA ($C_{22:6}$)
Menhaden	11	19	10	3	15	1	11	8
Mink	4	16	18	2	42	18	—	—
Orange roughy	1	1	12	—	56	2	—	—
Shark liver[a]	2	12	7	4	30	6	—	10
Tallow	4	28	5	23[b]	38[b]	2	—	—
Lard	2	24	3	13	44	9	—	—
Milk fats[c]	10	30	4	11	25	2	—	—

[a]Shark liver oil contains 5% linolenic acid.
[b]Extremely variable amounts; these are average values of wide ranges.
[c]Contains significant amounts of lower fatty acids (~15%).

3. *Marine Oils*

The marine oils are an interesting third class of triglycerides since they have value not only for their physicochemical properties as emollients, but also for their potential in aiding the biological functioning of the skin. This potential for biological activity is especially present in the highly polyunsaturated oils from cold-water fish, such as salmon, anchovy, and menhaden.

Although a full description of the biological activity of these polyunsaturated fatty acid (PUFA) containing triglyceride oils is beyond the scope of this chapter, some brief description of these oils and their component fatty acids may be useful as an introduction to the subject. Cold-water fish, like those mentioned above, are rich in triglyceride content. These triglycerides contain two very important PUFAs: eicosapentaenoic acid (EPA), and docosahexaenoic acid (DHA). Their structures are shown in Figure 4.

Both EPA ($C_{20:5}$) and DHA ($C_{22:6}$) play an important role in human nutrition and biophysiology because they are precursors to prostaglandins, a class of biologically active compounds which help to control a variety of metabolic functions. Prostaglandins are made in the "arachidonic acid cascade." There are a number of sources of valuable information concerning the metabolism of these fatty acids.

Marine oil use in topical systems has been limited due to their extreme sensitivity to oxidation, by virtue of their highly unsaturated nature. The triglycerides also represent a serious formulating challenge since most marine oils have a "fishy" odor. In addition, as these oils oxidize (rancidify), they become increasingly malodorous, with a more paintlike smell.

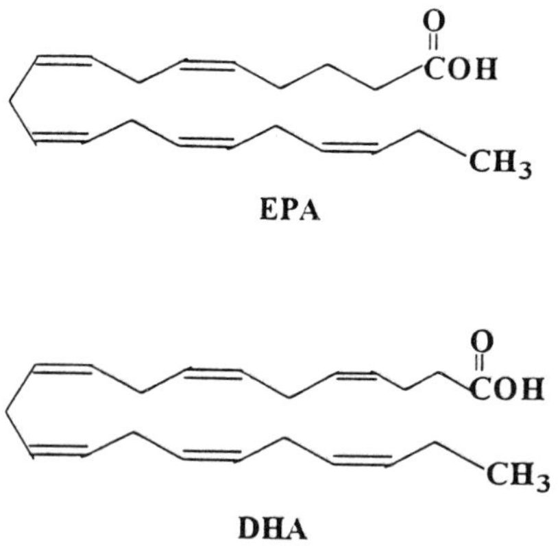

Figure 4 Structures of eicosapentaenoic and docosahexaenoic acids.

B. Lanolin and Lanolin Derivatives

Unlike triglycerides, where the alcohol fraction of the ester is glycerin, the alcohol fraction of the naturally occurring esters that comprise lanolin are very diverse. This alcohol fraction can be divided into two large classes of compounds: the aliphatic alcohols and the steroidal alcohols. Examples of the structures of these materials are shown in Figure 5 and Table 4. These various alcohols are present in their esterified form. The fatty acid portion of the esters also presents a diverse range, as described in Table 5.

Lanolin can be described as the purified form of wool wax, or wool grease. This is a naturally occurring lipid present in the fleece of sheep and typically obtained during the "scouring" of the wool to clean it prior to further processing. This lipid consists of a complex mixture of esters whose acid and alcohol fractions have been described in the previous figure and tables.

Lanolin has a long history of use as an emollient in cosmetics and pharmaceuticals; its first use dates to ancient times. Lanolin is a soft, tenacious solid with a melting point of approximately 40°C. Besides being useful as an emollient, lanolin has some beneficial properties as an auxiliary emulsifying agent

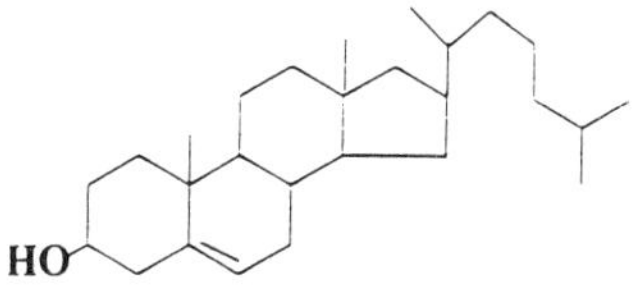

Cholesterol
$C_{27}H_{46}O$

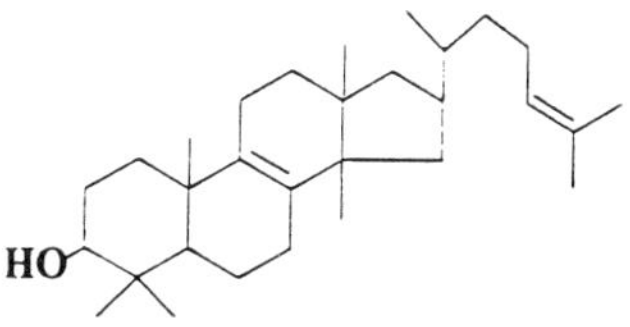

Lanosterol
$C_{30}H_{50}O$

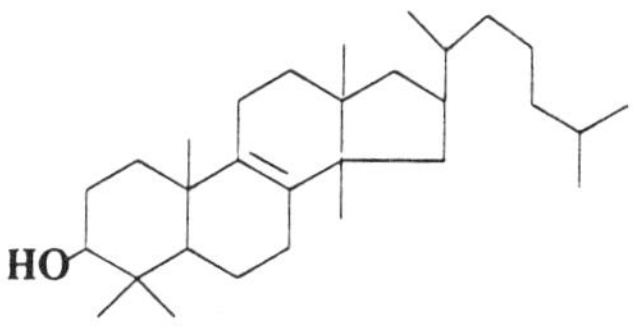

Dihydrolanosterol
$C_{30}H_{52}O$

Figure 5 Structures of cholesterol, lanosterol, and dihydrolanosterol.

Table 4 Alcoholic Constituents of Wool Wax or Lanolin

Constituents	Approximate %
Aliphatics	
n-alcohols: 7 members, octadecanol to triaceontanol	4
iso-alcohols: 5 members, 16-methylheptadecanol to 24-methylpentacosanol	6
anteiso-alcohols: 6 members, (+) 14-methylhexadecanol to (+) 24-methylhexacosanol	7
n-alkan-1,2-diols: 1 member, hexadecanediol	0.5
iso-alkan-1,2 diols: 4 members, *iso*-octadecanediol to *iso*tetracosanediol	3
Sterols: 5 members	
cholesterol	20
7-oxocholesterol	5
cholestane-3,5,6-triol	2
cholest-7-en-3-ol (and) cholesta-3,5-dien-7-one	2
Isocholesterol: 6 members	
lanostero	10
dihydrolanosterol	10
agnosterol	1
dihydroagnosterol	4
7,11-dioxolanost-8-en-3-ol (and) 7-oxolanost-8-en-3-ol	2
Hydrocarbons: number unknown structure unknown	1
Total	78
Unidentified residue	22

Source: Ref. 6.

of the water-in-oil type. The emollient properties of lanolin have been demonstrated by a number of workers.

1. Lanolin Oil

Lanolin itself can be physically separated into a fluid and a wax fraction. The resulting fluid fraction is known as lanolin oil. It is more lubricious than lanolin since it is a viscous fluid and not a soft solid at room temperature. The major advantage of lanolin oil is that it is clearly miscible with mineral oil in all proportions and does not form a precipitate. Thus, stable "absorption bases" composed of lanolin oil, lanolin alcohol, and mineral oil can be made which are useful formulating tools for emulsions.

Table 5 Acidic Constituents of Wool Wax or Lanolin

Constituents	Approximate %
n-acids: 9 members	
decanoic to hexacosanoic	7
iso-acids: 10 members	
8-methylnonanoic to 26-methylheptacosanoic	22
anteiso-acids: 12 members	
(+) 6-methyloctanoic acid to (+) 28-methyltriacontanoic	29
α-hydroxy-*n*-acids	
2-hydroxydodecanoic to 2-hydroxyoctadecanoic	25
α-hydroxy-*iso*-acids: 1 member	
2-hydroxy-16-methylheptadecanoic	3
Total	86
Unidentified residue (mostly unsaturated acids?)	14

Source: Ref. 6.

2. *Lanolin Derivatives*

A number of derivatives can be made from lanolin. Often these derivatives are made by first saponifying lanolin with sodium hydroxide and water thereby hydrolyzing the lanolin esters into their constituent acids and alcohols. After this, the constituents can be further reacted with a large number of compounds to alter their properties. Typically, these materials have altered physical forms as well as different solubility characteristics. The resulting esters will be described, with the other synthetic esters, in the next section.

C. Other Natural Esters

The manufacture of esters is one of the more elegant ways that an organism has to make compounds to protect itself and to store energy.

III. SIMPLE ESTERS

Emollient esters are made from a wide variety of acids and alcohols. Typically at least one of these fractions could be described as fatty in nature, meaning that it possesses a long hydrocarbon chain which would make that fraction soluble in oil and insoluble in water. This provides us with one useful way of categorizing these esters: namely, according to their constituent acid and alcohol fractions.

Of course these acids and alcohols could also possess more than one functional group. We could use polyhydric alcohols, polybasic acids, or hydroxy acids in our synthesis. This differentiation provides us with another organizational system. We can have esters made from one acid and one alcohol (simple esters) or those made from multifunctional acids or alcohols (complex esters).

As in the case of the triglycerides, many of the properties of the synthetic esters are directly related to the nature of the hydrocarbon chain contributed by the fatty acid portion of the molecule. In contrast, however, we now also need to concern ourselves with variations in the hydrocarbon chain of the alcohol fraction, since this will also contribute to the properties of the resulting ester. These parameters can be summarized as follows:

1. Hydrocarbon chain length of the component acid and alcohol
2. Degree of unsaturation of the component acid and alcohol
3. Presence of branching in either carbon chain
4. Presence of multiple acid or alcohol fractions
5. Presence of trace or minor components

The properties may be better understood if we examine the structure of a "typical" synthetic ester and its synthesis route. As an example we'll look at isopropyl myristate (IPM), a well-known ester used in personal care products. IPM is manufactured by combining isopropyl alcohol (IPA) and myristic acid. Typically this is done in the presence of an acid catalyst under conditions where the water produced by the esterification can be removed to allow the reaction to continue to completion. Figure 6 provides an illustration of this reaction. In this case, isopropyl alcohol quantities in excess of the stoichiometric amount needed for the reaction are frequently used since some IPA is removed with the water from the reaction.

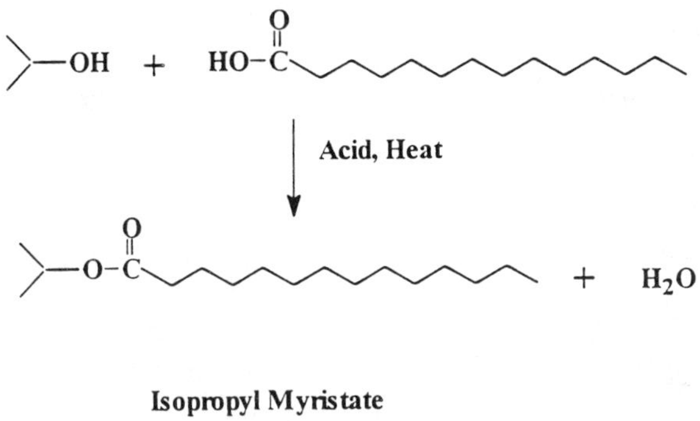

—OH + HO—C(=O)—

Acid, Heat

—O—C(=O)— + H_2O

Isopropyl Myristate

Figure 6 Reaction of isopropyl alcohol and myristic acid to produce isopropyl myristate.

A brief description of the structural parameters listed above will lead the reader to consider the infinite variety of esters that can be manufactured. Fortunately for the purposes of this review, there is one common characteristic of all the emollient esters used in personal care. They all possess at least one fatty chain (or similar long chain) contributed by either the acid fraction, the alcohol fraction, or both. This means that the fundamental nature of the synthetic ester revolves around the chemical modification of a fatty chain to achieve a desired set of properties.

Isopropyl myristate is perhaps the best known of the simple esters. We can engage in some reasonable speculation concerning the thought behind its development, based on our knowledge of the functionality of the triglycerides. The triglycerides, which function as fluid emollients, do so because of the presence of unsaturated fatty acid esters (typically oleic acid, $C_{18:1}$) in the triglyceride. While conferring fluidity and emollience, the oleate chains also contribute to the potential for oxidation, and hence instability.

In contrast, IPM is a fluid ester that does not rely on the unsaturation of the fatty component for its fluidity. Rather, it is due to the branched nature of the alcohol component. Branching helps to interrupt the orderly association of the long straight chain fatty components, which leads to the irregular packing of the molecules and hence fluidity. There are two important lessons to be learned from this: (1) changes in the alcohol fraction can "substitute" for variations in the acid fraction, and (2) branching can be a substitution for unsaturation with regard to fluidity.

In the case of IPM, the branched chain nature of the alcohol "substitutes" for the lack of unsaturation in the fatty acid portion of the molecule. We might refer to these two lessons as the "superposition principles" of ester development.

A. Straight-Chain Esters

This section will discuss the variety of physical properties and emollient ester characteristics that are obtainable from simple esters made from either straight hydrocarbon chain fatty acid or straight hydrocarbon chain fatty alcohols.

1. Acid Component Variations

Since we already have an understanding of the effect of variations in the fatty acid hydrochain for the triglycerides, it may be useful to review this information since it applies to the synthetic esters as well. As we increase the hydrocarbon chain length of the fatty acid, keeping the alcohol portion the same, we increase the melting point of the resulting ester. At the same time, we decrease the fluidity and increase the hydrophobicity. This will typically result in an oilier-feeling ester or, if the molecular weight is high enough, an ester with a waxier feel.

In our example of the simple ester IPM, we treated it like a single chemical entity. It must be stressed that this is typically not the case for cosmetic ingredients. Usually, an ester like IPM consists of isopropyl esters of a variety of fatty acids. IPM consists mainly, but not exclusively, of isopropyl esters myristic acid. Also present, depending on the fatty acid distribution of the starting material, will be quantities of isopropyl laurate, palmitate, and stearate. This fact is often overlooked in the comparisons of emollient esters since these comparisons are most often performed on the commercially available materials. Such comparisons are practically useful, but they often are filled with ambiguity, or at the least imprecision, when they are used for the scientific purpose of relating the structure of the ester to performance. We need to keep these limitations in mind when reviewing such information. These limitations, however, do not prevent us from making useful generalizations or interpretations based upon our understanding of the individual components.

The straight-chain, saturated fatty acids typically used in simple emollient esters range from lauric ($C_{12:0}$) acid to stearic ($C_{18:0}$) acid. Shorter hydrocarbon chains are not capable of delivering the nongreasy, lubricating emollient feel associated with cosmetic esters. Higher hydrocarbon chains are likely to produce esters with melting points well above skin temperature. It is difficult to consider these esters as emollients; they are probably better categorized as wax esters. Of course, the simple esters can also be based on unsaturated fatty acids such as oleic ($C_{18:1}$) and linoleic ($C_{18:2}$) acids, as well as the less common palmitoleic ($C_{16:1}$) acid.

Although specific comparative data for the isopropyl esters are not complete, based on the information available we can make some generalizations concerning these straight-chain fatty chains. As the molecular weight of the isopropyl esters increases, so do the solidification or melt point, viscosity, surface tension, and amount of ester that persists on the skin after 1 or 2 hours (Table 6).

With regard to the correlation of the skin feel of these esters in formulations to their chemical structure, the “Rosetta Stone” of such information is still the Goldemberg Article. In this paper, various esters are compared for their “initial slip” and “end feel” in a simple oil-in-water emulsion. Goldemberg notes that the isopropyl esters were often ranked the best in each series of esters studied. The data suggest that although the feel properties of the myristate and palmitate are very similar, the myristate scored slightly higher in initial slip but slightly lower in end feel. Perhaps the most interesting observation is that the isopropyl linoleate ($C_{18:2}$) ester is significantly poorer for both initial slip and end feel. Perhaps this is why the saturated-chain and branched-chain isopropyl esters have been preferred in actual use.

Table 6 Some Physical Data for Isopropyl Alcohol Esters

Fatty acid	MP (°C)	BP (°C)
Acetic	–73.4	90
Butyric	—	130-1
Caprylic	—	93.8
Capric	—	121
Lauric	—	196, 117
Myristic	—	192.6, 140.2, 167
Palmitic	13–14	160
Stearic	28	207
Oleic		223-4

Woodruff, in a more recent review of cosmetic esters, classifies all esters into three categories: (1) protective emollients, (2) nonocclusive emollients, and (3) dry and astringent emollients. Except for isopropyl isostearate, which is described as a protective emollient, the remaining isopropyl esters are described under the dry and astringent category. Woodruff's description of this category is that these esters are "used to reduce the greasy feel of vegetable oils in water in emulsions."

Briefly let us now consider the situation of esters of lower-molecular-weight alcohols. The methyl esters of fatty acids are certainly well-known compounds, but are not known as cosmetic ingredients. This could be because the hydrolysis products include methanol, which could cause irritation and could be toxic.

Ethyl esters are also well known compounds, but again, not as cosmetic ingredients. Ethyl oleate is used as a topical pharmaceutical agent, to enhance the skin penetration of lipophilic active ingredients. A pharmaceutical monograph exists describing the use of ethyl oleate for this purpose. Perhaps the enhanced skin penetration conferred by the unsaturated oleate is the reason the isopropyl esters of unsaturated acids (i.e., isopropyl linoleate) perform poorly in Goldemberg's subjective feel tests. They may penetrate into the skin and not remain at or near the surface to produce a subjective effect.

Since physiochemical data exist for the ethyl esters in the pure forms, it may be worthwhile for us to consider the variation in physiochemical properties with changes in the fatty acid component of the ester. These data are summarized in Table 7. It is worthwhile to point out the increase in melting point (MP) with the increase in carbon chain length of the fatty acid. The boiling point (BP) also increases with increasing molecular weight. As pointed out earlier, one of the great advantages of synthetic esters is that we are no longer limited to using the fatty acids typically found in natural triglycerides.

Table 7 Some Physical Data for Ethyl Alcohol Esters

Fatty acid	MP (°C)	BP (°C)
Acetic	–83.6	77
Butyric	–93.3	120
Caprylic	–43	208
Capric	–20	243
Lauric	1.8	273
Myristic	12.3	295
Palmitic	α:24, β:19.3	191
Stearic	31-3	199
Oleic	0.87	207
Linoleic	—	212

In addition to the fatty acids already discussed, we can use fatty acids that are far outside those occurring naturally—for example, benzoic acid. Examples of benzoate esters are illustrated in Figure 7, and some of the physical properties are summarized in Table 8.

As we would anticipate, we seen an increase in MP and BP with increasing moelcular weight. Benzoate esters tend to be quite stable both thermally and hydrolytically because they lack aliphatic beta hydrogens and therefore cannot form an enolic structure. Finally, benzoate esters have a very dry, nonoily feel

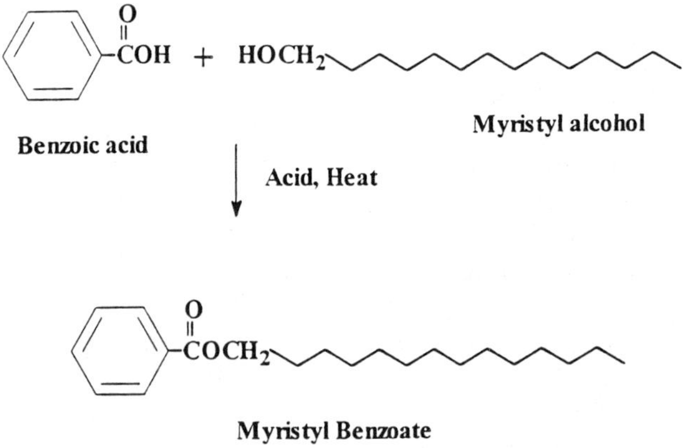

Benzoic acid + Myristyl alcohol → (Acid, Heat) Myristyl Benzoate

Figure 7 Reaction of benzoic acid with alcohols to produce benzoate esters.

Table 8 Some Physical Properties of Benzoate Esters

Fatty alcohol	MP (°C)	BP (°C)
Ethanol	–34.6	213.87
Propyl	–51.6	211
Isopropyl	—	218
Hexyl	113–117	272
Octyl	—	305–306
Cetyl	30	—

on skin and seem to be readily absorbed. Many formulators take advantage of this nonoiliness to reduce the oiliere feel of other esters and mineral oil.

2. *Fatty Alcohol Component Variations*

We began our discussion of simple esters by looking at the effects of variations in the acid components within the family isopropyl esters since these are popular, well-known cosmetic ingredients. We used them to illustrate how we might alter the properties of a simple ester by altering the nature of the fatty acid compound. Many of these generalizations will be applicable if we change the nature of the alcohol fraction of the ester.

Since we started with a low-molecular-weight, volatile alcohol (isopropyl alcohol), it seems logical to increase the molecular weight of the alcohol fraction of the ester and then determine if the generalizations regarding changes in the fatty acid component remain true. Also, we can determine how the properties of the esters are changed by increasing the molecular weight of the alcohol.

Referring to Table 9 one can see the variation in MP and BP caused by changing the fatty alcohol portion of esters of acetic acid. As is apparent from the table, the MP and BP both increase with increasing fatty alcohol hydrocarbon chain lengths.

As seen in Table 10, if the acid portion of an ester is small, i.e., formic or acetic acids, there are no dramatic variations in the melting point or boiling point of the esters, provided the acid fraction remains relatively small. This is because the alcohol fraction is large relative to the acid fraction and is predominantly responsible for the physical properties. This effect becomes more apparent as the size of the alcohol increases relative to the acid portion.

Hydrolytic stability is a major consideration for all esters. Possibly one of the reasons for the popularity of the isopropyl alcohol esters of fatty acids in preference to similar esters that can be made from a low-molecular-weight acid (such as propionic acid) and a fatty alcohol, is their improved hydrolytic

Table 9 Some Physical Data for Acetic Acid (Acetate) Esters

Fatty alcohol	MP (°C)	BP (°C)
Isopropyl	–73.4	90
Propyl	–95	101.6
Caprylic	–38.5	210, 112-3
Capric	–15	244, 125.8
Lauric	—	260–270
Cetyl	α: 18.5; β: 24.2	220–225
Stearyl	34.5	222–223

Table 10 Comparison of Some Physical Properties of Acetate, Formate, and Propionate Esters

Chemical compound	MP (°C)	BP (°C)
Ethyl formate	–80.5	54.5
Ethyl acetate	–83.6	77.06
Ethyl propionate	–73.9	99.1
Hexyl formate	–62.6	155.5
Hexyl acetate	–80.9	171.5
Hexyl propionate	–57.5	190
Octyl formate	–39.1	198.8
Octyl acetate	–38.5	210
Octyl propionate	–42.6	228

stability. It is important to consider that when an ester such as isopropyl myristate does hydrolyze, the resulting products are isopropyl alcohol and myristic acid. However, when an ester such as myristyl propionate hydrolyzes, the resulting components are myristyl alcohol and propionic acid. In this example, isopropyl alcohol would have a much more agreeable odor than propionic acid. Additionally, the propionic acid will lower the product pH possibly to a point where it will be detrimental to the product or consumer.

As with fatty acids, one must be careful of the exact chemical makeup of a particular ester when comparing physical properties, because the raw materials are not necessarily pure. The classic example is that of triple-pressed stearic acid which actually contains about 55% palmitic acid and about 45% stearic acid. The chemical composition of the fatty alcohols can vary just as markedly. Typically, stearyl alcohol will contain 65% stearyl alcohol and 35% cetyl alcohol. Esters made from this fatty alcohol will have the same distribution. This means that the resulting ester will be a mixture. Mixtures will generally have properties intermediate between the components. However, occasionally the mixtures will have "eutectic points" where the melting point of the mixture is much lower than that predicted arithmetically.

One major factor in the manufacture of esters is the cost of the respective fatty acids versus fatty alcohols. In most cases we find that the lauric, myristic, palmitic, and stearic fatty acids are significantly cheaper than their counterpart alcohols. This is because the fatty alcohols are usually produced by the reduction of the acid group to an alcohol. Therefore, it is generally less expensive to produce an ester from a low-molecular-weight alcohol and a fatty acid than from a higher-molecular-weight fatty alcohol and a small acid.

3. Modified-Chain Alcohols

Under the heading of simple esters, we have another type of ester which was developed to meet several specific needs. One of the needs was economic. As mentioned above, fatty alcohols can be significantly more expensive than fatty acids. One way to reduce this cost is to react the fatty alcohol with a compound of lower cost which will maintain the alcohol functionality, yet dilute the cost. Two materials that react in this manner are ethylene oxide and propylene oxide.

Ethyoxylation increases the water solubility of the molecule. Increasing the number of moles of ethylene oxide further increases the water solubility. Propylene oxide confers increasing alcohol solubility on the molecule with increasing moles of propylene oxide. In addition, propylene oxide also confers fluidity by virtue of the branched methyl group in the propylene oxide repeat unit.

When reacted with an acid, ethylene oxide adducts of fatty alcohols produce more polar esters. This is seen in terms of skin feel and adherence, i.e., lubricity enhancing and film-forming effects. If sufficient moles of ethylene oxide are used, the ester becomes water-soluble and functions as a surfactant. In the case of propylene oxide-modified fatty alcohols, one sees an increase in fluidity and a dramatic lowering of viscosity. Propoxylation can also change the solvency of the fatty acid esters. These effects can be seen in pigment wetting and in their ability to dissolve or solubilize other materials.

Another effect of alkoxylation is that the molecular weight of the fatty alcohol is greatly increased. For each repeat unit of ethylene oxide the molecular weight goes up by 44 daltons. The molecular weight goes up by 58 daltons for each propylene oxide repeat unit. This can produce a dramatic increase in the overall molecular weight. One of the effects of increasing molecular weight can be to reduce irritation. Generally, larger molecules are less likely to penetrate the skin to cause irritation. However, alkoxylation can change the solubilizing ability of an ester, making it more soluble in the skin and actually enhancing its penetration. Therefore, all esters must be carefully tested to determine their irritation potential.

Ethylene oxide addition can also be used to make "water-soluble esters." These materials are in fact surfactants, as they have both a water-soluble portion and an oil- or hydrocarbon-soluble portion. The ethoxylation can be carried out to such a degree that these esters become water-soluble to the point where they will actually foam. In use, it is felt that they confer a nonoily type of emolliency to a surfactant-based formulation such as a shampoo or bath gel.

While these are not truly modified alcohols, let us look at two "alcohols" that have been little explored commercially: hydroxy-terminated dimethyl polysiloxanes, and hydroxy-terminated perfluoroalkyls. These alcohols can be reacted with acids to make esters with unique properties. The advantages may not be readily apparent at first glance; however, the combination of a fatty acid and a perfluoro alcohol could impart water and oil repellency while giving the molecule some hydrocarbon compatibility to aid in formulation.

Similarly, a dimethylpolysiloxane alcohol esterified with a fatty acid provides a combination of silicone slip and feel with hydrocarbon compatibility. One may find unique applications for a silicone or perfluoroalkyl ester which could be more readily dispersed in a formulation. Silicone alkyl wax esters do exist commercially and provide interesting spreading and skin feel properties. These are easily formulated into the oil phase of an emulsion and help to make a cream that does not have a "waxy drag" when applied.

B. Branched-Chain Esters

As we have already seen, branching in the hydrocarbon chain of fatty acids causes the molecule to lose its ability to readily crystallize. This is because branching does not allow the close association of molecules, and therefore prevents crystallization. So, branched-chain molecules remain fluid over a much greater temperature range. The result of using a branched-chain fatty alcohol to make an ester is similar in that more liquidity is imparted to the ester. The melting point is reduced, the boiling point is reduced, and the esters are more fluid and have a higher spreading factor on skin. Also, the esters become less oily-feeling and can be used to reduce the oily feel of other esters or oils. Now, let us look at the effects of some specific branched-chain alcohols.

1. Short Branched-Chain Alcohols

The shortest of the branched short-chain alcohols is isopropyl alcohol (IPA). We have already discussed the properties of many of the esters of isopropyl alcohol (Table 6). One point to make is that IPA is a secondary alcohol as opposed to a primary alcohol. The esters of secondary alcohols are generally more stable to hydrolysis than the esters of primary alcohols. Therefore, one would expect to have formulations that are more stable with isopropyl myristate than if they were formulated with ethyl oleate. Indeed, IPM is used in

Table 11 Comparison of Some Physical Properties of 2-Ethylhexyl (Branched-Chain) and Octyl (Straight-Chain) Esters

Chemical compound	MP (°C)	BP (°C)
2-Ethylhexyl acetate	–93	199
Octyl acetate	–38.5	210
2-Ethylhexyl adipate	–67.8	214
Di(octyl) adipate	9.5–9.8	—

many different formulations without significant hydorlysis and must therefore be considered as providing adequate stability for cosmetic formulation purposes.

Now let us compare the properties of esters made from octyl alcohol versus 2-ethyl hexyl alcohol. Both alcohols contain eight carbons, but 2-ethyl hexyl alcohol is branched. From Table 11 it can be seen that the boiling points and melting points of the branched-chain alcohols are lower than the straight chains. This shows that the effects caused by branching of the alcohol versus branching of the acid are similar. So it doesn't really matter on which side of the ester the branching occurs. The point is that branching disrupts the ability of the molecules to associate closely, and this reduces its viscosity, its boiling point, and its melting point.

2. *Branched-Chain Fatty Alcohols*

Perhaps the best known of the branched chained fatty alcohols is isostearic alcohol. This is usually compared to stearic acid as an example of a branched-chain ester versus a linear ester. Esters made from linear fatty alcohols are higher in melting and boiling points and generally feel oilier or waxier than comparable esters made from a branched-chain alcohol.

Why would one choose a branched-chain alcohol when there are alternatives available that can reduce the oiliness and greasiness of esters? One answer to this question is reduced irritation potential. Irritation occurring with higher-molecular-weight fatty esters is generally much less than that seen with lower-molecular-weight fatty esters. Branching in an ester is one way to maintain a high molecular weight and confer luquidity upon the resulting ester. Hopefully, by this means one can make a fluid, low-viscosity, nonoily ester that is also low in irritation.

As you recall, liquidity can also be conferred by introducing a double bond into either the fatty acid or the fatty alcohol portion of an ester. When this is done, the resulting ester can suffer from reduced oxidative stability. This situation does not occur with branching. Thus, to make an ester that has improved

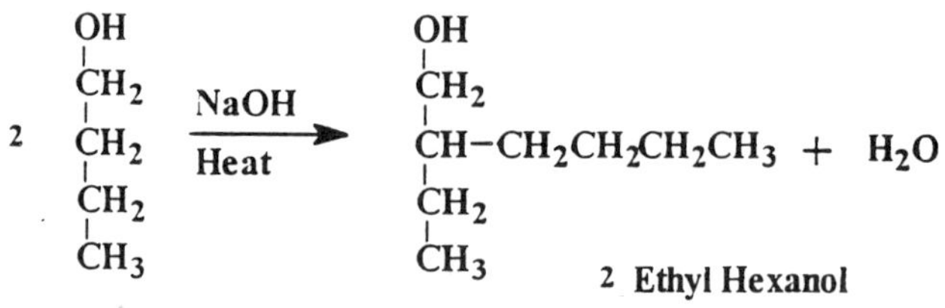

$$
2\;\begin{array}{c} OH \\ | \\ CH_2 \\ | \\ CH_2 \\ | \\ CH_2 \\ | \\ CH_3 \end{array} \xrightarrow[\text{Heat}]{\text{NaOH}} \begin{array}{l} OH \\ | \\ CH_2 \\ | \\ CH-CH_2CH_2CH_2CH_3 \\ | \\ CH_2 \\ | \\ CH_3 \end{array} + H_2O
$$

2 Ethyl Hexanol

Figure 8 Guerbet alcohol synthesis.

oxidative stability and yet remains liquid is another reason to choose a branched-chain alcohol ester.

In addition, branching can alter the skin feel of an ester. As an extreme example, when a Guerbet alcohol is used for the alcohol portion of an ester, the result is a liquid ester that has a high molecular weight and low irritation. Guerbet alcohols are the condensation product of 2 moles of alcohol in which the reaction proceeds under basic conditions through the elimination of a molecule of water. The reaction scheme is shown in Figure 8. As can be seen, the reaction of two moles of n-butanol results in one mole of 2-ethyl hexanol. Often, though, the two legs of the branched Guerbet alcohol are made fairly long such as by the condensation of n-octanol to make 2-hexyl decanol or n-decanol to make 2-octyl dodecanol; this dramatically affects the skin feel. The feel is described as "cushiony." This is a somewhat difficult concept to establish in words, but it describes the fact that the ester film is not readily absorbed by the skin and seems to act as a film or coating on the skin. This effect is seen with longer fatty branched-chain alcohols and with multifunctional alcohols or acids that produce di-, tri-, or tetrafunctional esters. These will be discussed in more depth below.

IV. COMPLEX ESTERS

The easiest way to describe complex esters is to compare them to simple esters. A simple ester is one that has a single ester group. Complex esters are made of either alcohols or acids, which have multifunctionality; therefore the resulting esters have more than one ester group.

A. Dibasic Acids

A great deal of work has been done by chemists in the esterification of dibasic acids. Much of this work was done prior to and during World War II to produce heavy-duty lubricating oils as replacements for the vegetable, animal, and petroleum oils that were then used as machine and motor lubricants (4). But, while much work has been done, not much of this work has been transferred to the cosmetics industry. At this point, we find ourselves with dioctyl sebecate, dioctyl adipate, and dioctyl maleate as the primary cosmetic esters of dibasic acids. The driving force for use of these esters is mostly economic. Adipic and sebacic acids are two of the least costly dibasic acids commercially available. In addition, 2-ethyl hexanol (octyl alcohol) is an inexpensive, readily available commercial alcohol. The dibasic acid esters can be made low in color (almost water white) and low in odor. The esters that are produced are very fluid, light-feeling and extremely nongreasy. Higher-molecular-weight alcohols produce more viscous and more oily products which may be more occlusive and lower in irritation. However, their use in cosmetic formulations is limited because of their oiliness.

Dibasic acid esters can enter into a formulation without one even being aware of it. If a cream or lotion is packaged in plastic bottles made from polyvinyl chloride (PVC) or polyethylene terephthallate (PETE), the plasticizer used in processing these polymers may be extracted from the plastic and into the product. The most common plasticizer used is dioctyl phthallate. This is an inexpensive plasticizer that is made in large quantities and used to soften (plasticize) PVC and PETE resins to allow them to be processed. You are probably familiar with dioctyl phthallate as the "new shower curtain" smell. As an odor, it is usually appreciated because it connotes "newness," but dioctyl phthallate in your formula is probably a contaminant. Its presence will usually go undetected unless it causes a problem. Normally, though, it is only noticed as a contaminant when gas chromatograms or high-pressure liquid chromatograms of the product are run.

So-called dimer acids are another type of dibasic acid. These are very high-molecular-weight acids usually made by dimerizing oleic or erucic acids. The resulting diacids therefore contain either 36 or 44 carbon atoms (Fig. 9). Esters made from these acids tend to be extremely oily and sticky due to their high molecular weight. But they have extreme tenacity to skin and form very occlusive waterproof films on skin.

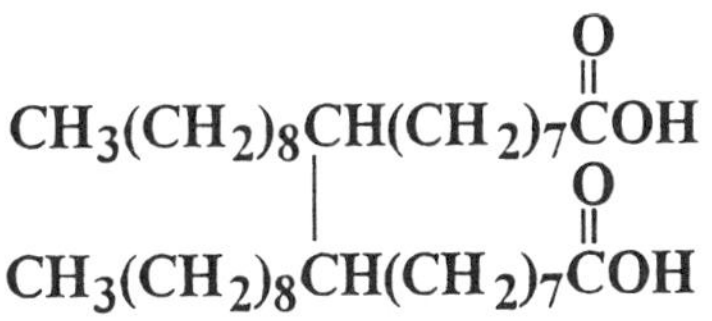

$$\begin{array}{l} CH_3(CH_2)_8CH(CH_2)_7\overset{O}{\overset{\|}{C}}OH \\ \qquad\qquad\quad | \\ CH_3(CH_2)_8CH(CH_2)_7\overset{O}{\overset{\|}{C}}OH \end{array}$$

Figure 9 Dimer acid structure.

Maleic acid is used to make a dioctyl ester, which has an extremely low viscosity with a high spreading coefficient. This ester finds use in spreading bath oils and in reducing the greasy feel of other esters. In addition, this ester is a good solvent for oxybenzone and therefore finds use in sunscreen formulations.

As the hydrogens on the center carbon are replaced by methyl groups, the molecular weight of the acid increases and therefore it will trend toward higher boiling points and melting points (4). Esters made from these acids show similar tendencies.

B. Polyhydric Alcohols

The simplest of the glycols is ethylene glycol. This material has been used for making esters and diesters for many years. The most common of these are the stearate esters. These are made by the reaction of stearic acid (usually triple-pressed stearic acid) and ethylene glycol in various ratios, to produce either a glycol monostearate ester or a glycol distearate ester. Variations on this theme have used diethylene glycol or triethylene glycol to add second or third ether linkages between the ester groups in order to increase the polarity and improve water dispersibility. These materials are used in surfactant products to produce opacity and pearlescence. They are also used in stick products to produce structure and in emulsions as the low HLB emulsifier component.

Ethoxylation of ethylene glycol or diethylene glycol is used to produce high-molecular-weight dihydroxy or polyether polymers of approximately 6000 daltons (or 150 oxyethylene repeat units). These materials are then esterified with stearic acid to produce viscosity-building agents. The well-known PEG-6000 distearate (PEG-150 distearate) is a material of this type. More recently, materials based on pentaerythritol ethoxylates have been made available. These are the tetrastearate esters of PEG-150 pentaerythritol. These materials have enhanced viscosity-building ability due to the three-dimensional structure conferred upon them by the central tetrahedral pentaerythritol molecule.

Similarly, esters can be made through the esterification of 1,2 or 1,3 propane diols. These materials are commercially produced as monostearate or distearate esters. Their uses are similar to those of ethylene glycol stearates in that they are waxy, fatty, solid esters used to provide structure in stick formulations and as pearlizing and opacifying agents in shampoos and detergent products. Caprylic/capric esters of propylene glycol are also available that find use in makeup formulations as emollients and as pigment-wetting agents.

Esters of neopentyl glycol (2,2 dimethyl-1,3 propane diol) and trimethanol propane are not in great evidence in the cosmetic industry, although these esters are used commercially as aircraft and automotive engine lubricants.

Table 12 Comparison of Some Physical Properties of Pentaerythritol Esters

Chemical compound	Specific gravity[a]	Surface tension[a]	Spreading factor[a]	Kinematic viscosity[a]
Pentaerythritol tetracaprylate/tetracaprate	0.950	24.7	4.7	56.0
Pentaerythritol tetraisostearate	0.922	30.4	2.9	305.4

[a]All measurements taken at 25°C.

The last member of the neopentylpolyol series is pentaerythritol. Esters of this material, as described before, are available as ethoxylates which are used as viscosity-building agents in surfactant systems. Also, esters of capric/caprylic acids and isostearic acids are available. The data for these two materials are presented in Table 12. The capric/caprylic esters are low-viscosity esters with a longlasting, lubricating skin feel. The capric/caprylic esters are used in makeup products due to their fairly low spreading ability, nongreasy feel, and ability to wet pigments. The tetraisostearate esters are much oilier because of their much higher molecular weight, but they are still liquid due to the branched natures of the isostearic acid and the pentaerythritol. They find use in sunscreens because of their low spreading ability.

Glycerin is one of the most common polyhydric alcohols and is the basic alcohol used by nature to make vegetable and animal oils. We have now returned to the triglycerides albeit from a synthetic direction in order to produce materials with new and potentially more useful physical properties. The first example of these new triglycerides is capric/caprylic triglyceride. These materials are C_8 and C_{10} straight-chain fatty acid triglycerides manufactured to produce a material with consistent properties and low color and odor. Capric/caprylic triglycerides have a moderate spreading ability and a very light, slightly oily feel.

As discussed previously, if the glycerin is ethoxylated prior to esterification, the resulting materials can be water-soluble. For example, the ester of capric/caprylic acids with PEG-6 glycerine or the ester of coconut fatty acids and PEG-7 glycerine are water-soluble and can be used as foaming agents and wetting agents that provide an emollient skin feel.

C. Multifunctional Acids and Alcohols

The combination of multifunctional acids and alcohols has not been used to any great degree for cosmetic purposes. These reactions produce polymers that are generally of high molecular weight and usually intractable. Modification of these polymers through the use of branched-chain fatty alcohols and fatty acids (or alkoxylation of these materials) has not resulted in commercially acceptable materials to date. However, as described earlier, polyhydric alcohols—multifunctional acids such as pimelic, tricarballic, and trimellitic—can be used to make cosmetic esters.

V. FUTURE CONSIDERATIONS

As already stated, much work has been done to synthesize esters and determine their properties. As we continue to explore new avenues of chemistry, product applications, and cosmetic skin uses, no doubt we will develop new products to meet the needs that are discovered. One of the major directions in which these products will go is toward safer, less irritating products. These will probably come from increased use of higher-molecular-weight, branched-chain alkyl groups, alkoxylation, or silicones, or possibly the use of fluorocarbons.

In addition, future efforts will be directed to better characterize the physical properties of esters. While there is much information available, it is widely dispersed and not readily available in one source. Current efforts are directed toward improving our understanding of the physical properties of esters and working to develop a more complete database.

As a further consideration, investigations are under way to develop measurements that will have greater use for the cosmetic industry. Specifically, these are an index to relate changes in viscosity to increases in temperature, and an index to compare skin spreading. The former could be envisioned as similar to the viscosity index used in the petroleum industry, which relates to decreases in viscosity to increases in temperature. The latter gives a measure of the ability of an ester (or another compound) to spread when applied to the skin. Comparative data for various esters and standardized methodologies to measure these indices are planned for future publication.

REFERENCES

1. Balsam M, Sagarin E. Cosmetics Science and Technology. 2nd Ed. Krieger, 1992.
2. The Merck Index. 11th Ed. Merck & Co., 1989.
3. Remington's Practice of Pharmacy. 11th Ed.
4. Synthetic Lubricants.
5. Swern D. Bailey's Industrial Oil and Fat Products. 4th Ed. John Wiley & Sons, 1979.
6. Truter EV. Woolwax Chemistry and Technology. Cleaver-Hume Press, 1956.

SUGGESTED READING

1. CRC Handbook of Chemistry and Physics, 72nd Ed., CRC Press, Boca Raton, FL, 1991–1992.
2. Heilbron IM. Dictionary of Organic Compounds, Vol. 1–3, Oxford University Press, 1943.
3. Leonard EC. The Dimer Acids, Humko Sheffield Chemical, 1975.
4. Chemfinder, camsoft.com-Internet
5. deNavarre MG. The Chemistry and Manufacture of Cosmetics, 2nd Ed., Continental Press, 1975.
6. Goldemberg RL and de la Rosa CP. Correlation of Skin Feel of Emollients to their Chemical Structure, J. Soc Cosmet Chem 1971; 22(10):634–654.

7
Proteins for Conditioning Hair and Skin

Gary A. Neudahl
Costec, Inc., Palatine, Illinois

I. INTRODUCTION

Proteins make up the majority of the dry weight of living cells (1), fulfilling both structural and functional roles. The building blocks of proteins are amino acids, monomers which are so named because each has, in addition to a hydrogen atom and an R group, an amino group (generally unsubstituted) and a carboxyl group attached to a central carbon atom. The identity of the R group defines the specific amino acid. At neutral pH this group may be anionic, cationic, polar, or nonpolar (Table 1). Regardless, the monomer is amphoteric. Twenty genetically encoded amino acids are commonly found in proteins. Additional amino acids, as well as non-amino acid components, may also be present as a result of posttranscriptional modification.

Proteins are polymers generated through the formation of peptide bonds between amino acids. These bonds are formed from amino and carboxyl groups with the consequential loss of a molecule of water and two charged groups per amino acid added. Thus, as polymerization proceeds, the growing molecule (polypeptide) becomes decreasingly hydrophilic. The number of amino acids constituting a protein subunit is typically 50 to 1000 (2). With multiple subunits, the overall molecular weight for a protein is typically between 5,000 and 7,000,000 daltons (Da) (1). The genetically determined order of attachment and the physicochemical environment yield the intra- and intermolecular covalent, ionic, hydrogen, and nonpolar bonding which produce a characteristic three-dimensional structure. When this structure matches that

Table 1 Amino Acid R-Group Classifications

Cationic	Polar	Nonpolar
Arginine	Asparagine	Alanine
Desmosine[a]	Cysteine	Glycine
Histidine	Glutamine	Isoleucine
Hydroxylysine[a]	Hydroxyproline[a]	Leucine
Isodesmosine[a]	Serine	Methionine
Lysine	Threonine	Phenylalanine
Anionic	Tyrosine	Proline[b]
Aspartic acid		Tryptophan
Glutamic acid		Valine

[a]Amino acids not genetically coded; synthesized via posttranscriptional modification.
[b]The only genetically coded imino acid (R group and amino group fused to form a ring).

found in nature (i.e., when a protein is in its native conformation), the protein's inherent biochemical properties, including any enzymatic activity, may be observed.

A. Types of Proteins

Fibrous proteins (protein chains arranged in parallel) include the water-swellable (but not soluble) structural proteins keratin, collagen, and elastin, major components of hair, skin, and connective tissue, respectively. Intermolecular crosslinks are common in these proteins, particularly those containing the monomer cysteine ($R = -CH_2-SH$). When oxidized, two cysteines form one cystine linkage ($-CH_2-S-S-CH_2-$). Such bonds are responsible for hair's strength as well as its ability to be reconfigured, or "permanent waved."

Globular proteins (protein chains compactly folded) include many which are water soluble, such as serum albumin, hemoglobin, and catalase. As a general rule, these proteins are more susceptible to denaturation (permanent loss of native conformation and hence biochemical function) at high temperature, low concentration, extremes of pH, and air/solution interfaces (i.e., in foams) than are fibrous proteins (2).

Some proteins, such as myosin and fibrinogen, contain elements of both fibrous and globular proteins (long length coupled with water solubility). Others have various prosthetic groups, e.g., the various lipids in the lipoproteins of blood plasma, the iron protoporphyrin of hemoglobin and catalase, the sugars and their derivatives of gamma-globulins, and the phosphated serine residues of casein.

B. Isolation of Proteins

Purification of proteins requires their separation from other organic materials and from one another. Among the methodologies which may be utilized, depending on the purity required, are fractional crystallization from salt solution, precipitation at the isoelectric (isoionic) point (the pH at which the protein has no net charge), isoelectric focusing, gel filtration, electrophoresis, ultracentrifugation, and various forms of chromatography (2).

In many cases, partial hydrolysis is required to separate protein subunits from one another or from matrix material and to promote water solubility. Hydrolysis enhances water solubility at near neutral pH, in part by increasing the number of charged groups per unit chain length (each peptide bond cleaved yields an amino group and a carboxylate group). Limited hydrolysis often reduces or eliminates enzymatic activity, and as the extent of hydrolysis increases (yielding polypeptides and, ultimately, amino acids) all enzymatic activity is lost. What remains is functionality related to specific amino acids present in the polypeptides, or, in the event of total hydrolysis, the functionality of the individual amino acids.

II. HISTORICAL DEVELOPMENT

Perhaps the earliest purported application of unisolated protein in personal care was the story of Cleopatra bathing in milk (3), with the milk proteins (in part due to their phosphated serine residues) functioning as effective emulsion stabilizers, buffers, and conditioning agents. More recently, prior to the widespread use of high-purity soaps and synthetic surfactants, eggs were utilized as shampoos. Their use in hard water yielded improved hair luster while avoiding the dullness which accompanied soap scum.

A. Early Production

Production of a partial hydrolyzate of collagen (gelatin) was reported as early as the late seventeenth century in Holland, with the first commercial production occurring in the United States in 1808 (4). These accomplishments preceded an understanding of the chemical composition of the material: the amino acid glycine was purified from gelatin in 1820 by H. Braconnot (1), but gelatin's physical and chemical structure were not elucidated until the early twentieth century. Some typical gelatin production processes are shown in Figure 1.

The vast majority of early protein hydrolyzates were derived from animals, and for good reason: significant portions of the animals slaughtered for food were unsuitable for this or other purposes unless modified and so would

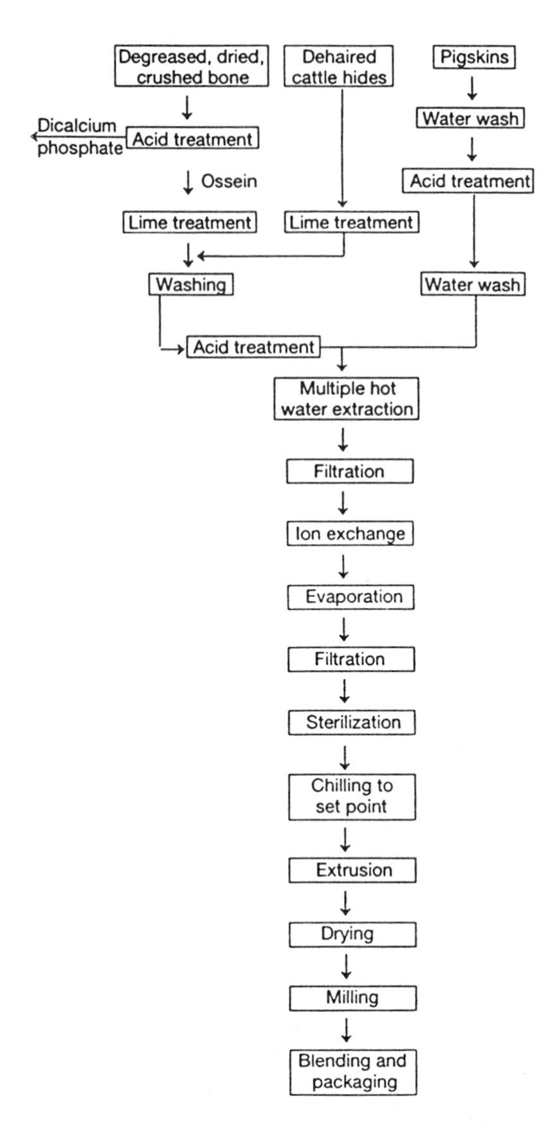

Figure 1 Some typical gelatin production processes. (From Ref. 4.)

otherwise be discarded. So methods were developed to produce protein hydrolyzates of skin, connective tissue, and bone (e.g., gelatin and hydrolyzed elastin), concurrently providing an environmental benefit. Plant-derived protein hydrolyzates arose primarily to meet a new definition of environmental friendliness (5).

B. Use of Hydrolyzates in Personal Care Products

The first protein hydrolyzates developed for personal care applications were reported to have been prepared in the early 1900s by the German company Chemische Fabrik Gruenau Berlin (3). They arose from observation of the alkaline dyeing of wool (which, like human hair, consists primarily of keratin). Dye baths laden with protein hydrolyzate proved markedly less damaging to the wool. Subsequent testing confirmed similar dramatic benefits for collagen hydrolyzates at a 10% use level in reducing human hair damage during bleaching and permanent waving (6) and identified the preferred method of hydrolysis (enzymatic) and optimal molecular weight [1000 to 2000 Da (7,8)] and pH [near neutral for bleaching hair, slightly alkaline for virgin or permanent waved hair (9)] for maximum substantivity. Quantification of hydroxyproline (an amino acid found almost exclusively in collagen) content of collagen-treated materials proved an effective means of establishing substantivity (10). An early patent in this area references the use of gelatin on previously tinted hair to resist dye action during the coloring of new growth (11).

1. Protein Effect on Hair

Testing also affirmed that substantial quantities of at least some hydrolyzates penetrated through the cuticle (hair's outermost, shinglelike protective layer) into the cortex (the fibrillar, main structural component) and that the amount of hydrolyzate bound increased markedly with increasing damage (virgin < bleached ≪ bleached and waved) (12). These hydrolyzates were shown to reduce cuticle damage and fiber embrittlement when present during the bleaching and waving processes (6).

2. Protein Effect on Skin

Evaluation of hydrolyzates applied to human skin revealed that penetration was limited to the outer layers of the stratum corneum (4,8). Hydrolyzates thus function as moisturizers and, as a result of their film-forming properties, irritation mitigants in the presence of strongly anionic surfactants (13).

C. Improvements in Manufacturing

The collagen hydrolyzates used in the aforementioned studies were dark in color, high in ash (inorganic salts), and malodorous. The efforts of many producers gradually yielded lighter colors, lower ash, and less objectionable odor, allowing their application in a broader range of products. Further, with increasing interest in products from renewable (i.e., plant) sources, protein hydrolyzates from other sources—animal, plant, and microbial—have been commercialized and International Nomenclature Cosmetic Ingredient (INCI) names (14) assigned, yielding a wide range of materials and potential functionalities from which to choose.

Building on the development of hydrolyzates, a variety of condensation products were then generated and evaluated for application in personal care (4,8). Reaction of fatty acid chloride (RCOCl, R = C_8–C_{18}) with primary amino groups of the hydrolyzate in alkaline media attached the R group via an amide linkage to the hydrolyzate. These hydrolyzates thus became anionic surfactants. Depending on the length of the polypeptide chain and the identity of the R group, mild wetting, foaming, conditioning, plasticizing, and emulsifying properties were achieved. Neutralized C_{12} and C_{14} versions (e.g., TEA-lauroyl hydrolyzed collagen) behaved similarly to synthetic detergent but with much less eye and skin irritation. Their C_{16} and C_{18} counterparts (e.g., potassium stearoyl hydrolyzed collagen) behaved similarly to soaps.

Reaction of fatty tertiary amines with primary amino groups attached these moieties to the hydrolyzate. A quaternary nitrogen atom resulted, imparting cationic character to the hydrolyzate, which was maintained at high pH (>11). These condensates were thus more substantive to hair and skin than the unmodified hydrolyzates and imparted conditioning benefits to hair and skin. Indeed, quaternium-76 hydrolyzed collagen, the coconut-based quaternary hydrolyzate, was as effective as stearalkonium chloride at a 2% actives level in improving the wet combability of hair and was markedly less irritating to skin and eyes (15).

Additional modifications have been made since by a variety of manufacturers, and representative products will be described in the sections which follow.

III. GENERAL FORMULATING CONSIDERATIONS

Certain formulation guidelines apply regardless of intended area of application. Among the considerations are microbiological preservation (proteins, their hydrolyzates, and most derivatives are excellent nutrients), odor and color stability (free amino and sulfhydryl groups are reactive toward a number of commonly employed cosmetic ingredients and substrates), conformational stability (enzymes retain their activities only when in, or very close to, their native states), molecular weight (humectancy, film forming, and the potential for allergenicity vary with polymer size), amino acid composition (humectancy is greater with more charged amino acids), ash (salt) content, and net charge (substantivity is highly dependent on pH).

A. Preservation of Protein-Based Materials

Proteins are components of every living thing, attesting to the broad-based compatibility of their monomers, the amino acids. It should thus not be surprising that proteins and their hydrolyzates and derivatives are readily biodegradable. A suitable preservation system must thus be incorporated in products containing both water and protein-based material. The protein products sold as aqueous solutions utilize a variety of preservation systems. Most frequently employed are the parabens (methyl- and propylparaben, in particular). Finished products containing proteins in their native states (generally present to make use of their enzymatic activities) require preservatives which have very low reactivity toward the protein and so do not appreciably modify the conformation of the protein. A combination of parabens and phenoxyethanol at a level of up to 1% by weight of the formulation may be employed. When hydrolyzates and their derivatives are utilized, more reactive (and more cost-effective) preservatives can be utilized. Such agents may include quaternium-15, imidazolidinyl urea, diazolidinyl urea, DMDM hydantoin, and methyl[chloro]isothiazolinone, with or without parabens. Specific preservation systems are referenced by some manufacturers (16,17). Certain preservatives are incompatible with proteinaceous matter, particularly at high concentration. A striking example is the firm gel produced after overnight storage at room temperature of a solution of approximately equal proportions of 37% formaldehyde and 55% hydrolyzed collagen (2000 Da).

B. Protein Formulation Stability Concerns

Odor and color stability can be significant concerns when formulating with protein-based materials. As a start, proteins and hydrolyzates must be stored under cool, dry conditions, taking precautions to prevent contamination. However, even with proper storage, the free amino groups are potential reaction sites for carbonyl and aldehyde groups. As a result, formulation with fragrance oils containing these groups may result in reaction with the free amino groups to produce changes in odor and/or color of the product. Color changes are also an area of concern when formulating with sugars, as their aldehyde groups may react with free amino groups via the Maillard reaction to induce brown color formation.

1. Using Proteins Containing Sulfhydryl Groups

Use of protein-based material containing sulfhydryl groups presents its own unique challenges. The hydrolyzed proteins in particular have a sulfurous odor which requires some measure of masking. Further, these sulfhydryl groups (present as the amino acid cysteine, in polypeptides containing this amino acid or in derivatives thereof) are frequently employed due to their capability to undergo sulfhydryl exchange, particularly with the keratin of hair. Oxidation of these groups to cystine linkages or cysteic acid, which may occur upon exposure to air or oxidizing agents in the product, reduces the effectiveness of interchange and changes the odor of the product. Further, precipitation of oxidized ("dimerized") material often results. Auxiliary reducing agents are therefore often employed to maintain the sulfhydryl group's reduced state. When hydrolyzate odor must be minimized, use of the spray-dried version of a product may prove helpful, as a portion of the odor-producing molecules are volatilized and removed during the drying process.

2. *Maintaining Conformation Stability of Proteins*

Conformational stability is of concern for both structural proteins and enzymes, the former so that consistent binding and tactile properties are ensured, the latter so that enzymatic activity is maintained. As noted previously, permanent loss of native conformation is prevented by avoiding high temperature, low concentration, extremes of pH, and foaming. Unfortunately, the acceptable ranges differ widely from protein to protein and must be determined empirically (e.g., for soluble collagen, temperatures above 30°C should be avoided).

C. Effects of Molecular Weight on Protein Properties

Molecular weight primarily determines a number of properties of hydrolyzates. The molecular weights cited for hydrolyzates represent average (or median) values for relatively broad (and frequently nongaussian) distributions. Thus, a range of 1000 to 5000 Da for a hydrolyzate may be considered "narrow." Humectancy (hygroscopicity) is greatest for individual amino acids, which can absorb several times their weight in water at high (>80%) relative humidity. Humectancy diminishes exponentially as the size of the polypeptide increases. Nonocclusive, protective colloidal film-forming properties increase concurrently. At low molecular weight, hydrolyzates may be considered nontoxic (18) and hypoallergenic. The potential for toxicity or allergic response increases with molecular weight, however, with the threshold dependent on protein type, conformation, prosthetic groups present (19), and site of contact. Allergic reaction is most likely to occur when material is inhaled as a dust or aerosol (20).

D. Effects of Protein Amino Acid Distribution

Amino acid composition is important not only because of its effect on conformation, but also because of its effect on humectancy and solubility. Both humectancy and water solubility of proteins and hydrolyzates increase as the proportion of amino acids with charged R groups increases. Thus, depending on the particular attributes desired in the finished product, one protein source may be preferred over others, based on its amino acid content. The amino acid profiles for a number of proteins are presented in Table 2 (as numerical percentages, i.e., mol%), Table 3 (animal and yeast proteins as weight percentages, i.e., %wt), and Table 4 (plant proteins, as wt%, i.e., g/100 g). Asparagine and glutamine typically assay as aspartic acid and glutamic acid, respectively, as the amide bonds in their R groups are hydrolyzed when utilizing the classical assay procedure. Source-dependent variations in apparent composition are evident in each table.

E. Effects of Other Protein Components

Inorganic salts (typically measured as ash) may remain in a protein-based product through its isolation, be added in the form of a buffer, or result from the alkaline or acid hydrolysis of a protein or its subsequent derivatization. For many formulations, inorganic salts present no concern as long as their level is consistent from lot to lot. On occasion, however, they may adversely effect emulsion stability in creams and lotions or yield cleansing systems with viscosity already over the salt curve (i.e., salt content higher than required to achieve maximum viscosity in the system). In such cases, differently processed material may resolve the problem.

F. Importance of Protein Net Charge

The net charge of protein-based products in finished formulations is important for several reasons. First, substantivity is higher when hydrolyzates have a net positive charge (i.e., are below their isoionic points), as both skin and hair have numerous anionic sites available for binding under typical conditions. Second, the solubilities (and frequently stabilities) of native proteins are typically lowest at their isoionic points. Thus, manufacturing procedures for finished products are best designed to preclude fluctuation of batch pH back and forth across the isoionic point of a purified protein. Third, many proteins consist of several subunits which assemble properly only within a certain pH range. Further, even many single-subunit proteins require a certain pH range to take on the correct conformation to exhibit enzymatic activity.

IV. HAIR CONDITIONING

Consumers know, whether through folklore or experience, that proteins (actually, protein hydrolyzates) and amino acids work to condition and moisturize hair effectively (21). Nonetheless, concerns over what constitutes "protein" in products resulted in a legal definition in the United States (22). Selected underivatized hydrolyzates with demonstrated performance attributes follow.

Table 2 Amino Acid Profiles (mol%) for Selected Protein Sources

Amino acid	Collagen	Wool keratin	Human hair	Silk [S] fibroin	Silk [B] fibroin	Pearl conchiolin	Milk casein	Wheat protein	Wheat gluten	Soya [B] bean	Soya [S] bean	Oat protein
Alanine	11.6	6.3	4.7	30.6	29.82	14.0	4.5	4.2	3.4	3.7	7.2	8.46
Arginine	3.6	6.9	6.3	0.1	—	1.4	2.8	0.1	2.3	7.3	1.6	4.78
Aspartic acid[a]	4.6	6.7	5.4	2.1	1.17	9.2	7.3	3.0	3.1	11.4	15.3	15.42
Cystine/2	—	7.8[b]	16.0[b]	Trace	—	—	0.4	2.2	0.9	1.3	0.3	1.03
Glutamic acid[a]	7.6	14.9	12.9	1.5	0.41	3.0	20.6	39.6	36.9	20.8	23.6	24.34
Glycine	33.8	7.0	6.1	42.9	45.99	42.9	3.2	6.8	5.8	3.9	9.5	11.65
Histidine	0.4	1.0	1.0	Trace	—	0.2	2.8	2.2	1.6	2.7	3.5	1.39
Hydroxylysine	0.2	—	—	—	—	—	—	—	—	—	—	—
Hydroxyproline	10.6	—	—	—	—	—	—	—	—	—	—	—
Isoleucine	1.3	3.5	3.3	1.0	0.33	3.2	4.4	1.6	3.3	4.0	3.8	2.10
Leucine	2.7	8.1	6.2	0.6	0.17	6.3	7.2	6.5	6.8	6.3	7.9	6.19
Lysine	2.5	2.8	2.8	0.5	—	1.2	7.3	1.1	1.3	2.7	4.4	3.16

Methionine	0.3	0.9	0.6	Trace	—	0.4	2.4	1.4	1.1	1.3	1.0	0.98
Phenylalanine	1.3	2.3	1.8	2.3	5.69	1.2	3.0	4.5	4.6	3.9	3.5	1.09
Proline	11.5	8.7	7.0	0.3	—	2.7	12.3	14.4	14.0	5.6	7.2	5.20
Serine	3.1	7.9	10.5	9.7	11.45	7.5	7.7	6.1	5.6	4.7	3.2	6.75
Threonine	1.8	5.5	7.2	0.9	1.12	2.3	5.1	2.1	2.6	3.7	1.2	3.26
Tyrosine	0.1	2.7	2.3	4.9	1.66	Trace	3.3	1.5	2.4	1.2	1.6	1.05
Valine	3.0	6.8	5.8	2.6	2.19	4.5	5.7	2.7	4.3	4.2	5.2	3.13

[a] See comments in text regarding asparagine and glutamine content.
[b] Cysteic acid, an oxidative reaction product of cyst[e]ine, was present at 0.2 mol% in wool keratin and 0.1 mol% in human hair.

Sources:
Collagen, silk [S] fibroin, pearl conchiolin, milk casein, wheat protein and soya [S] bean from Promois Digest, Seiwa Kasei Co., Ltd., May 1995.
Wool keratin and human hair from Promois WK-Q technical data sheet EKPHOO, Seiwa Kasei Co., Ltd., Apr. 26, 1996.
Silk fibroin [B] from Solu-Silk Protein Data 1301/Rev1, Brooks Industries.
Wheat gluten from WPH-DGF 1191 Product Information, DGF Stoess, Feb. 1992.
Soya [B] bean from Solu-Soy EN-25 Technical Data 1400/1, Feb. 1990, Brooks Industries.
Oat from Hydrolyzed Oat Protein HOPA Product Information, Canamino, Inc.

Table 3 Amino Acid Profiles (g/100 g) for Selected Animal and Yeast Protein Sources

Amino acid	Pork skin gelatin (type A)	Calf skin gelatin (type B)	Bone gelatin (type B)	Collagen [H]	Fish collagen	Cattle hide collagen	Collagen [C]	Bovine collagen	Human hair	Hair keratin	Bovine elastin	Yeast
Alanine	8.6–10.7	9.3–11.0	10.1–14.2	9.3	8.6	9.4	8.0–11.0	9.6	4.5	5.5	18.2	7.16
Arginine	8.3–9.1	8.55–8.8	5.0–9.0	8.3	16.7	7.4	7.8–9.0	7.3	9.1	11.0	1.5	9.31
Aspartic acid[a]	6.2–6.7	6.6–6.9	4.6–6.7	6.3	3.9	5.5	5.7–9.0	5.6	9.0	9.4	2.0	9.21
Cystine/2	0.05	Trace	Trace	—	—	—	0.0–0.9	—	1.3	1.3	—	—
Glutamic acid[a]	11.3–11.7	11.1–11.4	8.5–11.6	9.6	4.2	9.5	10.0–11.7	10.6	18.1	20.0	3.9	9.38
Glycine	26.4–30.5	26.9–27.5	24.5–28.8	24.6	25.6	23.0	20.0–30.5	23.1	5.6	7.0	23.1	14.02
Histidine	0.85–1.0	0.74–0.78	0.4–0.7	0.9	1.7	0.7	0.7–1.0	1.0	1.3	1.5	2.8	1.30
Hydroxylysine	1.04	0.91–1.2	0.7–0.9	—	—	0.8	0.7–1.2	2.3	—	—	—	—
Hydroxyproline	13.5	14.0–14.5	11.9–13.4	13.9	0.8	12.0	12.1–14.5	11.2	—	—	2.7	—
Isoleucine	1.36	1.7–1.8	1.3–1.54	1.6	1.2	1.4	1.3–1.8	1.4	2.4	0.9	3.2	1.70
Leucine	3.1–3.34	3.1–3.4	2.8–3.45	3.3	2.2	2.8	2.8–3.5	3.1	4.6	1.4	7.5	2.05
Lysine	4.1–5.2	4.5–4.6	2.1–4.36	3.8	—	3.9	3.9–5.2	3.8	3.6	4.3	1.4	4.07
Methionine	0.8–0.92	0.8–0.9	0.0–0.6	1.0	1.0	0.7	0.7–0.9	1.8	0.8	0.2	—	1.17
Phenylalanine	2.1–2.56	2.2–2.5	1.3–2.49	2.3	—	1.9	1.1–2.6	1.9	2.3	1.6	5.1	3.87
Proline	16.2–18.0	14.8–16.35	13.5–15.5	13.6	8.0	13.8	13.7–18.0	12.3	9.3	9.5	11.9	—

Serine	2.9–4.13	3.2–4.2	3.4–3.8	3.0	4.7	—	2.9–4.1	2.6	13.1	13.4	1.2	5.68
Threonine	2.2	2.2	2.0–2.4	2.0	2.1	—	1.8–2.6	1.7	9.1	8.1	1.2	4.08
Tryptophan	—	—	—	—	—	—	—	—	—	—	—	3.25
Tyrosine	0.44–0.91	0.2–1.0	0.0–0.23	0.3	—	0.2	0.2–1.0	—	0.8	0.5	—	1.28
Valine	2.5–2.8	2.6–3.4	2.4–3.0	3.2	1.7	2.2	2.1–3.4	2.0	5.2	4.3	14.4	4.24

[a]See comments in text regarding asparagine and glutamine content.

Sources:

Gelatins from Gelatin, New York: Gelatin Manufacturers Institute of America, 1993 (Table 1, p. 8).
Collagen [H] from Amino Acid Distribution of Peptein, Hormel Foods Specialty Products. Unpublished.
Fish collagen from Solu-Mar EN-30 Technical Data 1090/1, Brooks Industries, Feb. 1990.
Cattle hide collagen from Hydrocoll Technical Data, Issue 2, Brooks Industries, Nov. 1990.
Collagen [C] from Crodata: Crotein SPA, SPO and SPC, Croda, Inc., June 14, 1985.
Bovine collagen, hair keratin, and bovine elastin from Products for Applications in Cosmetics, Diamalt, Inc.
Human hair from Crodata: Crotein HKP—Keratin Amino Acids, Croda, Inc., Jan. 23, 1982.
Yeast from VEGP/2: Plant Proteins, Brooks Industries, Spring 1992.

Table 4 Amino Acid Profiles (g/100 g) for Selected Plant Protein Sources

Amino acid	Soy [H]	Soya [B] bean	Soy [D]	Soy [C]	Corn zein	Potato	Golden pea	Wheat [C]	Wheat [B] gliadin	Wheat [D] gluten	Wheat [H] gluten	Rice gluten	Oat protein
Alanine	5.5	3.50	4.1	4.5	12.84	5.43	4.47	2.7	1.60	2.1	5.2	5.9	6.4
Arginine	8.0	8.42	7.9	7.2	4.15	4.40	8.80	3.2	2.01	3.4	3.5	8.5	6.1
Aspartic acid[a]	11.6	13.56	12.4	12.3	6.77	14.69	11.86	3.1	2.24	3.3	2.9	9.2	8.6
Cystine	2.0	1.23	—	1.0	2.35	0.73	0.90	1.8	1.24	—	1.7	—	2.4
Glutamic acid[a]	19.6	26.82	21.9	20.1	18.68	13.86	19.93	36.8	22.91	40.9	38.3	20.4	24.3
Glycine	4.3	2.10	4.0	4.5	5.35	8.92	4.33	3.5	42.37	3.1	3.8	4.6	8.8
Histidine	2.3	3.21	2.9	2.6	2.50	1.70	2.66	2.2	1.07	2.3	2.2	3.4	2.2
Isoleucine	4.9	4.18	4.9	3.8	3.78	5.08	4.58	3.4	1.87	4.1	4.8	5.0	2.6
Leucine	6.7	6.59	7.9	7.4	11.81	10.22	8.46	7.3	4.04	6.8	7.2	9.1	7.5
Lysine	6.4	2.90	6.5	6.4	3.20	6.04	7.58	1.7	1.12	1.5	1.7	3.7	4.1
Methionine	2.3	1.23	1.0	1.3	1.71	1.81	0.98	1.5	0.79	1.3	2.1	2.1	1.8
Phenylalanine	4.3	5.87	5.2	4.8	0.62	4.74	5.14	5.4	3.11	5.2	4.8	5.9	4.3
Proline	5.3	4.61	4.9	5.3	11.19	5.82	4.93	12.0	8.28	13.0	10.8	4.4	3.9
Serine	4.4	4.98	4.7	5.4	5.61	4.67	3.68	5.7	3.20	4.1	3.8	4.9	5.1
Threonine	4.0	3.92	3.6	4.1	4.50	4.94	2.94	2.9	1.60	2.4	2.4	3.7	3.4
Tryptophan	0.9	1.44	—	1.2	—	0.20	0.51	—	0.06	—	0.7	—	1.5
Tyrosine	3.0	1.44	2.3	3.6	0.20	0.20	3.31	2.9	0.35	1.3	3.3	2.0	2.9
Valine	4.4	3.97	5.1	4.7	5.59	6.52	4.92	4.1	2.12	4.2	4.2	6.9	3.9

[a]See comments in text regarding asparagine and glutamine content.

Sources:

Soy [H] from Peptein VgS, Document 09-020-5790 Revision C, Hormel Foods Specialty Products, Dec. 1995.

Soya [B] bean, wheat [B] gliadin, corn zein, potato, and golden pea from VEGP/2: Plant Proteins, Brooks Industries, Spring 1992.

Soy [D], wheat gluten [D], and rice gluten from Products for Applications in Cosmetics, Diamalt, Inc.

Soy [C] from Crodata: Hydrosoy 2000/SF, Croda, Inc., July 14, 1984.

Wheat [C] from Crodata DS-23: Hydrotriticum 2000, Croda, Inc., Aug. 17, 1994.

Wheat [H] gluten from Peptein VgW, Document 09-019-5790 Revision B, Hormel Foods Specialty Products, Dec. 1995.

Oat protein (whole oat) from Ceapro, Inc., unpublished.

A. Hydrolyzed Collagen

Hydrolyzed collagen, the first broadly utilized functional proteinaceous material for hair care, remains in widespread use today. Collagen is the most common protein in the human body, comprising 30% of total protein content. More than a dozen variants are found in various tissues, including tendon, cartilage, bone, and skin (23). Types 1 and 3, produced primarily by young skin, have the highest water-binding capacity. With advancing age, crosslinking of collagen increases and water-binding capacity decreases, contributing to an increase in the number of fine lines and wrinkles.

Collagen presents an excellent case study regarding the importance of processing in relation to an end product's properties and performance. The gelatin from which hydrolyzed collagen is formed is classified as either type A or type B, depending on whether its precursor is acid (A) or alkali (base, B) treated. The significance of this treatment lies in the isoionic point of the gelatin, and subsequent hydrolyzates, which result: pH 7–9 for type A and pH 4.7–5.4 for type B (4). Thus, type A gelatin and its hydrolyzates are typically much more substantive to both skin and hair (anionic substrates) at the neutral to slightly acidic pH of most cosmetic products, as they carry a net positive charge in this range. Type B products must be quaternized to achieve such substantivity because of their greater net negative charge, which may be attributed to conversion of glutamine and asparagine to glutamic acid and aspartic acid, respectively, during processing. This replaces polar R groups with anionic ones (unless adjusted to very low pH).

The gelatin is then further hydrolyzed using acid, base, or enzymes to form collagen hydrolyzates. Chemical hydrolysis necessarily adds to ash (salt) content, which can destabilize emulsions. Enzymatic hydrolysis does not contribute to ash content and has the further benefits of generally producing a lighter color and a product with more consistent molecular weight (24). Benefits of collagen hydrolyzates are typically and most readily observed in finished products when protein hydrolyzate solids are present at 2% or more in leave-in products and at 5% or more in rinse-out products. Hydrolyzates are stable in the range of pH 3–12 (16).

Hydrolyzed collagens of 5000 to 15,000 Da are excellent moisturizing film formers. Their relatively high molecular weight minimizes absorption, depositing most of the collagen on the hair's cuticle. Added body, resiliency, shine, and manageability, as well as reduced static charging, is the result, particularly from leave-in products.

Hydrolyzed collagen of 2000 Da has demonstrated substantivity to hair, increasing with the extent of damage to the hair (bleached and permanent waved > bleached > neither). Further, the more damaged the hair, the greater the difference in substantivity between type A- and type B-derived material is observed. The substantivity is the result of both adsorption (coating) and absorption (penetration) of the hydrolyzate. Surface adsorption contributes to shine, combability, and volume (24), while penetration of the cortex yields increased tensile strength and elasticity. Temporary mending of split ends when applied as a leave-in product (25) at 2.5% solids level (above this level, damaged hair picks up little additional hydrolyzate) (24) and moisturization of the hair also result, improving both the hair's manageability and appearance. Type A gelatin-derived hydrolyzed collagen has appreciable substantivity to hair from pH 4 to 10, with the greatest substantivity from pH 6 to 9.

Hydrolyzed collagen is available as a powder and in preserved aqueous solution form. The powder can be added with mixing to water for ready dissolution. Hydrolyzed collagen (and other hydrolyzates and small ionic compounds) are best added to an emulsion-based product after the emulsion has been formed, as salts and similar ionic materials may adversely affect emulsion formation. In surfactant systems such as shampoos, hydrolyzed collagen acts as a foam stabilizer and booster when used at one-fourth the level of active foamer on a solids basis. The hydrolyzate's colloidal, film-forming nature strengthens the bubble film, prolonging the time until breakage (16).

B. Elastin

Elastin is the second most common connective tissue protein after collagen, constituting 60–80% of the dry weight of blood vessels and ligaments. Its unique amino acids are desmosine and isodesmosine. In skin, elastin fibers enmeshed in collagen provide skin with its elastic strength (23). Excessive sun exposure leads to abnormal, disoriented elastin fibers and folding of the epidermis (i.e., wrinkles) (26). Like hydrolyzed collagen, hydrolyzed elastin of 2000 to 5000 Da is a good film former, but is much less hygroscopic due to its much lower polar and charged amino acid content (see Tables 2 and 3). As a result it has much higher hydroalcoholic and polyol solubility and can reduce swelling of hair during permanent waving and coloring processes. Like other low-polarity protein hydrolyzates, such as silk (17), it may be preferable to collagen (and other more hydrophilic hydrolyzates) where humidity resistance is desirable.

C. Keratin

Keratin, as the primary structural protein of hair (and nails as well), provides hair its strength. Keratin hydrolyzates are notable primarily because of their high cystine content (see Tables 2 and 3), which, in a reducing environment, allows for sulfhydryl interchange with cysteine residues in hair. High-molecular-weight hydrolyzed keratin (125,000 Da) demonstrated long-term substantivity and conditioning effects when applied to reduced hair (i.e., in the midst of permanent waving) at a 1.5–5% solids level (27). Patents related to alkaline permanent waving and the better afterfeel which resulted from incorporation of these hydrolyzates were issued in the early 1980s (28,29).

D. Vegetable Proteins

1. *Soy and Wheat Proteins*

Soy and wheat protein hydrolyzates produce benefits similar to those noted for collagen, but also have the added potential for hair protection and conditioning through bonding via disulfide interchange. Collagen does not have this capacity, because of its very low cystine content (the presence of cystine in collagen is an indication of impurity). Hydrolyzed wheat protein in a peroxide-based hair bleach at 2% solids reduced formation of cysteic acid (an undesired oxidative by-product) by a third (30).

2. *Hydrolyzed Sweet Almond Protein*

Hydrolyzed sweet almond protein, contrary to its INCI name, actually contains both polypeptides and oligosaccharides as major components. A significant carbohydrate content is typical of plant- (but not animal-)derived proteins. The result is conditioning by both penetration (of the polypeptides) and coating (by the oligosaccharides). In a leave-in product, enhancement of moisture retention and shine would thus be expected (31). Observable strengthening of damaged hair and protection of hair fibers would be expected at a use level similar to that required for hydrolyzed collagen.

3. *Hydrolyzed Wheat Proteins*

Hydrolyzed wheat protein (and) hydrolyzed wheat starch, similar to hydrolyzed sweet almond protein, conditions by both penetrating and coating. Typical of collagen and soy hydrolyzates, substantivity to normal hair increases with concentration, rapidly to 2% actives and plateauing above 5% actives. Studies of stress relaxation and elasticity on hair damaged with a 2% sodium hydroxide cream for 15 min suggested a reduction in brittleness at low relative humidity (RH) and a reduction in limpness at high RH for hair pretreated with 5% hydrolyzate solids (32). Other, more readily apparent benefits to be expected from protein/oligosaccharide combinations in leave-in products are improved body and shine (33).

4. *Hydrolyzed Oat Protein*

Hydrolyzed oat protein can also provide substantivity, shine, split-end repair, and a protective barrier to hair. Aspartic and glutamic acids, the two amino acids with carboxyl-containing R groups, together account for 40% of the amino acid content of the protein (34), making for improved humectancy. Exceptionally light color allows greater formulation latitude. A 10% use level is recommended for irritancy mitigation in shampoos, and a 2% level in formulated leave-in and rinse-out conditioners.

E. Amino Acids

Wheat amino acids are derived from wheat protein, fully hydrolyzed to its constituent amino acids. Like oat protein, 40% of the amino acids have carboxyl-containing R groups (see Tables 2 and 4). The small size of amino acids allows absorption and retention in the hair fibers (35). Their hygroscopic nature allows them to function as humectants when applied directly to hair at a low 0.2% active solution (36). In rinse-out conditioners, a use level of 0.5% actives is recommended to enhance hair's equilibrium moisture content.

Collagen "amino acids" covers hydrolyzates which vary by manufacturer from actual amino acids to short polypeptides (<700 Da). Regardless, they are highly hygroscopic and so excellent humectants with very efficient moisture-binding properties for hair. Substantivity is somewhat greater than for the higher-molecular-weight hydrolyzates (24). Penetration results in more pliable hair with a soft, nontacky feel, particularly at high relative humidity.

Certain amino acids find application in specific areas. The predominantly anionic amino acids glutamic acid and aspartic acid are particularly effective humectants. Taurine has been patented for use as a "protein magnet," to enhance deposition of positively charged ingredients in damaged hair (37). Cysteine and derivatives have found application in permanent waving compositions, contributing to the reduction of disulfide bonds necessary as the first step in reconfiguration of the hair.

F. Quaternized Protein Hydrolyzates

Quaternization of protein hydrolyzates typically raises their isoionic points to pH 10 or higher, regardless of their original isoionic points. It thus tends to equalize (and improve upon) the performance of collagen hydrolyzates, whether made from type A or type B gelatin. Like the underivatized hydrolyzates, substantivity increases with the extent of damage to the hair. A slight increase in moisture binding has also been observed (for wheat and soy quaternaries). The increased cationic character yields enhanced substantivity, even under the alkaline conditions encountered in permanent waving, oxidative coloring, and straightening of hair. Hydroxypropyltrimonium- and propyltrimonium-substituted protein hydrolyzates thus extend the benefits of the underivatized hydrolyzates to alkaline pH, allowing effective protection and conditioning during "chemical processing" (covalent modification) of hair. Further, such products in typical cleansing systems yield denser and more stable foam. Certain products can markedly reduce the irritation potential of these systems as well. Five parts sodium lauryl sulfate to one part of one manufacturer's propyltrimonium hydrolyzed collagen reduced irritation ratings from severe to nonirritating, and its addition to a wide array of commercial surfactant-based products yielded irritation reduction (38).

Incorporating a long alkyl chain in the quaternary imparts the conditioning characteristics expected of traditional quaternaries such as stearalkonium chloride, but with much lesser irritation and much greater compatibility with anionic surfactants. As such, they may be utilized in clear conditioning shampoos, conditioners, and styling mousses, enhancing wet and dry combability, luster, feel, and static control. Quaternium-79 hydrolyzed proteins added to 10% sodium lauryl sulfate improved combability to at least 0.4% (solids) and feel to 0.1%, while substantivity increased to 2% with only a small decrease in foam generation (39). Typical use levels for fatty alkyl quaternary protein hydrolyzates are 0.4–4% (solids basis).

Fatty quaternary hydrolyzates of proteins containing cyst[e]ine residues, such as wheat, may be particularly appropriate for permanent waving solutions, as they offer the potential for "permanent conditioning," i.e., grafting fatty moieties covalently to the hair via disulfide interchange. While substantivity is essentially independent of fatty moiety chain length (C_{10}–C_{18}), it decreases by a factor of 4 between pH 4.5 and 8.5 (40).

G. Other Chemically Reacted Proteins

Reaction of fatty acid chloride with primary amino groups of hydrolyzates yields anionic surfactants. Unneutralized, they may be incorporated in pomades, functioning as dispersants, to assist in pomade removal during shampooing. When amine neutralized (e.g., AMP-isostearoyl hydrolyzed collagen) they have excellent alcohol solubility and may be incorporated in hairspray formulations as resin plasticizers. Potassium cocoyl hydrolyzed collagen is a low-irritation, relatively high-foaming surfactant which can be utilized, generally with other surfactants, to produce very mild (e.g., baby) shampoos. It is also an effective dye leveler in hair coloring applications. Triethanolamine-neutralized material combined with sorbitol yields greater compatibility in traditional shampoo systems, allowing formulations to pH as low as 5.5 (41).

Direct esterification can also be employed to impart alcohol solubility. Ethyl ester of hydrolyzed collagen (or other protein hydrolyzates) as a plasticizer in hairsprays will provide a harder, more lustrous film than the amine-neutralized fatty derivatives. As is typical for resin plasticizers, use level should not exceed 10% of resin level.

Silicone grafted protein hydrolyzates can impart improvements in wet combing and feel from shampoos (42) and enhance deposition over that of the parent hydrolyzate as a result of peculiar solubility profiles (43).

V. SKIN CONDITIONING

A. Enzymes Used in Skin Products

There is interest internationally in cosmeceutical-type skin care products containing enzymes (44). Superoxide dismutase, a superoxide radical scavenger which may thereby function as an antiinflammatory agent, has been protected in gelatin microspheres, allowing the majority of the enzyme to be delivered in active form from the preparations. This mode of entrapment improved stability toward pH and temperature extremes as well as providing increased resistance to protease attack (45). Similar stabilization can be expected with other enzymes. Papain, a protease, has been immobilized on polyacrylic acid for use as an alternative to AHAs as an exfoliant (46). Papain is active at pH 5–7, allowing formulations at less extreme pH than is required for AHA-based formulas. A composite of enzymes involved in the protection of various plants from the effects of ultraviolet irradiation has been assembled for use as UV absorbers and attenuators in topical skin care applications (47).

B. Collagens

The natural protein choice for skin care applications is still collagen, as it is the major structural protein in skin. Soluble collagen, a triple helix of 250,000 to 300,000 Da, has strong moisture-binding properties without the tackiness that can be associated with the use of amino acids as humectants. Incorporation in moisturizing lotions and creams provides an elegant, silky, cushioned feel to the skin and leaves the skin smooth, soft, and perceptibly moisturized. Applied medicinally, soluble collagen appreciably shortens the time for burns and wounds to heal.

To minimize degradation during storage, soluble collagen is best kept under refrigeration. To prevent denaturation during incorporation, it should be added to finished products after emulsion formation and at no greater than 30°C. Greatest stability is at near-neutral pH. Soluble collagen loses its effectiveness in formulations containing high levels of salt, alcohol, or surfactant, and urea, tannins, chloroacetic acid, formaldehyde, and other denaturing or crosslinking agents. Low-salt (less buffered) versions are available for use in emulsions which are extremely salt sensitive.

1. Gelatin

Gelatin is a collagen hydrolyzate of about 150,000 Da. Such high molecular weight is possible despite a collagen monomer molecular weight of only 95,000 Da because mature collagen has covalent crosslinks between monomers which are not broken during heat-induced hydrolysis of the collagen. Gelatin finds use as a natural viscosity builder and thickening agent. It also forms a protective film on the skin and so acts as an antiirritant.

Gelatin and its incomplete hydrolyzates (>700 Da) applied directly to the skin feel tacky and so require a balanced formulation to produce desirable esthetics. Formulations best employ a coarse mesh (6–10 mesh on U.S. sieves), as finer meshes wet out quickly, clumping easily to form "fish eyes" (dry gelatin cores encased in partially hydrated lumps of gelatin). To keep this from happening with finer meshes, high shear is required, often resulting in aeration of the batch. The optimal hydration temperature is thus 170–180°F (77–82°C), a compromise between clumping from too rapid wetting at higher temperature and excessively long hydration times at lower temperatures. Shortly after hydration is complete, the batch temperature should be reduced to retard thermally induced degradation of the gelatin.

2. Hydrolyzed Collagen

Hydrolyzed collagen of 5000 to 15,000 Da in lotions and creams can function as auxiliary emulsifiers, stabilizing emulsions, as well as enhancing the water-binding characteristics of the finished product, thereby slowing transepidermal water loss (TEWL) upon application to the skin as a result of the protective colloidal film which forms.

Hydrolyzed collagen of 2000 Da in lotions and creams can form moisture retentive films on the skin. Further, in surfactant systems it complexes with monomeric anionic surfactant molecules, effectively raising the critical micelle concentration of such systems, thereby reducing irritancy (13). At a 2–3% level in hand dishwashing liquids, a reduction in skin tautness following usage is observed as well (16). These hydrolyzates are also said to combat the drying effects of detergents on skin via their protective colloidal film-forming properties (48,49). Soy and wheat protein hydrolyzates can be expected to perform comparably.

C. Hydrolyzed Elastin

Hydrolyzed elastin of 2000 to 5000 Da is also a good film former, but is much less hygroscopic due to its low polar and charged amino acid content. As a result, it is effective in reducing TEWL and may be preferable to collagen (and other more hydrophilic hydrolyzates) where moisture-retentive and humidity-resistant features are desirable. Its high solubility in hydroalcoholic systems allows its use in facial toners and aftershave liquids.

D. Amino Acids

Collagen "amino acids" (amino acids to 700-Da polypeptides, depending on the manufacturer) are highly hygroscopic and substantive to the skin, enhancing moisture binding of products in which they are incorporated and so slowing TEWL. These low-molecular-weight hydrolyzates do not have the tacky feel of higher-molecular-weight hydrolyzates. Emulsions should be formed prior to the addition of collagen amino acids because of their high ionic strength. Wheat amino acids are of small size, allowing penetration of the skin's outer layer, in a manner similar to that observed for hydrolyzed collagen, for moisturization from within owing to their hygroscopic nature (21).

Individual amino acids and derivatives are used for specific effects. Tyrosine and derivatives find application in sun care products because of their involvement in skin coloration processes, synthetic and natural. Gelatin glycine enriched with lysine was reported to reduce the irritancy of emulsions containing 10% glycolic acid while enhancing recovery of skin elasticity and depigmentation of age spots (50). Use of acetyl cysteine was recently patented by Procter and Gamble as an alternative to alpha-hydroxy acids (AHAs) for the removal of dead skin. The method utilizes sulfhydryl compounds such as this to improve skin suppleness and smoothness and to treat acne (45).

E. Other Modified Proteins

Quaternization of protein hydrolyzates increases their isoionic points, enhancing substantivity and reducing the irritation of anionic surfactants in cleansing formulations (13). For example, polytrimonium gelatin (gelatin hydroxypropyltrimonium chloride) (51) reduced both the eye and skin irritation of a chloroxylenol-based antiseptic cleanser and propyltrimonium hydrolyzed collagen reduced irritation ratings from severe to nonirritating when one part was added to five parts sodium lauryl sulfate (38).

Anionic hydrolyzates, produced from the grafting of fatty acid residues on primary amino groups, remain film formers. They are readily incorporated into solvent-based systems such as nail polishes and polish removers, and may also be used in tanning oils to assist in their removal during subsequent skin cleansing. Neutralized forms (e.g., TEA-cocoyl hydrolyzed collagen) are low-irritation, relatively high-foaming surfactants (17) that are useful in mild surfactant-based skin cleansing compositions such as facial washes and bubble baths.

Further testing has been performed on specific "lipoamino acids" (anionic hydrolyzates). Their amphophilic character facilitates transport across biomembranes. They thus penetrate the skin to an extent and in a manner which underivatized amino acids cannot, penetrating the intercellular cement and contributing the regulation of water content of the skin from within. Toxicity and antigenicity are low, and some have the capacity to inhibit proteolytic enzymes such as elastase, urokinase, and plasmine (52). Palmitoyl hydrolyzed wheat protein was demonstrated to stimulate protein synthesis in human epidermal keratocytes in vitro at 0.1%, an effect which may have been nutritive in origin. Human dermal fibroblast cultures treated at 0.1 μg/L with this compound produced greater extracellular matrix collagen fiber crosslinking, similar to that observed with vitamin C, which would present itself in vivo as an improvement in the skin's elasticity and firmness. Using a gelatine cells method, dose-dependent reduction in TEWL across an emulsion film containing this material was demonstrated. An in-vivo soothing effect (a slight reduction in skin redness versus a control) was demonstrated on abraded skin as well.

VI. FUTURE TRENDS

In an era of formulation which has been defined by the concepts of renewable resources, natural derivation, environmental friendliness, biodegradability, irritation reduction, natural (plant) sourcing, volatile organic compound (VOC) reduction, and ecological soundness, nature-identical proteins and their hydrolyzates and derivatives are poised for increasing utility. The extent to which their potential is achieved will depend largely on the level of basic and applied research performed to generate and/or elucidate the properties desired in the next generation of products.

Selection of plant-derived protein products over animal-derived products is likely to continue. While initially driven by the greater marketing appeal (and occasionally different performance properties) of plant-derived materials, the shift has been accelerated by concerns related to bovine spongiform encephalopathy (BSE) in the United Kingdom (53,54). Further, globalization of raw material supplies and formulations requires accommodation of a significant portion of the world's population who find certain (or all) animal species, their biological components, and/or their by-products objectionable for use in foodstuffs or elsewhere.

Biotechnology, fermentation science, and genetic engineering are likely to yield highly functional materials for tomorrow. In hair care, proteins of customized amino acid content and configuration could be the polymers providing flexible hold over a broad humidity range from a low-VOC matrix. Immobilized enzymes without antigenic potential may be applied to damaged hair to remove frayed cuticle, thereby enhancing shine, much as cellulases are currently utilized to remove fuzz from cotton clothing during laundering and thereby brighten colors. Modified amino acids could penetrate the cortex and convert cysteic acid residues to moieties capable of forming disulfide bonds, thereby allowing strengthening of hair. Selected amino acids or polypeptides or their derivatives may enhance the protective effects observed with whole protein hydrolyzates and may do so at a much lower use level, allowing application in permanent waving or relaxing systems which currently are intolerant of the ionic strength. Already a blend of fruit enzymes has been proposed as a total replacement for traditional surfactants in shampoo and bath applications (55), and whey protein, with its activated cytokines [immunological regulators, signaling and controlling molecules in cell-regulating pathways (45)] has been shown in in-vivo human testing to improve skin firmness, touch, and smoothness and to increase skin elasticity and thickness (56).

Indeed, skin is likely to be the primary beneficiary of the next generation of protein-based products, despite regulatory roadblocks which may result from their druglike effects. Perhaps the greatest untapped resource in the advance of skin care lies in the delivery of functional enzymes through the epidermal barrier for the treatment of both "cosmetic" and physiological deficiencies. Advances in stabilization and delivery systems for macromolecules will make what was once an impossibility a reality.

The first step, delivery of functional and stable enzymes to the skin's surface, has already been achieved through the use of two-component packaging: polyol-stabilized serine protease phase has been co-dispensed with an aqueous activation phase to provide skin smoothing effects (20). Other proposed combinations include a tocopheryl acetate phase co-dispensed with an esterase to generate free tocopherol and a magnesium ascorbyl phosphate phase co-dispensed with a phosphatase phase to delivery free ascorbic acid (vitamin C). These proposals were given noting the potential for allergenicity and recommended attachment of enzymes to substrates of sufficient size to preclude inhalation, as well as proper safety evaluation (as respiratory allergenicity testing of formulations is not typical). Enzyme technology has also been referenced in the mitigation and prevention of adult acne (57).

The next step will be the intentional delivery of active principles (which may include enzymes, hormones, and specific polypeptides or amino acids, as well as many other materials) to living skin cells for cosmetic purposes, with the aim of providing noticeably "improved" skin texture and topography. In practice this occurs unintentionally when products which have been tested for performance on "normal" skin are utilized on excessively dry, burned, abraded, or otherwise compromised skin. Defects in the stratum corneum (and deeper layers of the skin) allow substances to pass directly to the living cells and exert an effect. It also occurs intentionally when drug products are applied to relieve pain, kill embedded microorganisms, fight acne, or otherwise promote the skin's well-being. Just as alpha- and beta-hydroxy acids and standardized botanical extracts have pushed the limits of cosmetic versus drug, so these new protein-based products will. Indeed, the deciding factor will likely be the claims made, not the results delivered.

Envisioned is the use of liposomal or micellar vesicles to deliver actives through intact skin and the development and marketing of products for "sensitive" skin, including compromised skin. A cosmetic manufacturer's "oil-control hydrator" has already featured an exclusive enzyme technology to help normalize skin to produce measurably less oil (57). Its intensive lifting creme includes an undisclosed restorative enzyme encapsulated in liposomes, and other manufacturers' products feature enzyme complexes for cell energy and glycoprotein to stimulate natural regeneration of the skin (58).

Because products will rely on intimate contact with living cells to achieve the visible and tactile improvements desired, the microbiological and biochemical purity of proteins throughout isolation or manufacture will be crucial. If deleterious agents such as endotoxins (lipopolysaccharide–protein complexes of the outer membrane of Gram-negative bacteria which produce release of pyrogens in mammals, resulting in fever) are present, a detrimental response may be expected for the user (59). On the other hand, direct delivery of a desired therapeutic material may have as rapid and pronounced a positive effect.

ACKNOWLEDGMENTS

Thanks are expressed to each manufacturer's representative who kindly responded to information requests, and particularly to Suellen Bennett (Brooks Industries), Michael Birman (Croda, Inc.), Steve Bell (Hormel Foods), and Anna Howe (Inolex Chemical) for extensive literature made available. Thanks are also expressed to Alberto Culver USA (and Jo Rathgeber, Senior Research Librarian) for making library resources available, and to Ed McKeown (Costec, Inc.) for heartily encouraging and supporting this effort.

REFERENCES

1. Lehninger AL. Proteins and their biological functions: a bird's eye view. In: Biochemistry, 2d ed. New York: Worth, 1977:57–70.
2. Perlmann GE, Manning JM. Proteins. In: McGraw-Hill Encyclopedia of Science and Technology. 6th ed. New York: McGraw-Hill, 1977, 14:405–409.
3. Johnsen VL. Innovations in protein products and technology. Cosmet Toilet 1977; 92 (Dec.):29–36.
4. Gelatin. New York: Gelatin Manufacturers Institute of America, 1993.
5. Burmeister F, Brooks GJ, O'Brien KP. Vegetable/plant proteins in shampoos. Cosmet Toilet 1991 (Apr.); 106(4):41–46.
6. Bouthilet RJ, Karler A. Cosmetic effects of substantive proteins. Proceedings of the Scientific Section of the Toilet Goods Association 1965 (Dec); 44:27–31.

7. Stern ES, Johnsen VL. Studies on the molecular weight distribution of cosmetic protein hydrolyzates. J Soc Cosmet Chem 1977; 28:447–455.
8. Johnsen VL. Proteins in cosmetics and toiletries. Technical Services Report #8. Philadelphia: Inolex Chemical Company.
9. The effect of pH on the sorption of collagen-derived peptides by hair. Technical Services Report #19. Philadelphia: Inolex Chemical Company.
10. Karjala SA, Williamson JE, Karler A. Studies on the substantivity of collagen-derived polypeptides to human hair. J Soc Cosmet Chem 1966; 17:513–524.
11. Frowde HL. Hair Dyeing Method. U.S. Patent 3,193,465 (1965).
12. Stern-Cooperman ES, Johnsen VL. Penetration of protein hydrolyzates into human hair strands. Cosmet Perf 1973 (July):88–92.
13. Tavss EA, Eigen E, Temnikow V, Kligman AM. Effect of protein cationicity on inhibition of in vitro epidermal curling by alkylbenzene sulfonate. J Am Oil Chem Soc 1986; 63(4):574–579.
14. Wenninger JA, McEwen GN Jr. International Cosmetic Ingredient Handbook. 3d ed. Washington, DC: The Cosmetic, Toiletry, and Fragrance Association, 1995.
15. Stern ES, Johnsen VL. Cosmetic proteins: a new generation. Cosmet Toilet 1983; 98(5):76–84.
16. Croda Cosmetic and Pharmaceutical Formulary, Chapter 19 (Crotein), Parsippany, NJ: Croda, Inc., 1977.
17. Promois Digest. Osaka, Japan: Seiwa Kasei Co., Ltd. 5th printing. May 1995.
18. Cosmetic Ingredient Review, final report: safety assessment of hydrolyzed collagen, J Am Coll Toxicol 1985; 4(5):199–221.
19. Verbent A. The fascinating challenge to mimic nature in producing recombinant glycoproteins. Chimica Oggi/Chemistry Today 1996 (Oct.); 14(10):9–14.
20. Edens L, van der Heijden E. Delivery of enzymes and high actives in cosmetics. Cosmet Toilet Manufacture Worldwide 1997:189–193.
21. Poppe CJ. Enriching the life of hair. Soap/Cosmet/Chem Specialties 1996 (Oct.); 72:28, 30, 32, 34, 36, 40.
22. Fed Reg 42:17109, March 31, 1977 (FTC Ruling).
23. Gelita-Proteins: Application in Skin-Care Products. Eberbach, Germany: DGF Stoess, July 1995.
24. Hormel Foods Specialty Products Hydrolyzed Collagen. Brochure SP3812, Austin, MN, 1994.
25. Peptein 2000. Soap, Perf Cosmet 1996 (Dec.); 69(12):9.
26. CO_2 laser vanquishes skin wrinkles (temporarily). Biophotonics Int 1996 (Nov./Dec.); 3(6):31–32.
27. Crodata: Kerasol (A Unique Keratin Protein). Parsippany, NJ: Croda, Inc., July 9, 1987.
28. U.S. Patent 4,279,996 (1981), Keratin hydrolyzate useful as hair fixatives, assigned to Seiwa Kasei.
29. U.S. Patent 4,390,525 (1981), Keratin hydrolyzate useful as hair fixatives, assigned to Seiwa Kasei.
30. Plantasol W 20 Product Information. PIW20-8E. Eberbach, Germany: DGF Stoess. Oct. 1995.

31. The compounder: sweet almond derivative helps produce healthy hair. Drug Cosmet Ind 1996 (Aug.); 159(2):55.
32. Crodata DS-31R-1: Cropeptide W. Parsippany, NJ: Croda, Inc., July 6, 1994.
33. Peptein WPO. Document 09-058-5790. Austin, MN: Hormel Foods Specialty Products, Oct. 1996.
34. Hydrolyzed Oat Protein HOPA Data Sheets. Saskatoon, Saskatchewan, Canada: Canamino, Inc., 1996.
35. Hydrotriticum WAA. Drug Cosmet Ind 1996 (Aug.); 159(2):5.
36. Crodata DS-27R-1: Hydrotriticum WAA. Parsippany, NJ: Croda, Inc., June 1, 1994.
37. U.S. Patent 4,793,992 (1988), Hair treatment composition, assigned to Redken Laboratories, Inc.
38. Protectein Data Sheet. Document 09-008-5790 Revision D. Austin, MN: Hormel Foods Specialty Products, Dec. 1995.
39. A Guide to Formulating Protein Hair Care Products. MGL6003. University Park, IL: McIntyre Group Ltd. Aug. 1995.
40. Crodata DS-19R-1: Hydrotriticum QL, QM and QS. Parsippany, NJ: Croda, Inc. June 10, 1994.
41. Foam-Colls Technical Data 1070/2. South Plainfield, NJ: Brooks Industries. Sept. 1990.
42. Promois SIG Series Technical Data Sheet. Osaka, Japan: Seiwa Kasei Co., Ltd., Apr. 22, 1995.
43. Crodata DS-39R-2: Crodasone W. Parsippany, NJ: Croda, Inc., Sept. 10, 1996.
44. Shaw A. International Cosmetics Report. Soap Cosmet Chem Specialties 1996 (Oct.); 72:46.
45. Fox C. Technically speaking. Cosmet Toilet 1996 (Sept.); 111(9):15, 16, 19, 20.
46. New chemicals for specialties: AHA alternative. Soap Cosmet Chem Specialties 1997 (Feb.); 73(2):107.
47. Campo Research Plants' UV protectant active principle(s) "UVzymes." Household Personal Prod Ind 1997 (May); 34(5):30.
48. U.S. Patent 3,548,056 (1970), Eigen E et al., assigned to Colgate-Palmolive Company.
49. U.S. Patent 4,195,077 (1980), Detergent Compositions Comprising Modified Proteins, assigned to The Procter & Gamble Company.
50. Goldemberg RL. The compounder/compounder's corner. Drug Cosmet Ind 1997 (May); 160(5):66–67.
51. Data 1045: Quat-Coll IP-10 (30%). South Plainfield, NJ: Brooks Industries, Inc., Fall 1992.
52. Michel N. Stoltz C. The interest of amino acid biovectors. Drug Cosmet Ind 1996 (Sept.); 159(3):36–38, 40, 42, 104.
53. Shepherd T. Rendered products market post March 96. Oils Fats Int. 1996; 12(6): 36.
54. Regulations: a new EU ban on British gelatin? Household Personal Prod Ind 1997 (May); 34(5):40.
55. Natural ingredients listing, Campo Research. Household Personal Prod Ind 1997 (June); 34(6):70, 72.

56. MPC—milk peptide complex. Soap Perf Cosmet 1996 (Dec.); 69(12):26.
57. Colwell SM. Clearly in demend. Soap Cosmet Chem Specialties 1997 (Feb.); 73(2): 32–34, 36, 38.
58. Dunn CA. The skin care market. Household Personal Prod Ind 1997 (May); 34(5): 79, 80, 82, 84.
59. Held DD, Mehigh RJ, Wooge CH, Crump SP, Kappel WK. Endotoxin reduction in macromolecular solutions: two case studies. Pharmaceut Technol 1997 (Apr.); 21(4):32, 34, 36, 38.

8
Organo-Modified Siloxane Polymers for Conditioning Skin and Hair

Eric S. Abrutyn
The Andrew Jergens Company, Cincinnati, Ohio

I. BASICS OF ORGANO-MODIFIED SILOXANE POLYMERS

A. Historial Perspective

In general, the term *silicone* is used to describe organo-modified siloxanes—mentioned by Kipping in 1901 and based on the generic formula R_2SiO—where R is CH_3 (polydimethylsiloxane, PDMS) and R_2SiO is referred to as a *siloxy group*. Two well-known types of silicones are cyclomethicone and dimethicone. Organic groups attached to a siloxy group (e.g., alkyl, phenyl, polyether, trifluoropropyl, vinyl), in turn create organo-siloxane polymers which have physical properties of the "inorganic" siloxane backbone (silicon–oxygen–silicon, Si–O–Si) and the "organic" group attached to it. These organo-modified siloxane polymers have unique properties that make them very useful in personal care products.

The basic raw material from which silicones are formed is quartz, i.e., silica or silicon dioxide (SiO_2). In the form of crystals or fine grains, quartz is the main constituent of white sand. In 1824, Jöns-Jacob Berzelius, a Swedish chemist, was successful in liberating elemental silicon (Si) from quartz by reduction of potassium fluorosilicate with potassium. Alkylation of elemental silicon to prepare alkyl silanes was done initially by Friedel and Crafts (1863) using zinc compounds, by Kipping (1904) using organo-magnesium compounds (Grignard reaction), and independently in the 1930s by Hyde (Corning Glass Works) and Rochow (General Electric) using methyl chloride. These scientists synthesized the silicon–carbon bond—one of the most important steps in the history of organo-siloxane polymer development (1,2). The silicon–oxygen–silicon backbone was synthesized by Ladenburg in 1871 by hydrolyzing diethyldiethoxysilane in the presence of a dilute acid to form an oil (silicone). Between 1899 and 1944, Kipping published 54 papers on the subject of silicon chemistry, describing the first systematic study in the field. This work helped Hyde and Rochow develop a commercial process—"the direct process"—using elemental silicon and methyl chloride to produce organo-silicon compounds. Current reviews of the synthesis of organo-siloxane polymers have been written by Colas (3) and Rhone Poulenc (4).

B. Polymer Manufacturing

The great diversity of functional siloxanes, which can be obtained from a relatively few silicon-containing monomers, may best be understood by considering the chemistry of these monomers and their many possible combinations. As mentioned earlier, this process begins with high-grade quartzite rock that is heated with charcoal in an electric furnace to produce elemental silicon at a purity of 98–99%; with heat, the carbon combines with the oxygen from the silica. Extracted silicon is ground to a fine powder and induced to react with methyl chloride gas at 250–300°C in what is called the *direct process* (5), producing several important chlorosilane monomer building blocks—*mono*chlorotrimethyll silane, *di*chlorodimethyl silane, *tri*chloromethyl silane and *tetra*chloro silane. These compounds are fractionally distilled as purified chlorosilane monomers.

Organo-modified siloxane polymers are derived from chlorosilane monomers via hydrolysis and polymerization and/or polycondensation. The hydrolysis and condensation product of dichlorodimethyl silane is of particular importance [$(Me_2SiO)_x$], as it is the major component of most organo-siloxane polymer fluids and elastomers. When two or more types of functional chlorosilanes are hydrolyzed together, the process is called *co-hydrolysis*. An example is co-hydrolysis of monochlorotrimethyl silane and dichlorodimethyl silane to form trimethyl siloxy end-cap polydimethylsiloxane chains. The resultant polymers from co-hydrolysis may be fluid, gel, crystalline solid, or resin, depending on the average functionality of the monomer—$[(CH_3)_nSiO_{(4-n)/2}]_x$. If $n > 2$, the product is a fluid; if n equals 2, the product is a cyclic or linear siloxane fluid (a gum, if a high-molecular-weight siloxane); and if $n < 2$, the product is a resin.

C. Polymer Nomenclature

Today's nomenclature system (6) for organo-silicon polymers is based on Sauer's (7) recommendations. These recommendations were subsequently developed and adopted by the American Chemical Society (8), the *Journal of the Chemical Society* (9), and IUPAC (10).

In general, organo-silicon nomenclature is applied to any structure containing at least one silicon atom. Silanes (R_4Si) are silicon-containing compounds with one silicon atom and four directly bonded groups. Silicones, containing alternating silicon and oxygen atoms, are cyclic, linear, branched, caged, or three-dimensional polymers of the monomeric siloxy group. The prefix to these siloxane polymers designates the number of silicon atoms in the polymer—that is, disiloxane has two silicon atoms, while trisiloxane has three silicon atoms, etc. Siloxanes and silanes are named similarly; the root describes whether it is a siloxane (Si–O–Si backbone) or silane (only one Si atom), and the organofunctional portion describes the type and amount of substitution: hexamethyldisiloxane, decamethylpentacyclosiloxane, Tris(trimethylsiloxy)silane, and (poly)dimethylsiloxane.

Shown below is a shorthand nomenclature of the four major organo-siloxy chain units—M, D, T, Q. Each Si–O bond has another Si attached through an oxygen linkage. Therefore, the mono-, di-, etc., refers to the number of polymer propagating oxygen bonds on Si.

Me	Me	Me	$O_{1/2}$
$MeSiO_{1/2}$	$O_{1/2}SiO_{1/2}$	$O_{1/2}SiO_{1/2}$	$O_{1/2}SiO_{1/2}$
Me	Me	$O_{1/2}$	$O_{1/2}$
M (Mono-)	D (Di-)	T (Tri-)	Q (Quad-)
$Me_3SiO_{1/2}$	$Me_2SiO_{2/2}$	$MeSiO_{3/2}$	$SiO_{4/2}$

As an example using this shorthand nomenclature system, (poly)dimethylsiloxane, containing 10 repeating dimethylsiloxy groups and terminated with trimethylsiloxy groups, would have a shorthand description of $MD_{10}M$. Additionally, the T structure above represents silsesquioxane (e.g., the simplest form is M_3T) and the Q structure represents silicate (e.g., the simplest form is M_4Q).

D. Chemical-Physical Properties

The unique structure of the basic organo-modified siloxane polymer accounts for the fundamental properties and resulting benefits to the personal care industry (11a, 11b). A few fundamental chemical properties (12) that make organo-modified siloxane polymers versatile are:

1. Si–C bond—low energy (75 kcal/mol) and long length (1.88 Å), Si–O bond—low energy (106 kcal/mol) and long length (1.63 Å) (13) means reduced steric interactions.
2. Flatter Si–O–Si bond angle (130–150°) versus C–O–C (105–115°) (14).
3. Lower energy of rotation that results from Si–O longer bond length and flatter bond angles.
4. Lower silicon electronegativity (1.8) than carbon (2.5), which leads to a very polarized Si–O bond that is highly ionic.
5. Low intramolecular and intermolecular (van der Waals) forces.

These chemical properties, in turn, help explain some of the physical properties (e.g., easy diffusion of organic molecules through organo-modified siloxane films, low-temperature stability, water repellency, and low surface tension) of organo-modified siloxane fluids.

For an understanding of the physical properties of organo-modified siloxane compounds, one needs to understand the molecular architecture (15) of silicon–oxygen and silicon–carbon compounds compared to organic carbon–carbon and carbon–oxygen compounds. Low bond energy and low bond rotational energy contribute to a high degree of rotation of a Si–O–Si backbone. This freedom of rotation leads to a unique flexibility of a siloxane molecule. A siloxane molecule can be compared to a spring, Slinky, or accordion as its relates to back-and-forth and twisting motion. This bond flexibility is an important aspect of organo-siloxane polymers. It is believed that the freedom of rotation of a Si–O–Si backbone allows for an organo-cloud orientation [see Figure 1 (16)] that facilitates an effective spatial orientation/alignment at the interface of the surface to which an organo-siloxane polymer is exposed (lines A and B in Figure 1 represent interfaces). This freedom of rotation allows for maximizing surface activity, aligning the inorganic backbone (high-surface-energy Si–O–Si backbone) to high-polarity surfaces (line B in Figure 1) and organo-groups (low-surface-energy CH_3 groups) to low-polarity surfaces (line A in Figure 1).

The spatial orientation of nonpolar methyl groups (as represented by CH_3 in Figure 1), with their low surface energy—measured as lower interface

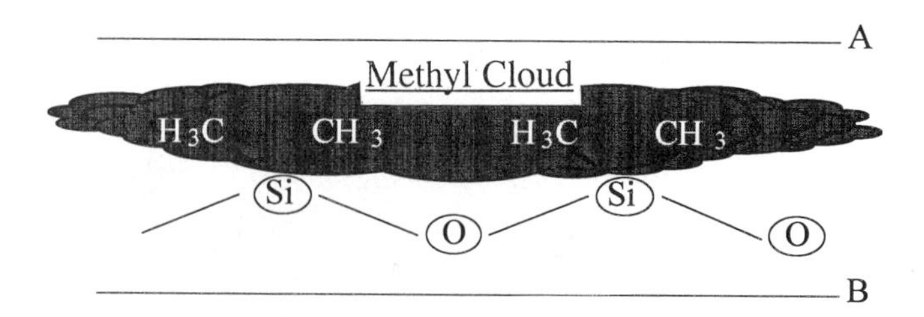

Figure 1 Organo-cloud orientation of a (poly)dimethylsiloxane molecule.

tension—yields an interface which can be characterized as organic and hydrocarbon-like. These methyl groups, with low intermolecular forces and surface energy, are inert and hydrophobic. The flexibility of the Si–O–Si backbone allows for the maximization of hydrophobicity and surface activity attributes of the methyl groups. Due to the packing of methyl groups at a surface, the surface tension is very low. The low surface tension of PDMS fluids (PDMS, 20 dynes/cm; cf. benzene, 28.9, and water, 72) permits them to spread easily over surfaces, which, coupled with other properties (e.g., low coefficient of friction), translates to silky, smooth, nontacky esthetic qualities.

The intra/intermolecular forces are low between siloxane molecule(s), giving rise to low resistance to flow when stressed, creating the ability to engineer high-molecular-weight polymers with relatively low viscosities. The glass transition temperature (T_g) of (poly)dimethylsiloxane is low, at ~–120°C (flowable liquid down to the T_g). Polymers with molecular weight (M_N) as high as ½ million are still fluids—a unique distinguishing point compared to similar molecular-weight organic polymers. At the same time, changing the structure of organo-modified siloxane polymers allows for physical form differences, e.g., from waterlike-viscosity volatile siloxanes to rubberlike, and even glasslike siloxane resins. The resultant physical property data on organo-modified siloxane compounds have been referenced in many sources (17).

Siloxane chain entanglement also plays an important role in solvation and flow. Ferry (18) characterized the relationship of viscosity and molecular weight ($M_n \propto \eta$). Below 1000 cP viscosity for PDMS, viscosity and molecular weight are more nearly proportional—thus, viscosity is increasing at a corresponding rate to M_n. Above 1000 cP, viscosity is not proportional to molecular weight—viscosity increases nearly 3.5% faster than M_n. This can be explained using the "reptation" model (molecules are moving through the mass by pulling in and out through tunnels). If one thinks of a mass of worms (each worm—strand—is a PDMS molecule), it takes time for the strands to slide in and out. With branching or crosslinking, the constraint on flow is dramatically slowed and viscoelastic properties play a stronger role. Thus, with crosslinked or high-molecular-weight PDMS, solvents can swell but not dissolve the molecules, time and energy being required to solvate and disentangle the molecules.

E. Classification

The silicone family of products is extensive. The silicon atom has the ability to accept many different substituents, and the siloxane backbone can take on different structures, which then create the opportunity to change the physical and, as a result, the performance characteristics of functional silicones from:

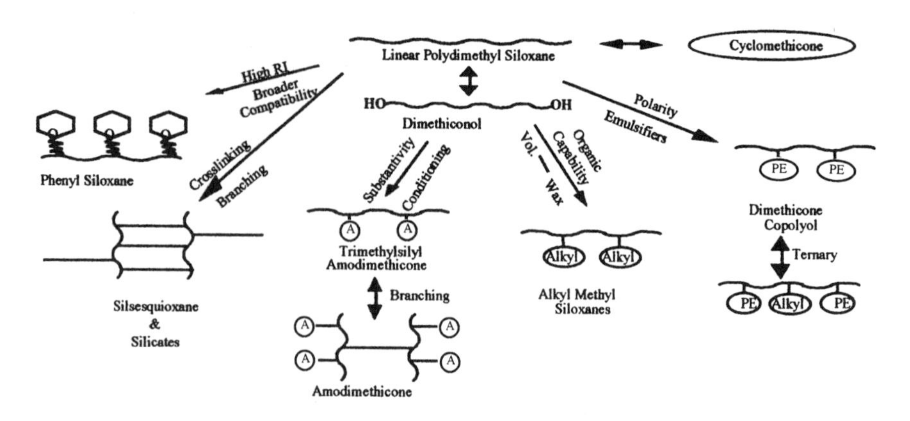

Figure 2 Pictorial of silicone family of products.

Volatile → range of viscosity → high-melting-point wax
Defoamer → profoamer → emulsifier
Fluid → elastomeric → resinous powder.

Figure 2 shows a pictorial representation of the silicone family of products—dimethyl, alkyl, aromatic, polyethers, amino, and three-dimensional—that are commercially available and commonly used in the personal care industry.

What follows is a more in-depth look at some key functional siloxane product families.

1. (Poly)dimethylcyclosiloxanes

The most common (poly)dimethylcyclosiloxanes [International Nomenclature Chemical Ingredient (INCI): Cyclomethicones] possess 6, 8, 10, 12, or 14 atoms in their ring structures (Figure 3) with 3 siloxy bonds (INCI: Cyclotrisiloxane), 4 siloxy bonds (INCI: Cyclotetrasiloxane), 5 siloxy bonds (INCI: Cyclopentasiloxane), 6 siloxy bonds (INCI: Cyclohexasiloxane), and 7 siloxy bonds (INCI:

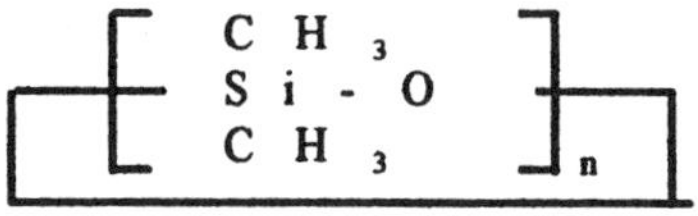

$$\left[\begin{array}{c} CH_3 \\ Si-O \\ CH_3 \end{array} \right]_n$$

where n = 3, 4, 5, 6, etc.

Figure 3 Empirical structure of C-PDMS (cyclomethicone).

Table 1 Heat of Vaporization of Common Personal Care Carriers

Material	Heat of vaporization (cal/g)
Water	539
Ethanol	210
Cyclotetrasiloxane	32
Cyclopentasiloxane	32

Cycloheptasiloxane), respectively. Volatile (poly)dimethylcyclosiloxanes are considered non-Volatile Organic Components (VOC) as regulated by the U.S. federal (19) and state Environmental Protection Agencies (EPA). Cyclomethicone fluids have unique physical properties that translate into benefits for personal care applications. These include volatility, low viscosity, and nonresidual transient skin feel. Cyclomethicone fluids have low heat of vaporization (Table 1), they are noncooling (i.e., cooling during evaporation from skin is minimal), and are nonstinging—a combination rare for personal care carriers. Other attributes of cyclomethicone fluids include: colorless, odorless, nonstaining, good spreading, detackification, water sheeting, and transient emolliency properties. The level of use of cyclomethicone in personal care applications is typically from 0.1% to 85%; the low end represents use levels which enhance skin feel in creams (e.g., hand and body moisturizers), and the high-end use levels are to carry actives to skin and hair (e.g., antiperspirant salts, hair cuticle coat). Another aspect of cyclomethicone fluids that make them functional in the personal care industry is that they have broad compatibility with most personal care ingredients, including ethanol, mineral oil, and fatty acid esters.

Another key attribute ot cyclomethicones is their ability to provide appealing sensorial esthetics to personal care products. Key sensorial esthetics typically associated with (poly)dimethylcyclosiloxanes, of which cyclopentasiloxane is a good example, are low residue, low stickiness, low tackiness, low greasiness, and low waxiness. Using a protocol developed by Dow Corning Corporation (20) in Figure 4 one sees that cyclopentasiloxane has sensory properties normally associated with silicones. The scale in Figure 4 that is used to rank each attribute is based on 0 to 10, with 10 representing the most and 0 representing the least for the attribute measured. As the ring size of (poly)dimethyl-cyclosiloxane changes, there are minor changes in sensory effects. Larger-ring-size cyclomethicone fluids are slower to volatilize, and they reside on the skin longer to give a longer-lasting feel until they totally evaporate.

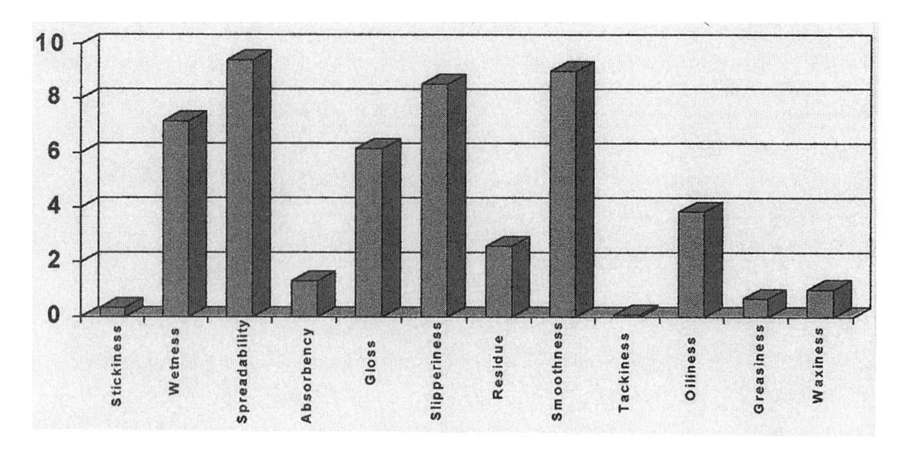

Figure 4 Cyclotetrasiloxane sensory profile.

2. *Linear (Poly)dimethylsiloxanes (Dimethicone and Dimethiconol)*

Linear (poly)dimethylsiloxane (INCI: Dimethicone)—see Figure 5—having trimethyl end blocking, can have from two to thousands of repeating dimethylsiloxy groups (Figure 5a), with viscosities ranging from 0.65 to >100,000 centistokes (cS) (Table 2). When linear (poly)dimethylsiloxanes are hydroxy terminated, they are called dimethiconols (INCI), which have viscosities as high as 30,000,000 cS (Figure 5b).

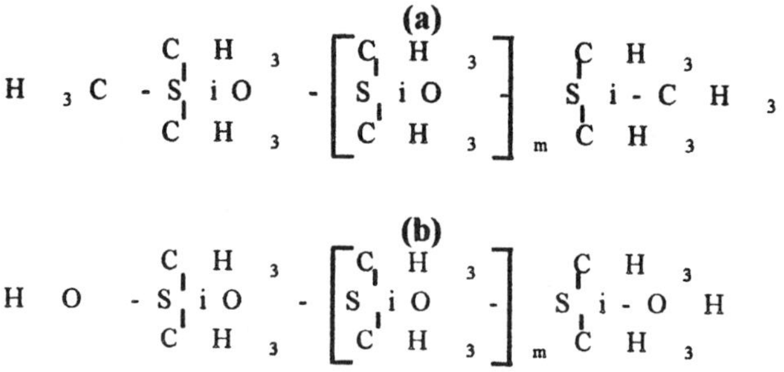

(a)

$$H_3C-\underset{CH_3}{\overset{CH_3}{Si}}O-\left[\underset{CH_3}{\overset{CH_3}{Si}}O\right]_m-\underset{CH_3}{\overset{CH_3}{Si}}-CH_3$$

(b)

$$HO-\underset{CH_3}{\overset{CH_3}{Si}}O-\left[\underset{CH_3}{\overset{CH_3}{Si}}O\right]_m-\underset{CH_3}{\overset{CH_3}{Si}}-OH$$

where m is greater than 0

Figure 5 Empirical structure of (a) linear poly(dimethylsiloxane) and (b) dimethiconol.

Table 2 Comparative Molecular Weights for Different Viscosity (Poly)dimethylsiloxanes

Average viscosity (cS)	Approximate molecular weight	No. of subunits (MD_xM)	Example of similar viscosity
5	800	9	Water
20	2,000	27	Cooking oil
50	3,800	50	
100	6,000	80	Paint
200	9,400	125	Shampoo
350	13,700	185	Mineral oil
500	17,300	230	Ketchup
1,000	28,000	375	Motor oil
5,000	49,300	665	Corn syrup
10,000	62,700	845	Molasses
12,500	67,700	910	
30,000	91,700	1,235	Honey
60,000	116,500	1,570	Hot tar
100,000	139,000	1.875	
300,000	204,000	2,750	
600,000	260,000	3,510	Thick milkshake
4,000,000	400,000	5,400	Petroleum jelly
20,000,000	550,000	7,430	Silly putty

As a class of personal care additives, linear PDMS is considered a good skin emollient and lubricant. Linear PDMS generally acts as (a) skin-feel modifiers, (b) water barrier protectants, (c) defoamers, (d) desoapers (i.e., eliminators of creamy whitening of a cosmetic formulation during the initial rubbing onto skin or hair), and (e) providers of conditioning and emolliency.

Dimethicone is covered in the "Skin Protectant Monograph" as an over-the-counter drug (21) and is listed in the National Formulary compendium (22). Because dimethicone can be recognized as a drug, when sold under a compendia name with skin protectant claims it must meet the requirements as outlined in the compendium, and the substance's name must be distinguished clearly on the label. The tentative final monograph identifies which active can be used at use levels of 1–30% as an active ingredient in OTC products with label claims such as "helps prevent and temporarily protects chafed, chapped, cracked, or windburned skin and lips."

Linear PDMS is noted for its thermal and oxidative stability over a wide range of temperature. When held in contact with air at 150°C, linear PDMS is stable for long periods. These compounds oxidize at higher tempertaures, well above those typically associated with operating temperatures used in the personal care industry. In the absence of oxygen and at temperatures above 250°C, linear PDMS can be broken down to smaller polymers. This breakdown process is known as depolymerization (a phenomenon similar to the cracking of petroleum). Additionally, the low glass transition temperature ($T_g \approx -120°C$) of PDMS means that the polymers are flowable liquids down to that T_g.

It is the compatibility (solubility) characteristics of linear PDMS that contributes to its unique esthetics and film-forming properties. Because linear PDMS has limited compatibility in typical cosmetic ingredients, it tends to migrate to the surface (away from the skin or hair). The compatibility of linear PDMS varies with its viscosity—with the low-viscosity fluids having compatibility with more materials than higher-viscosity fluids. Linear PDMS is insoluble in water, but soluble in aliphatic hydrocarbons, aromatics, and chlorinate solvents (e.g., hexane, toluene, methylene chloride, and chloroform). Furthermore, linear PDMS is incompatible in mineral oils and solvents such as isopropanol; lower-viscosity PDMS has greater miscibility, especially in mineral oils and solvents such as isopropanol.

Volatility and vapor pressure decrease while viscosity and molecular weight increase for linear PDMS. Hexamethyldisiloxane (0.65 cS) possesses the highest volatility of linear or cyclic (poly)dimethylsiloxane fluids—similar to ethanol. PDMS starts to lose its volatility at about 5 cS (around 800 molecular weight). Volatile PDMSs are classified as non-Volatile Organic Components (VOC) as regulated by the U.S. Environmental Protection Agency (23) and individual state governments.

When linear PDMS fluids are incorporated with cosmetic ingredients to make a personal care product, sensory attributes may be slightly different compared to the neat PDMS fluid. The ease of spread and lubricity of a linear PDMS fluid produces the characteristic velvetlike feel typically associated with silicones. As the molecular weight increases for linear PDMS, an increase in residue, oiliness, and smoothness can be observed. Figure 6 denotes the sensory attributes for a medium-viscosity linear PDMS (350 cS)—sensorial attributes are evaluated by trained sensory panelists. In Figure 6 one sees an increase in residue, oiliness, greasiness, and waxiness of 350-cS dimethicone, a linear PDMS, compared to cyclomethicone (reference Figure 4). Dimethicone, 350 cS, shows different attributes than cyclomethicone, because it does not volatilize and thus remains on the skin.

To effect a perceived sensorial change in a personal care product, one would usually need to use between 0.1 and 5% of a linear PDMS fluid. Figure 7 shows the minimum threshold levels required for various PDMS polymers before they are judged as perceptibly different. As the molecular weight of

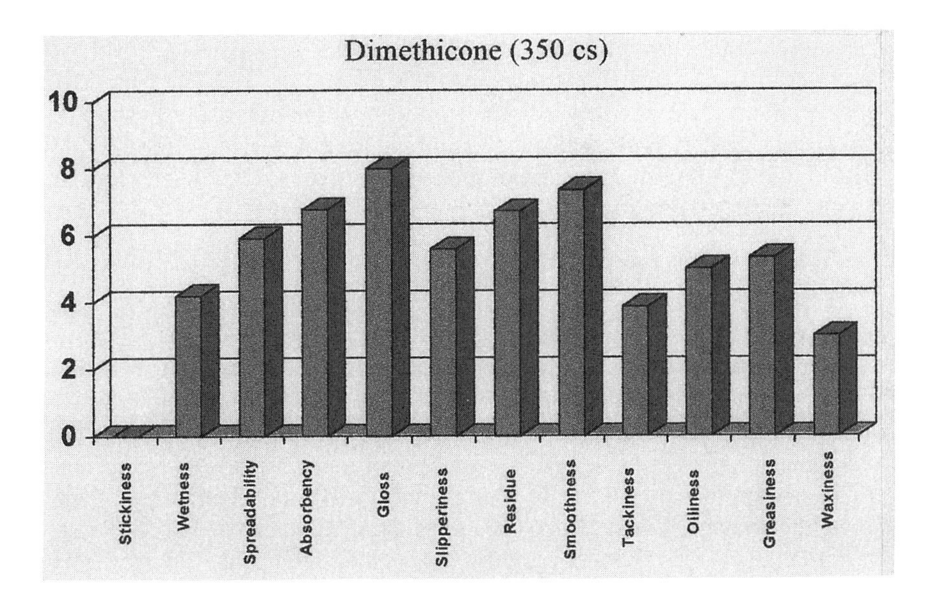

Figure 6 Sensory profile of 350 cS linear (poly)dimethylsiloxane.

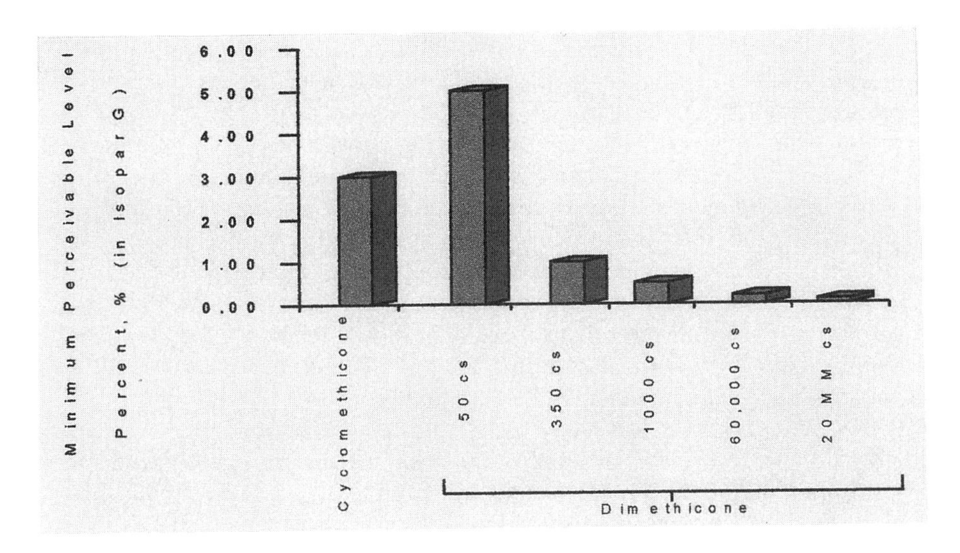

Figure 7 Threshold level of various PDMSs.

linear PDMS increases, a lower weight percent will be needed in formulations—the hypothesis is that a thicker film is produced—to effect perceived sensory change, in particular, residue, oiliness, and smoothness. What can be concluded is that, to elicit a sensory perception, (a) typically less linear PDMS (dimethicone) is required than cyclomethicone; and (b) as viscosity (molecular weight) increases, less linear PDMS is required.

3. Other Functional Siloxanes

a. Polyether Functional (24). Polyether functional siloxanes (SPE, siloxane polyether) are created by replacing some of the methyl groups on the siloxy backbone with polyether (polyoxyalkylated) substituents (see Figure 9, below)—primarily ethylene oxide and propylene oxide. Substitution of a polyoxyalky substituent for a methyl group on the PDMS backbone results in a change in the hydrophile/lipophile balance (HLB) of a SPE polymer. Addition of polyether groups allows for modification of compatibility of SPE polymers in polar and nonpolar solvents. Not only do polyether groups effect solubility, but also interfacial tension characteristics of polyether functional siloxanes are different from those of comparable-molecular-weight PDMS.

Polyether functional siloxanes can range from water soluble to oleophilic soluble, thus having applications in both water and oil phases of cosmetic formulations. Water and polar solvent solubility increases when the mole percent and molecular weight of the polyoxyethylene moiety is increased. Complete water solubility occurs when the mass ratio of polyoxyethylene to dimethylsiloxy groups is in the range of 2–4. They can be used to detackify water and oil phases of cosmetic formulations. High-HLB polyether functional siloxanes are used for detackification of water, and low-HLB polyether functional siloxanes are used for oil phases.

Polyether functional siloxanes are playing an ever-increasing role in the personal care industry as emulsifiers [Silicone Formulation Aids® (25)] for water-in-poly(dimethylcyclosiloxane) and water-in-oil emulsion. Their uniqueness as siloxane emulsifiers is due to flexibility of the siloxane backbone, which in turn provides for reduction of interfacial tension, more effective surface coverage, and a thick phase between the water and oil phases without temperature or shear as the driving force (26). The stability of oil-continuous-phase (water-in-oil) emulsions is the result of a film of high viscoelasticity at the water–oil phase boundary (27) of the emulsion.

Polyether functional siloxanes provide advantages in producing water-in-poly(dimethylcyclosiloxane) and water-in-oil emulsions: (1) low levels of emulsifier—less than 2%, but usually less than 1%; (2) creation of emulsions that vary in viscosity from pumpable fluids to extrudable gels to sticks (using phase volume to control viscosity), and from opaque to crystal clear (using refractive index matching of the water and oil phase)—without the need for

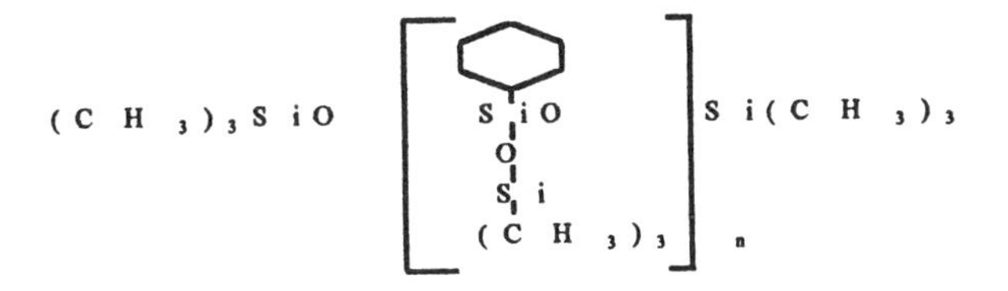

Figure 8 Empirical structure of phenyl trimethicone.

making significant modifications in phase compositions; and (3) nonirritation or drying of skin. In addition to being efficient water-in-oil or water-in-(poly)dimethylcyclosiloxane emulsifiers, SPE polymers can enhance perceptible skin feel in personal wash applications [e.g., bar soaps, shower gels (28)], act as good hairspray resin modifiers (29), and mild conditioners for clear shampoo. Also, it has been reported that certain water-soluble polyether functional siloxanes (30) can reduce eye irritation in shampoos.

b. Phenyl Functional. Replacement of some methyl groups on a PDMS siloxane backbone with phenol substituents results in an increase in the refractive index (imparting shine or sheen), and broadened compatibility in comon personal care systems. Phenyl groups also enhance thermal stability, oxidative resistance, and nonwhitening and nonresidue attributes (31). Figure 8 shows the empirical structure of a phenyl functional siloxane, Phenyl Trimethicone (INCI).

c. Alkyl Functional (32). Substituting alkyl groups (six or more carbons) for some methyl groups on a PDMS siloxane backbone results in an alkyl functional siloxane (more specifically, alkyl methyl siloxanes or AMS). Alkyl functionality of these siloxanes improves occlusivity (33) (decrease in water permeability) and compatibility in common personal care systems. AMS polymers have product forms ranging from fluids to very-high-melting-point waxes (above 70°C). Alkyl substitution results in the modification of sensory characteristics of a PDMS siloxane, providing esthetics approaching both hydrocarbons and silicones—that is, a waxier and greasier feel during rub-out and a "silicone" smooth after-feel. AMS waxes have been reported (34) to provide bodying and cushioning to all types of emulsions, with moisturization comparable to petrolatum.

d. Amino Functional. The addition of aminoalkyl groups to a PDMS siloxane backbone increases substantivity (35,36) of silicones. The most widely used amine group is based on the ethylene diamine structure. Amine functional siloxanes can be trimethyl siloxy or dimethyl hydroxyl siloxy end-

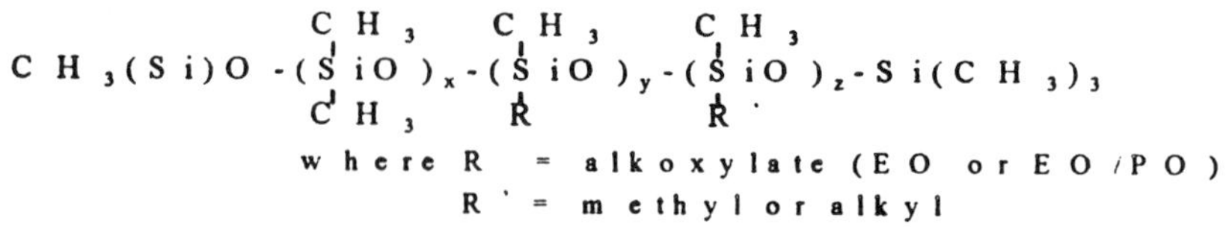

$$CH_3(Si)O-(\underset{CH_3}{\overset{CH_3}{Si}}O)_x-(\underset{R}{\overset{CH_3}{Si}}O)_y-(\underset{R'}{\overset{CH_3}{Si}}O)_z-Si(CH_3)_3$$

where R = alkoxylate (EO or EO/PO)
R' = methyl or alkyl

Figure 9 Empirical formula of siloxane polyether.

blocked. Typical INCI names assigned to this class of materials are Amodimethicone, Amodimethiconol, and Trimethylsilylamodimethicone (see Figure 9). Amine functional siloxanes can differ in length of siloxane backbone, linear versus branching of the siloxane backbone, substitution amount and type of amino groups on the backbone, and the linking group on the amino moiety. The addition of an amine group to a siloxane backbone allows for excellent hair conditioning for both leave-in and wash-off conditioners. Amino-functional PDMS polymers can potentially cause skin irritation; therefore, caution is required when using them as skin conditioners.

e. Crosslinked Poly(dimethylsiloxanes). This group of PDMS polymers is based on crosslinking of siloxane backbones. INCI names are Silicates ("Q" structures), Silsesquioxane ("T" structure), Crosspolymer, and Copolymer. Changing the degree of crosslinking of the PDMS backbone results in a change in the physical form from a soft gel to a glasslike solids. Recent patents have been issued outlining the ability to modify rheology of functional siloxanes with crosslinked PDMS polymers (37–40). These patents show how to improve appearance, bodying, cushioning of cosmetic products, and set retention/styling of hair. Also, it has been speculated that crosslinking of the PDMS backbone could improve substantivity without adding heaviness to resultant films.

f. Beyond Classical PDMS. How functional siloxanes are delivered to skin and hair is as critical as the structure/properties of the specific functional siloxane. As an example, functional siloxanes can be added directly to a formulation and delivered to specific skin or hair sites as a solution or dispersion of the formula—requiring transport through the formula matrix to reach the desired location. Functional siloxanes can be predispersed as colloidal systems (e.g., emulsions and microemulsions) for ease of incorporation. Engineering the PDMS colloidal systems—particle size distribution, polymer rheology, and surfactant type—can affect how other functional conditioners are delivered to hair and skin. Further, blending of functional siloxanes can provide the opportunity for synergistic enhancement of performance.

The above functional siloxane classifications and benefits for skin and hair conditioning are based on a simplistic view of the organo-modified siloxane backbone. One can visualize the potential for significant changes in properties and the resultant benefits when the backbone is further modified (e.g., combination of functionality on the backbone, or reaction product of active sites on the functional attached moiety to the PDMS backbone. New, evolving multifunctional siloxanes are covered by O'Lenick in Chapter 9 in this book.

II. ORGANO-FUNCTIONAL SILOXANES FOR SKIN CONDITIONING

Silicones have been used to condition skin since the 1950s. *Conditioning* in this context means beautification and esthetic improvement of skin, perceived by consumers as resulting in healthy skin. Silicones provide these two benefits by nature of the chemical and physical properties of the siloxane backbone. Silicones act as excellent emollients (41) to provide perceived smoothness, softness, and lubrication to the skin as delivered from personal care topical skin applications. Film-forming characteristics of PDMS polymers provide uniform deposition, self-healing films, and substantivity to skin.

Increasing the compatibility of a blend of PDMS polymers and an organic cosmetic ingredient may decrease spreading and film forming of the blend. This is because molecules can more easily co-mingle, and are less likely to migrate to the interfacial surface zone, where they would provide lubrication.

Organo-functional siloxane polymers are used in a variety of facial and body cleansing applications (42). Volatile (poly)dimethylcyclosiloxane fluids are used extensively in nonrinsable makeup removers and cleansers, as they are good solvents for organic-based oils, leaving skin with a dry, smooth, nongreasy feeling. Also, they are mild and nonirritating to the skin. (Poly)dimethylsiloxane fluids, such as dimethicone, polyether functional siloxanes, and amino-functional siloxanes, have been used in bar soaps to aid mold release and improve foam quality. As with 2-in-1 conditioning shampoos, PDMS polymers are emerging in the shower gel market because consumers perceive them to have a residual smooth after-feel.

New developments in water-soluble (poly)dimethylsiloxane polyether technology enable formulation of facial washes with improved stable foams and after-feel characteristics, along with potential for reduced eye and skin irritation. By reducing the surface tension of water, they become easier to spread. Also, polyether functional siloxanes, especially those with relatively high HLB (greater than 5), have benefits in surfactant-based systems by leaving a perceivable softness to the skin.

A. Formulation Examples

The following formulas exemplify the use of silicones in personal care applications as conditioning agents.

1. *Daytime Hand and Body Skin Cream*

Formula 1 demonstrates the use of cyclomethicone and dimethiconol (a blend of 13% high-molecular-weight PDMS polymer gum in cyclomethicone) to enhance the non-oil feel during application of cream to the skin. After the formula dries, resultant smooth skin feel is associated with residual dimethiconol. Also, dimethiconol at a very small level (0.45%) provides a smoother and drier feel during application of the cream on the hand and body.

Formula 1 Hand and Body Skin Cream (43)

Water	84.7%
Carbomer 934	0.1
Glycerin	3.0
Triethanolamine	0.9
Cetyl Alcohol	1.0
Stearic Acid	0.8
Glyceryl Stearate	1.5
Cyclomethicone (and) Dimethiconol	3.0
Isopropyl Myristate	1.5
Diisopropyl Adipate	1.5
Mineral Oil	2.0
Preservative/Fragrance	q.s.

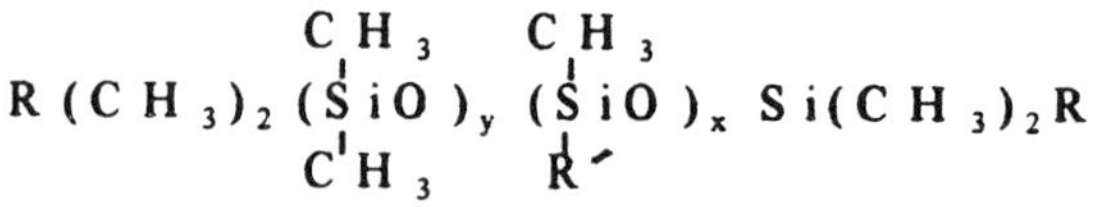

$$R(CH_3)_2\,(\underset{\displaystyle CH_3}{\overset{\displaystyle CH_3}{|}}\!\!\!\!SiO)_y\,(\underset{\displaystyle R'}{\overset{\displaystyle CH_3}{|}}\!\!\!\!SiO)_x\,Si(CH_3)_2R$$

where: $R = OH$ or (CH_3)

$R' = (CH_2)_a\,NHCH_2CH_2NH_2$

Figure 10 Empirical structure of amodimethicone.

2. Nighttime Skin Lotion

Formula 2 demonstrates how silicone emulsifiers (dimethicone copolyol, Figure 10) can be utilized to make simple water-in-cyclomethicone emulsion lotions that have an elegant after-feel (associated with dimethiconol). Even though polyether functional siloxane emulsifiers are not noted as conditioners of hair and skin, they provide unique and efficient delivery systems for other conditioning agents.

Formula 2 Nighttime Skin Lotion (44)	
Cyclomethicone (and) Dimethicone Copolyol	10.0%
Cyclomethicone (and) Dimethiconol	10.0
Cyclomethicone	10.0
PPG-3 Myristyl Ether	0.5
Water	62.5
Glycerine	5.0
Sodium Chloride	2.0

3. Contemporary Cold Cream

Formula 3 demonstrates how silicone emulsifiers can be utilized to make simple water-in-cyclomethicone emulsion creams that can remove makeup (due to cyclomethicone) and have an elegant, low-residue skin feel.

Formula 3 Contemporary Cold Cream (45)	
Cyclomethicone (and) Dimethicone Copolyol	10.0%
Cyclomethicone	10.0
C12–15 Alcohols Benzoate	5.0
PPG-3 Myristyl Ether	1.5
Sodium Chloride	1.5
Water	q.s. to 100%
Preservative	q.s.

4. Facial Cleanser

Formula 4 demonstrates the use of polyether functional siloxane (Dimethicone Copolyol) to add a soft feel during application and after. Choosing the right functional siloxane can also contribute to foam boosting and elegance of the foam structure as is demonstrated in this formulation.

Formula 4 Facial Cleanser (46)	
Water	60.5%
Disodium Cocoamphiodiacetate	12.0
Sodium Lauroyl Succinate	10.0
Cocamidopropyl Betaine	8.0
Lauramide DEA	4.0
Citric Acid (50% aq.)	0.5
Dimethicone Copolyol (Aq. Soluble)	5.0
Preservative	q.s.

5. *Shower Gel*

Formula 5 demonstrates the use of cyclomethicone and a high-molecular-weight PDMS (dimethiconol) polymer gum to provide a perceptible after-feel when the surfactants are washed off the skin.

Formula 5 Shower Gel (47)	
Water	29.35%
Sodium Lauryl Sulfate (30% active)	30.0
Cocamide MIPA	4.0
Cocamidopropyl Betaine	6.65
TEA-Cocohydrolyzed Animal Protein	6.0
Acrylates/C10–30 Alkyl Acrylates Crosspolymer (2%)	20.0
Cyclomethicone (and) Dimethiconol	2.0
Glycol Distearate	2.0
Preservative	q.s.

6. *Another Shower Gel*

Formula 6 is another example of the use of a polyether functional siloxane (copolymer of alkyl and polyether functionality) in a personal wash system. Again, the functional siloxane is left on the skin after the surfactant is washed off.

Formula 6 Shower Gel (48)	
Water	54.5%
Cocoamide MIPA	2.0
Laurylmethicone Copolyol	2.0
Sodium Lauryl Sulfate (30%)	30.0

Decyl Glucoside	6.0
Cocamidopropyl Hydroxy-sultaine	3.0
PEG-7 Glyceryl Cocoate (and) Polyol Alkyoxy Ester	2.5
Preservative	q.s.

7. Sunscreen Lotion

Formula 7 is an example of the use of a silicone emulsifier to provide sunscreen actives to the skin from the cyclomethicone continuous phase for improved uniform application. The alkyl methyl siloxane wax (Stearyl Dimethicone) is used to enhance the SPF (Sun Protection Factor) (49).

Formula 7 Sunscreen Lotion

Cyclomethicone	6.0%
Stearyl Dimethicone	5.0
Cyclomethicone (and) Dimethicone Copolyol	9.5
Octyl Dimethyl PABA	4.0
Laureth-7	0.5
Water	74.5
Sodium Chloride	0.5
Fragrance	q.s.

8. Low-Residue Solid Antiperspirant

Though underarm products are designed to reduce odor and wetness, there is a need to add ingredients that provide a mildness and esthetic that can be construed as conditioning. Cyclomethicone, dimethicone, and phenyltrimethicone have played critical roles in enhancing the esthetics and uniform delivery of antiperspirant active salts. Cyclomethicone acts as a transient, non-oily carrier of actives; dimethicone provides good lubrication and skin feel; and phenyltrimethicone reduces whitening and residue of antiperspirant salts.

Formula 8 demonstrates how a high-refractive-index phenyl functional siloxane (R.I. = 1.46) can enhance the appearance of an underarm antiperspirant, minimizing the whitening effect when the cyclomethicone (transient, nonstinging carrier fluid for antiperspirant actives) evaporates.

Formula 8 Low-Residue Solid Antiperspirant

Cyclomethicone	42.0%
Phenyl Trimethicone	10.0
Stearyl Alcohol	20.0

Hydrogenated Castor Oil (M.P. 80°C)	4.0
PEG-6 Distearate	4.0
Aluminum-Zirconium Tetrachlorhydrex-GLY powder	20.0

B. Skin Conditioner Trends

Current trends continue to move toward non-oily, non-greasy, light-feeling products, all of which silicones enhance. A number of patents have been issued over the past few years that highlight the benefits outlined above for the use of silicones in skin care applications:

1. U.S. Patent 5,013,763 (Tubesing et al.), issued in 1991, discusses the use of substantive silicones and quarternium ammonium compounds to leave the skin feeling moist, soft, and smooth even after it has been washed with soap and water.
2. U.S. Patents 5,021,405 and 5,210,102 (Klemish), issued in 1991, discusses the use of amido-functional organo-siloxane mixtures to make skin care products longer lasting and more esthetically acceptable.
3. W.O. 9,209,263 (Alban et al.), published in 1992, discusses the use of polyalkyl, polyaryl, polyalkylaryl, and/or polyether siloxanes in cosmetic gels to control sebum distribution on facial skin and provide improved skin feel, moisturization, rub-in, and absorption characteristics.
4. W.O. 9,307,856 (Decker et al.), published in 1993, discusses the use of alkyl-functional, alkyl/aryl, and/or polyether-functional siloxanes to improve skin feel and residue characteristics together with moisturizing, emolliency, rub-in, and absorption characteristics.
5. U.S. Patent 5,326,557 (Glover et al.), issued in 1994, discusses the use of water-soluble polyether functional siloxanes to aid in humectancy for facial cleansers, moisturizers, and aqueous conditioning gels.
6. W.O. 9,417,774, issued to Procter & Gamble in 1995, discusses the use of silicone gums with a molecular weight of 200,000–400,000 in oil-in-water dispersion to provide improved skin feel and reduced greasiness.
7. W.O. 9,614,054 (Simmons et al.), published in 1996, discusses the use of water-insoluble silicone emollients as emulsions of average particle size 5–4000 μm to give clear, surfactant-free, low-tack, excellent skin feel, and low irritating hydrogel thickened conditioning composition.

III. ORGANO-FUNCTIONAL SILOXANES FOR HAIR CONDITIONING

Functional siloxanes are used in conditioners, shampoos, mousses, hairsprays, hair colorings, permanent wave applications, and setting lotions. Generally, functional siloxanes help condition the hair to improve wet and dry combing, help retain moisture, heighten luster and sheen, improve manageability, and provide an elegant tactile feel. The term *conditioner*, in relation to hair grooming preparations, refers to ingredients that aid in enhancing manageability, appearance, and feel of hair. Such ingredients should be capable of acting as a lubricant—reducing combing resistance of wet and dry hair, minimizing tangling, making hair softer and smoother, and improving set retention when styled. These ingredients should also act as antistatic agents to reduce or eliminate flyaway hair (especially from dried-out hair) without producing buildup upon repeated application, as this would cause hair to become lank and dull.

Organo-modified siloxane polymers demonstrate functionality as effective ingredients for conditioning hair at low use levels (as low as 1% in most conditioners and shampoos) (50). Furthermore, blended combinations of an organo-modified siloxane and an organic conditioner has proven effective for conditioning hair. In particular, organo-modified siloxane polymers help damaged hair to appear and feel healthier—without greasy buildup. Besides reducing combing force, functional siloxanes—particularly the amine-functional varieties—can play a unique role in helping to prevent heat damage to hair. Incorporated in cream rinses, they speed drying time by capillary displacement of water held between hair fibers.

The low surface energies of functional siloxanes allow them to deposit on the hair surface and spread uniformly as a thin coating on the hair shaft. As a result of spreading characteristics of functional siloxanes, these organo-siloxane films have improved durability of the resultant hair coating, are self-healing, uniform, and assist other formulation ingredients to spread on the hair shaft. Durability is proportional to molecular weight or type of functional group attached to the siloxy group. This translates to reduction in combing forces of the hair shaft, leaving a soft, manageable feel to hair.

Conditioners that reduce combing force can have a significant impact on controlling damage from mechanical stress on hair. Tangle-free combing is important to both wet and dry hair, since hair is weakest when saturated with water. Also, shampoos with conditioning agents that remain on the hair help counteract sebum stripping.

A. Application Examples

Silicone technology has advanced dramatically over the past decade, but the basic (poly)dimethylsiloxane molecule still meets most requirements in conditioning the hair for a healthier-looking appearance. As a result, hair care formulators, with a selection of new materials with which to develop highly differentiated hair care products, find themselves asking, "How does one choose the right functionality of siloxane to achieve desired performance results?" A selection guide (51), as a rational approach to selecting the right functional siloxane, has been published to assist the formulating chemist in matching the right functional siloxane to the intended performance benefit required.

Depending on how a formulator desires a product to perform, he or she might choose a (poly)dimethylsiloxane fluid, polyether-functional siloxane, an amino-functional siloxane, a functional siloxane emulsion, or a clear formulation.

1. *(Poly)dimethylsiloxane Fluids*

(Poly)dimethylcyclosiloxane fluids offer temporary (transient) conditioning, improved wet combing, decreased drying time, resin plasticization, hair functional actives carrier, and detangling. Also, as (poly)dimethylcyclosiloxane fluids evaporate, there remains no residual deposit on hair. These cyclical fluids have applications in conditioner (52), 2-in-1 shampoo, and styling products.

Linear (poly)dimethylsiloxane fluids and emulsions are typically associated with improved conditioning (wet/dry combing) benefits with a nice sensory feel—the higher the molecular weight of PDMS fluid, the more effective (lower use levels) are the conditioning effects. Additional benefits for this class of functional siloxanes are reduce flyaway, improved shine, softness, and humidity resistance. These linear fluids have application in 2-in-1 shampoos, conditioners, styling, and cuticle coat treatment products (53).

(Poly)dimethylsiloxanes—both cyclical and linear—are used to aid in conditioning of shampoo bases. Patents have been issued that discuss how to incorporate water-insoluble siloxane into a surfactant system, while other patents focus on the conditioning effects and deposition-aid ingredients. Higher molecular-weight (poly)dimethylsiloxane fluids improve the potential for increased conditioning and could require less for equal conditioning effects.

a. Pearlescent Conditioning Shampoo. Formula 9 demonstrates the use of dimethicone (350 cS) to aid wet and dry combing performance. Hydroxylpropyl methylcellulose is used to disperse and stabilize dimethicone with respect to settling.

Formula 9 Pearlescent Conditioning Shampoo

Ammonium Lauryl Sulfate (30%)	30.0%
Water	61.0
Hydroxylpropyl Methylcellulose	2.0
Coconut Diethanolamide	3.0
Dimethicone, 350 cS	4.0
Citric Acid	q.s. to pH 6.0
Ammonium Chloride	q.s.

b. High-Shine Cuticle Coat. Formula 10 is an example of the use of high molecular-weight (poly)dimethylsiloxane polymer gum (dimethiconol) to help damaged cuticle. Phenyl Trimethicone is also used to help shine or sheen.

Formula 10 High-Shine Cuticle Coat (54)	
Cyclomethicone (and) Dimethiconol	90.0%
Cyclomethicone	8.0
Phenyl Trimethicone	2.0

2. *Polyether Functional Siloxanes*

Polyether functional siloxanes are typically associated with profoaming, light conditioning, improved feel (softness), resin plasticization, and anti-irritancy. These polyether fluids have applications in shampoo, conditioner, styling, and mousse products.

Polyether functional siloxane fluids provide light to medium conditioning, especially when clarity is required in a formulation. Polyether functional siloxanes can be incorporated into shampoos to give conditioning benefits, aid in foaming efficiency, and reduce the potential for eye irritation caused by anionic surfactants used in shampoos. In Figure 11 (55), 3% sodium lauryl sulfate (SLS) + dimethicone copolyol (solid line) is compared to 3% SLS (broken line) to show the irritation reduction associated with Dimethicone Copolyol.

a. Detangling Conditioner. Formula 11 uses a Dimethicone Copolyol to provide water-soluble, mild conditioning. Amodimethicone (as a silicone-in-water emulsion for ease of incorporation) is also used to aid synergistically in dry combing.

Formula 11 Detangling Conditioner	
Water	81.0%
Glycerin	11.0
Polysorbate 20	1.5
Quaternium-7	2.0
Lanolin (and) Hydrolyzed Animal Protein	1.0
Panthanol	0.2
Imidazolidinyl Urea	0.2
Dimethicone Copolyol	1.0
Amodimethicone (and) Cetrimonium Chloride (and) Trideceth-12	2.0
Methyl Paraben	0.1

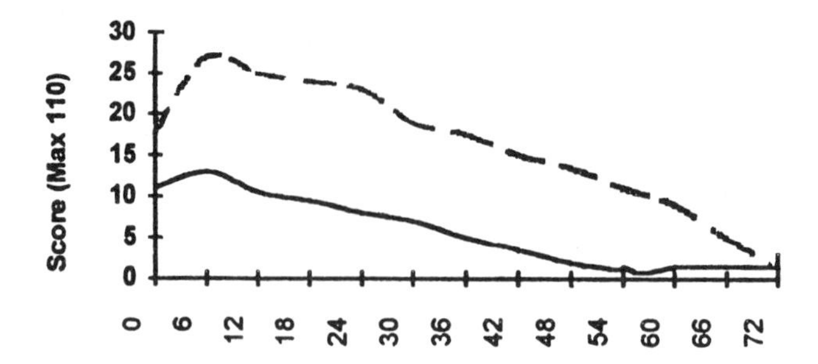

Figure 11 Draize mean eye irritation scores.

b. Superhold Hair Spray. Formula 12 demonstrates the resin plasticization properties of Dimethicone Copolyol, aiding in modification of the spray pattern and improved resin performance.

Formula 12 Superhold Hair Spray

Dimethicone Copolyol	0.5%
PVM/MA Copolymer, Ethylester	10.0
Aminomethyl Propanol	0.2
SD Ethanol 40	89.3

3. *Amino-Functional Siloxanes*

Amino-functional siloxanes are typically associated with all types of conditioners and some 2-in-1 shampoo products, to improve wet/dry combing, softness, durability, and set retention.

Amino-functional siloxane fluids and emulsions provide robust conditioning to hair because of their substantivity (56). Polar amine groups have a profound effect on siloxane deposition properties, giving affinity to the proteinaceous surface of hair. This translates to strong affinity to a hair's surface; and, depending on a formulation, can stay on hair through multiple washings. What is remarkable is that after several treatments, amino-functional siloxane fluids do not build up. In addition to the above, they are easily incorporated into clear and opaque hair products. Using 2.5% of an amino-functional siloxane emulsion and 0.9% of a polyether-functional siloxane in a rinse-out conditioner can help combat the effect of sebum stripping caused by excessive shampooing. By reducing friction and detangling, amino-functional siloxane at 1% can help prevent damage to hair from combing, without buildup on the hair shaft.

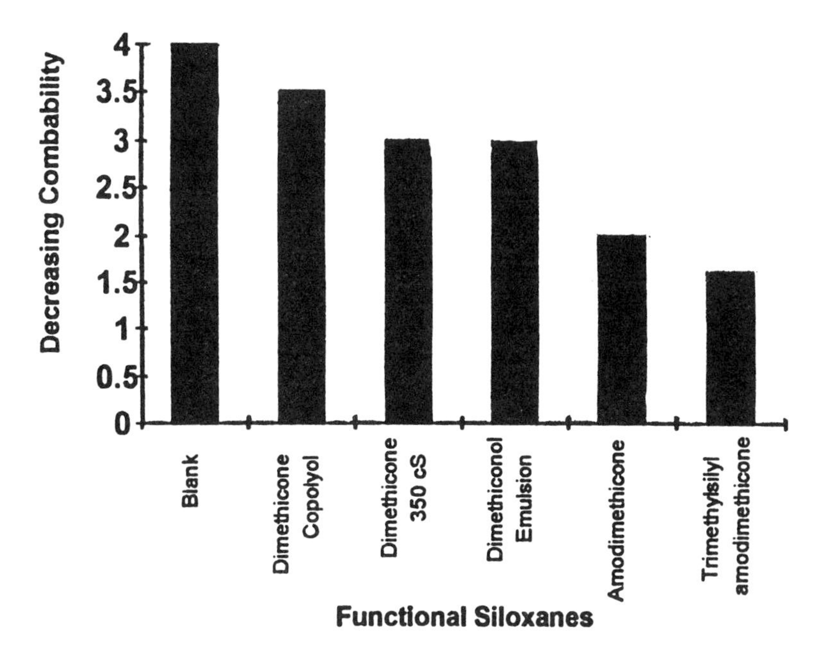

Figure 12 Wet combing data for functional siloxanes in shampoo base.

In Figure 12, a graph depicts wet combing data collected from shampoo bases in which there are different functional siloxanes at equivalent weight percent. The data show significant decrease in force during wet combing when amino-functional siloxanes are used. Therefore, one can provide a premium conditioner without buildup or affect on curl retention by using a very small amount of an amino-functional siloxane.

a. Durable Conditioning Cream Rinse. Formula 13 illustrates the use of an amino-functional siloxane (incorporated as a pre-made emulsion for ease of use) to improve wet and dry combing for perceivably better conditioning.

Formula 13 Durable Conditioning Cream Rinse

Water	86.5%
Sodium Chloride	0.5
Hydroxypropylmethyl Cellulose	2.0
Stearalkonium Chloride	5.0
Cetyl Alcohol	1.0
Amodimethicone (and) Cetrimonium Chloride (and) Trideceth-12	5.0

b. Intensive Conditioner. Formula 14 is an example of another formula incorporating a ready-to-use amino-functional siloxane emulsion to improve wet and dry comb, and provide soft, silky, and glossy esthetics to hair.

Formula 14 Intensive Conditioner

Hydroxyethyl Cellulose	1.5%
Water	91.5
Cetyl Alcohol	2.0
Amodimethicone (and) Cetrimonium Chloride (and) Trideceth-12	5.0

4. Functional Siloxane Emulsions

Both linear PDMS fluids and amino-functional siloxane fluids (57) can be supplied as silicone-in-water emulsions. As micro- and macroemulsions, linear PDMS fluids and amino-functional siloxanes are easier to incorporate in aqueous surfactant hair preparations. These emulsions can have anionic, cationic, or nonionic charges, which allows incorporation into most aqueous hair preparations—without incompatibility problems.

a. Conditioning Shampoo

Formula 15 Conditioning Shampoo

Ammonium Lauryl Sulfate (30%)	33.0%
Cocoamide DEA	3.0
Water	57.0
Dimethicone (and) Laureth-3 (and) Laureth-23 Emulsion (50% active of 60,000-cS Dimethicone)	6.0
Ammonium Chloride	1.0
Citric Acid	q.s. to pH 6.0
Fragrance	q.s.
Preservative	q.s.

b. Leave-in Conditioner

Formula 16 Leave-in Conditioner

Trimethylsilylamodimethicone (and) Octoxynol-40 (and) Isolaureth-6 (and) Glycol Emulsion	1.8%
Dimethicone Copolyol	1.8

Water	96.2
Quaternium-15	0.2
Color/Fragrance/Preservative	q.s.

c. Rinse-out Conditioner for Damaged Hair

Formula 17 Rinse-out Conditioner for Damaged Hair

Cetearyl Alcohol (and) Ceteth-20	5.0%
Sodium Lauryl Sulfate	0.1
Perfume	0.4
Trimethylsilylamodimethicone (and) Octoxynol-40 (and) Isolaureth-6 (and) Glycol Emulsion	6.8
D.I. Water	87.5
Quaternium-15	0.2

5. *Clear Formulations*

Since some hair preparations require clarity of formulation, silicones must be sufficiently compatible with other formula ingredients to maintain clarity of formulation while providing conditioning. Following are some functional siloxanes that can be used in clear hair preparations.

Microemulsions of linear PDMS fluids can be clear and maintain clarity in hair preparations. Particle size of a linear PDMS droplet is typically less than 50 nm, which results in an optically clear system. Microemulsions are typically used at less than 5% in formulas, representing less than 2.5% of active PDMS. (Poly)diorganosiloxane microemulsions can provide enhanced dilution and formulation stability for easier delivery of functional benefits.

Polyether functional siloxanes are water soluble when there is a sufficient level of polyoxyethylene attached to the siloxane backbone (typically associated with high calculated HLB). Also, they can reduce the viscosity of aqueous-based surfactant formulas; thus, when they are used in these formulas, addition of salt or a viscosity booster is recommended. Polyether functional siloxanes are typically used in formulas at 5% to provide light conditioning of hair.

a. Clear Conditioning Shampoo. Formula 18 illustrates the use of a high-calculated-HLB surfactant (12–14), which is soluble in the surfactant system, providing mild conditioning. These polyether functional siloxanes typically have an inverse cloud point (losing solubility as temperature rises). They cloud out of shampoo formula around normal hot bath water temperature (cloud points ranging from 40 to 58°C) and deposit on the hair—not easily rinsed off with a shampoo surfactant system.

Formula 18 Clear Conditioning Shampoo

Dimethiconol Copolyol	2.0%
Sodium Lauryl Ether Sulfate (30%)	30.0
Cocoamide DEA	2.0
PEG-120 Methyl Glucose Dioleate	1.5
Water	q.s. to 100%
Citric Acid	q.s.
Sodium Chloride	q.s.
Fragrance/Preservative	q.s.

Amino-functional siloxanes can be incorporated directly or via micro- or macroemulsion into clear shampoos and aid in strong conditioning and faster drying of hair. Amino-functional siloxanes also have been noted to reduce viscosity of a formulation; thus, when they are used in a surfactant-based formula, addition of salt or a viscosity booster is recommended. They are typically used in hair preparations at less than 1%. Work reported by Jachowicz and Berthiaume (58) suggests the ability of amino-functional siloxane microemulsions to penetrate into the interior of a hair fiber.

b. Conditioning Setting Gel. Formula 19 incorporates a cyclomethicone/dimethicone copolyol to aid in drying time and mild conditioning of hair. Trimethylsiloxyamodimethicone is used to give a durable dry-combing conditioning effect, even after a few shampoo washings, and to provide softness and a noticeably reduced drying time.

Formula 19 Conditioning Setting Gel

Vinylcaprolactam PVP/ Dimethylaminoethylmethacrylate Copolymer	10.0%
Trimethylsilyamodimethicone (and) Octoxynol-40 (and) Isolaureth-6 (and) Glycol Emulsion	0.5
Cyclomethicone (and) Dimethicone Copolyol	0.5
Carbomer 940	1.0
Sodium Hydroxide (20%)	1.4
D.I. Water	86.6
Preservative/Color/Fragrance	q.s.

B. Hair Conditioner Trends

Several authors have studied the fundamentals of hair conditioning using silicones, so that the phenomenon is reasonably well understood (59). Materials and delivery-system innovations have continued, as evidenced by a number of patents which have been issued over the past few years that highlight the benefits outlined above from the use of silicones in hair conditioning applications. Selected references are noted in the areas of silicone microemulsions, T/Q-type organosilicon compounds, further derivatization of organo-functional siloxanes, enhanced performance/delivery, and new formulations:

1. G.B. 9,117,740 (Birtwistle) discusses the use of microemulsions of functional siloxanes (particle size less than 1 μm) and cationic deposition polymers in various types of surfactant systems to be compatible in 2-in-1 optically clear shampoos to impart good conditioning benefits to hair and skin.
2. E.P. 0,268,982 (Harashima et al.) discusses the use of dimethylsiloxane microemulsions in cosmetic formulations.
3. U.S. 5,661,215 (Gee et al.) discusses gel-free crosslinked polydimethylsiloxane polymer microemulsions for hair conditioning applications that require lubrication properties without excess tackiness.
4. U.S. 5,244,598 (Merrifield et al.) discusses the preparation of aminosilicone microemulsions for hair conditioning.
5. U.S. 5,085,859 (Halloran et al.) discusses the use of T-type organosilicon compounds for curl retention of hair.
6. U.S. 5,135,742 (Halloran et al.) discusses the use of organo-modified T-type organosilicon compounds for imparting curl retention to hair.
7. J 94,282,138 (assigned to Shiseido Co., Ltd.) discusses the use of organic silicone resins that have good setting power and gives good glossiness and smoothness to hair.
8. GB 2,229,775 (Berthiaume et al.) discusses the use of low-viscosity organo-functional siloxysilicate resins to improve shine, volume, body, and curl retention in hair, and reduced combing force.
9. U.S. 5,226,923 (O'Lenick) discusses the use of silicone fatty acid esters to provide softening and lubrication when applied to hair and skin. They also afford antistatic properties.
10. E.P. 0,723,770 (Cretois et al.) discusses the use of modified oxyalkylene-substituted silicones in a guar gum surfactant system that oversomes undesirable residue left on hair and poor fixing power.
11. U.S. 5,409,695 (Abrutyn et al.) discusses the use of a polymeric lattice network to deliver functional siloxanes to the hair in a controlled manner.

12. D.E. 4,324,358 (Hilliard et al.) discusses the use of (poly)dimethylsiloxane (1,000–120,000 cS viscosity) in bar soaps at 1–6% to provide softer, glossier hair and skin.

IV. CONCLUSIONS

Formulators have come to trust the safety of silicones—a wide range of studies document a low order of toxicity (60). For this reason, the (poly)dimethylsiloxane backbone has become a key conditioning ingredient for both skin care and hair care applications. It was not too long ago that silicones were a strange and unique cosmetic additive, but today more than 50% of all new cosmetic and toiletry products have at least one silicone incorporated to aid in esthetics, improve delivery, or improve application ease to the skin and hair, or to provide a functionalized performance (e.g., nonwhitening/nonresidue, emulsification, enhanced substantivity, enhanced sun protection). Polyether functional siloxanes and amino-functional siloxanes will continue to play an important role in conditioning of hair, but branching and crosslinking of the basic (poly)dimethylsiloxane molecule will emerge as an important classification to improved deposition and measurable performance. Alkyl methyl siloxanes and copolymers will emerge as important classifications for improved deposition on skin. Polyether functional siloxanes (and copolymers) will continue their growth as important water-in-oil/silicone emulsifiers.

There is still a need to understand further how to deposit functional siloxanes more efficiently to hair and skin and then demonstrate the durability of the resultant silicone film. Learning more about the effect of other formulation ingredients will assist in understanding deposition effectiveness.

REFERENCES

1. Fearon FWG. In: Seymour R, Kirsenbaum G, High Polymers. Elsevier, 1986.
2. Rochow EG. Silicon and Silicones. Springer-Verlag, 1987.
3. Colas A. Silicone chemistry overview. Chemie Nouvelle 1990; 8(30):847.
4. Schorsch G. Silicones: Production and Applications. Techno-Nathan Publishing, 1988.
5. Rochow EG. Silicon and Silicones. Springer-Verlag, 1987.
6. Hardman B. Silicones. Encyclopedia of Polymer Science & Engineering. Vol. 15. 1989:204.
7. Sauer RO. Nomenclature of organosilicon compounds. J Chem Educ 1944; 21: 303.
8. American Chemical Society Nomenclature, Spelling and Pronunciation Committee. Chem Eng News 1952; 30:4517.
9. Editorial report on nomenclature. J Chem Soc 1952:5064.
10. IUPAC, CR 15th Conference 1949:127–132; Information Bulletin No. 31, 1973:87.

11a. Wendel SR. Use of silicones in cosmetics and toiletries. Perfums, Cosmetic Aromes 1984:59, 67.
11b. Disapio AJ. Silicones in personal care: An ingredient revolution. Drug Cosmet Ind 1994 (May).
12. Stark FO. Silicones. In: Comprehensive Organometallic Chemistry. Vol. 2. Pergamon Press, 1982:305.
13. Stark FO, et coll. Silicones. In: Encyclopedia of Polymer Science & Engineering. Vol. 15. 1989:204.
14. Allred L, Rochow EG, Stone FGA. The nature of the silicon-oxygen bond. J Inorg Nuclear Chem 1956; 2:416.
15. Corey NY. Historical overview and comparison of silicon with carbon. In: Patai S, Rappoport Z, eds. The Chemistry of Organic Silicon Compounds. Wiley, 1989: chap 1.
16. Disapio AJ. Silicones in personal care: an ingredient revolution. Drug Cosmet Ind 1994 (May).
17a. Noll W. Chemistry and Technology of Silicones. New York: Academic Press, 1968.
17b. Dean JA, ed. Lange's Handbook of Chemistry. 13th ed. New York: McGraw-Hill, 1985.
18. Ferry J. Visco-Elastic Properties of Polymer. 3d ed. New York: Wiley, 1980.
19a. Carter W. Pierce J, Malkina I, Luo D. Investigation of the ozone formation potential of selected volatile silicone compounds. Final report from Statewide Air Pollution Research Center, University of California, Riverside, CA, to Dow Corning Corporation, Oct. 1992.
19b. 59 Fed Reg 192, Oct. 5, 1994, pp. 50693–50696.
20a. Urrutia A, Glover DA, Oldinski RL, Rizwan BM. Sensory evaluation of silicones for personal care applications. XIII Congreso Latino-Americano e lb'erico de Quimicos Cosme'ticos, Acapulco, Mexico, Sept. 21–26, 1997.
20b. IFSCC Monograph No. 1.
21a. 21 CFR Part 347, 25204–25232 (Vol. 55, No. 119), June 29, 1990. Proposed "Skin Protectant Drug Products for Over-the-Counter Human Use."
21b. 21 CFR Part 347, Docket No. 78N-0021 (48 Fed Reg 6820), Feb. 15, 1983. Notice of Proposed rulemaking "Skin Protectant Drug Products for Over-the-Counter Human Use; Tentative Final Monograph."
22. Natl Formulary 18, pp. 2242–2243 (1995).
23. 59 Fed Reg No. 192, Oct. 5, 1994, pp. 50693–50696.
24. Disapio AJ, Fridd P. Silicone glycols for cosmetic and toiletry applications. IFSCC, London, Oct. 1988.
25. Tradename registered to Dow Corning.
26a. Kasprzak KA. Emulsion techniques using silicone formulation aids. Drug Cosmet 1996 (May).
26b. Gruening S, Hemeyer P, Weitemeyer C. New types of emulsions containing organo-modified silicone copolymers as emulsifiers. Tenside Surfactants Deterg 1992; 29(2):78–83.
27. Zombeck A, Dahms G. Novel formulations based on nonaqueous emulsions of polyols in silicones. 19th IFSCC Congress, Sydney, October 22–25, 1996.

28. Dow Corning Corporation. Shower: Formulation Guide. Dow Corning, Midland, MI.
29a. Handt CM. Hair fixative benefits from the physical and chemical properties of silicones. Soap Cosmet Chem Spec 1987.
29b. Personal communications from National Starch.
30a. Cosmetic Ingredient Review. Final report on the safety assessment of dimethicone copolyol. J Am Coll Toxicol 1991; 10(1).
30b. Dow Corning Corporation. Internal report.
31. Smith JM, Madore LM, Fuson SM. Attacking residue in antiperspirant—alternative to the clear stick. Drug Cosmet 1995; (Dec.):46–52.
32. Disapio AJ, LeGrow GE, Oldinski RI. Correlation of physical and sensory properties of alkylmethylsiloxanes with their composition and structure. IFSCC, Japan, 1994.
33. Van Reeth I, Wilson A. Understanding of factors influencing the permeability of silicones and their derivatives. Cosmet Toilet 1994 (July).
34. Glover DA. Unique alkyl methyl siloxane waxes for personal care. Cosmet Toilet. In press.
35. Starch MS. Silicones in hair care products. Drug Cosmet Ind 1984 (June):38–44.
36. Starch MS. Silicones for conditioning of damaged hair. Soap Cosmet Chem Spec 1986 (Apr.).
37. Schulz WJ, Zhang S. Silicone Oils and Solvents Thickened by Silicone Elastomers. U.S. Patent 5,654,362, 1997.
38. U.S. Patent 4,987,169, U.S. Patent 5,412,004, and U.S. Patent 5,236,986.
39. Stepniewski GJ, et al. Stable Water-in-Oil Emulsion Systems. U.S. Patent 5,599,533, 1997.
40a. Halloran DJ, et al. Hair Treatment for Curl Retention—Using Fixative Resin Film-Forming Non-Polar Silsesquioxane. U.S. Patent 5075103, 1992.
40b. Halloran DJ, et al. Imparting Curl Retention to Hair—By Using Pre-Hydrolyzed Organo-Functional Silane with Silsesquioxane Properties as Film Former. U.S. Patent 5,135,742, 1992.
41. Woodruff J. Manufact Chem 1996; 67(6):27–31. Defining the term emollient and key physical and perception characteristics.
42. Blakely JM. The benefits of silicones in facial and body cleansing products. IFSCC, Japan, 1994.
43. Dow Corning Corporation (Midland, MI) Formulation #E6845-128B.
44. Dow Corning Corporation (Midland, MI) Formulation #23/887.
45. Dow Corning Corporation (Midland, MI) Formulation #7/691.
46. Dow Corning Corporation (Midland, MI) Formulation #25-334-92.
47. Dow Corning Corporation (Midland, MI) Formulation #140/7.
48. Dow Corning Corporation (Midland, MI) Formulation #12/194.
49. Van Reeth IM, Dahman F, Hannington J. Alkyl methyl siloxanes as SPF enhancers: relationship between effects and physio-chemical properties. IFSCC, Sydney, Oct. 1996.
50. DeSmedt A, Van Reeth IM, Machiorette S, Glover DA, Nard J. Measurements of silicone deposition on hair by various analytical methods. Cosmet Toilet 1997 (Feb.); 112(2).

51. Halloran DJ. A silicone selection guide for developing conditioning shampoos. Soap Cosmet Chem Spec 1992 (Mar.):22–26.
52. Wagman J, et al. Aqueous Hair Conditioning Compositions. U.S. Patent 4,777,037.
53a. Geen H. Nontangling Shampoo. U.S. Patent 2,826,551.
53b. Pader M. Freely-Pourable, Stable, Homogenous, Shampoo Composition. U.S. Patent 4,364,837.
54c. Fieler et al. Process for Making a Silicone Containing Shampoo. U.S. Patent 4,728,457.
53d. Bolich R, et al. Hair Care Composition Containing a Rigid Silicone Polymer. U.S. Patent 4,902,499.
53e. Conditioning Shampoo Comprising a Surfactant, Non-volatile Silicone Oil, Guar, Hydroxypropyl Trimonium Chloride, Cationic Conditioning Polymer. U.S. Patent 5,151,210.
54. Thomson B, Vincent J, Halloran D. Anhydrous hair conditioners: silicone-in-silicone delivery systems. Soap Cosmet Chem Spec 1992 (Oct.); 68(10):25–28.
55. Disapio A, Fridd P. Silicones: use of substantive properties on skin and hair. Int J Cosmet Sci 1988; 10:75–89.
56a. Halloran D. Silicones in shampoos. HAPPI 1992 (Nov.):60–64.
56b. Sejpka J. Silicones in hair care products. Seifen, Oele, Fette, Wachse 1992; 118 (17):1065–1070.
56c. Demarco et al. Aqueous Hair Conditioning Compositions. U.S. Patent 4,529,586.
56d. Halloran D, et al. Clear Shampoo Compositions. U.S. Patent 5,326,483.
56e. Madrange et al. Detergent Cosmetic Composition. U.S. Patent 4,710,314.
56f. Starch MS. Hair Conditioners Containing Siloxane and Freeze-Thaw Stabilizers. U.S. Patent 4,563,347.
56g. Optically Clear Hair Care Compositions Containing Silicone Microemulsions. E.P. 514,934 A1.
56h. Chandra et al. Conditioning Shampoos Containing Amine-Functional Polydiorganosiloxanes. U.S. Patent 4,559,227.
56i. Krzysik D, Gatto S. Black hair care products: new formulating concepts with silicones. Drug Cosmet Ind 1987 (Nov.).
56j. Traver F, et al. Dialkylaminoalkyl Siloxy Terminated Polydiorganosiloxane Compounds. U.S. Patent 5,132,443.
57. Gee. A Method of Preparing Polyorganosiloxanes Having Small Particle Size. U.S. Patent 4,620,878.
58. Jachowicz J, Berthiaume M. Microemulsions vs. macroemulsions in hair care products. Cosmet Toilet 1993; 108:65–72.
59a. Nanavati S, Hami A. A preliminary investigation of the interaction of a quaternium with silicones and its conditioning benefits on hair. J Soc Cosmet Chem 1994 (May–June); 45:135–148.
59b. Robbins CR, Reich C, Patel A. Adsorption to keratin surfaces: a continuum between a charge-driven and a hydrophobically driven process. J Soc Cosmet Chem 1994 (Mar.–Apr.); 45:84–94.
59c. Lockhead RY. Conditioning shampoos. Soap Cosmet Chem Spec 1992 (Oct.): 42–49.

59d. Jachowicz J, Berthiaume M. Heterocoaggulation of silicone emulsions on keratin fibers. J Colloid Interface Sci. 1989; 133:118–134.
60a. Chandra G, Disapio A, Frye C, Zellner D. Silicones for cosmetic and toiletries: an environmental update. 1994.
60b. Crowston E, Kolaitis L, Verbiedse N. Purity as a unique quality property of Dow Corning dimethicone and cyclomethicone product line in the cosmetic industry. 1996.
60c. Disapio AJ, Zellner AN. Silicones (polydimethylsiloxanes): a profile for safety and functional performance in cosmetics and toiletries. 1994.
60d. Verbiese N. Organosilicon compounds as cosmetic ingredients and European regulatory environmental trends. 1994.

9
Specialty Silicone Conditioning Agents

Anthony J. O'Lenick, Jr.
Lambent Technologies, Norcross, Georgia

I. INTRODUCTION

Silicone compounds have enjoyed growing acceptance in a variety of personal care applications. This is true in part due to advances in the formulation techniques used in the preparation of silicone-containing products. Equally important and potentially more exciting to the formulator, however, is the synthesis and commercial availability of a variety of new organofunctional silicone compounds which offer both ease of formulation and enhanced performance in formulations.

II. BACKGROUND

A. Historic Uses of Polysiloxanes

Silicone compounds have been known since 1860, but were of little commercial interest until the 1940s. Over the years, silicone compounds have received growing acceptance in many personal care applications. In fact, it has been said that four of ten new personal care products introduced in the 1990s have silicone in them. There has been considerable confusion related to nomenclature of silicone compounds. The term silicon (no "e") refers to the element silicon (atomic number 14) and is correctly used to describe simple, nonpolymeric compounds such as SiS_2 (silicon disulfide). The term "silicone" was coined by F. S. Kipping to designate an organosilicon oxide polymer. The first of these compounds were thought to be silicon-based ketones, hence the contraction silicone. Despite this error, the term is still widely used and accepted.

The vast majority of the volume of silicone compounds used are silicone fluids, also known as polysiloxanes. They are the oldest and perhaps the best understood of the silicone compounds in use today. There has been a recent explosion in the availability of chemically modified silicone compounds that provide (a) conditioning, (b) softening, (c) irritation mitigation, (d) barrier properties, and (e) emulsification. This is due to the fact that many of the desirable properties that formulators want in their products cannot be achieved using silicone fluids alone, hydrocarbons alone, or blends. Blends are ineffective because silicone fluids and hydrocarbons are insoluble in each other. Organosilicon compounds offer the possibility of combining the best properties of each type of compound in one hybrid molecule. The hybrid molecule that results can be tailored for its properties in specific formulations.

Silicone fluids, also called silicone oils, are sold by their viscosity and range from 0.65 cs to 1,000,000 cs. These materials are discussed at length in Chapter 8. Unless the silicone fluid is made by blending two different-viscosity silicone fluids, the molecular weight of the silicone fluid is dependent on the number of repeating units in the polymer. Perhaps the major recent development in the use of silicone fluids in personal care is the so-called two-in-one shampoo. This technology was pioneered by Procter and Gamble and is the topic of many of their patents (1–6). The basic concept in these systems is to disperse silicone fluids in a thickened shampoo base and deliver the silicone fluid to the hair during the shampoo process, providing both cleaning and conditioning. The silicone is delivered to hair or skin this way due to a hydrophobic interaction. When oil is placed into water, it disrupts the hydrogen bonding between the water molecules in the water solution. This disruption is accomplished only when the energy of mixing is sufficient to break the hydrogen bonds. When the mixing is stopped, the oil is forced out of the water by the re-formation of the hydrogen bonds between the water molecules. This phenomenon can be used to deliver oil to a surface. Much has been written about the desirability of using silicone oil as a replacement for mineral oil in many cosmetic applications (7). The choice of materials should be based solely on the cost effectiveness of achieving the desired properties in the formulation; some formulations use both mineral oil and silicone fluid.

B. Drawbacks of Polysiloxane

Despite the fact that silicone fluids function in many applications, there are a number of formulations in which silicone fluids cannot be used. Among their drawbacks are the following.

1. Silicone fluids defoam many formulations, limiting their utility.
2. Silicone fluids lack solubility in water and in many organics and therefore are very difficult to use in formulations.

3. Silicone fluids are greasy.
4. Silicone fluids need to be emulsified to be used in water-based systems.

C. Improving Silicone Properties by Modifying Organofunctionality

The above-mentioned negative effects can be reduced by modifying the solubility of the polysiloxane molecule. This can be accomplished by adding various organofunctional constituents to the silicone chain. These modifications allow the material to be delivered to the skin or hair in ways other than the hydrophobic interaction described above. One way to modify the solubility of polysiloxane is to insert into the silicone molecule groups that modify the solubility of the resultant compound. Such groups include hydrocarbon groups that effect organic solubility, and polyoxyalkylene groups that effect water solubility. Silicone compounds that have hydrocarbon groups and/or polyoxyalkylene groups are one class of compounds referred to as organofunctional silicone polymers. Silicone modified in this way may be adhered to the skin or hair by virtue of one or more of the following mechanisms.

Ionic interactions. The charge on the molecule will also have an effect on the delivery of the oil to the hair or skin. For example, if the oil has a cationic charge on the molecule, it will form ionic bonds with substrates which contain negative surface charges. The two opposite charges together form a so-called pair bond.

General adhesion. If an oil is delivered to the skin or hair penetrates and then polymerizes, an interlocking network of polymer will develop. Although it is not bonded directly to the substrate, this polymer network will adhere to the substrate.

Specific adhesion. If an oil is delivered to the skin or hair penetrates and then reacts with groups on the hair or skin, there will be a chemical bond between the polymer and the substrate. This is the strongest and most permanent of the adhesion mechanisms.

Silicone fluids react almost exclusively by the hydrophobic mechanism described above. To the extent that other mechanisms may be introduced, the more strongly and efficiently the conditioner can be delivered to substrate. Organofunctional silicone seek in large part to capitalize on these additional mechanisms to provide thorough and efficient conditioning for hair and skin.

III. TYPES OF ORGANOFUNCTIONALLY MODIFIED SILICONE POLYMERS

A. Dimethiconol

1. *Structure/Background*

An important class of reactive silicone compounds id dimethiconols. This class of compounds has an Si–OH group and is also known as silanols. Dimethiconol compounds conform to the following structure:

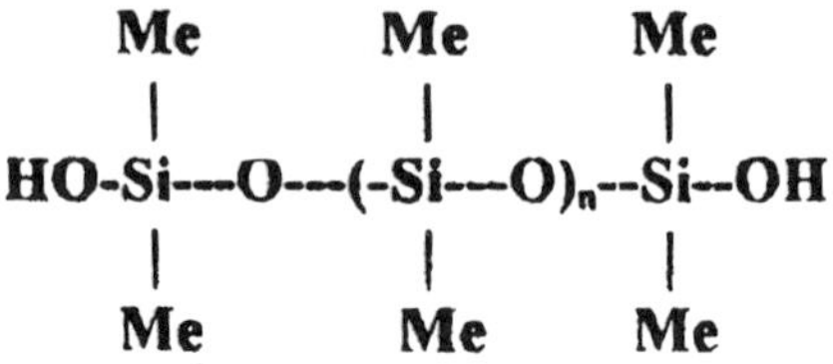

```
     Me        Me        Me
     |         |         |
HO-Si---O---(-Si---O)n--Si--OH
     |         |         |
     Me        Me        Me
```

where Me is methyl.

2. *Properties/Applications*

Like dimethicone compounds, dimethiconol compounds are sold by viscosity. Under certain conditions, dimethiconol compounds homopolymerize to give higher-molecular-weight species. The polymerization can be accomplished after the lower-molecular-weight polymer penetrates the hair or skin. This type of polymerization can result in an interlocking polymeric compound that adheres to the substrate. Dimethiconol can be supplied in emulsion form that can be applied to treated or colored hair to lock in the color or treatment. The initial mechanism of conditioning the substrate is hydrophobic interaction, but unlike silicone fluids, these materials can polymerize for added durability.

In addition to their usefulness in emulsions, another property of dimethiconol compounds is their ability to react to make organofunctional compounds. The reactivity of the dimethiconol group toward fatty acids to make esters has been compared to the reactivity of carbanols in fatty alcohols (8). Dimethiconols were found to be only slightly less reactive than the carbanols when esterified. This fact clearly illustrates why dimethiconols can be used in a variety of surfactant unit operations to give silicone surfactants. This reactivity of the dimethiconol group allows for the preparation of surfactants, in which silicone is a hydrophobe. The properties of the final surfactant depend on the silicone chosen and can be either hydrophobic and lipophobic. This is a result of the insolubility of the silicone in many hydrocarbons and in water. Silanols, or dimethiconols, are made by the polymerization of cyclic silicone compounds using acid catalysts and water.

B. Dimethicone Copolyol

1. Structure/Background

Another major class of silicone derivative is dimethicone copolyol. This material is used as a conditioning agent as well as to synthesize other derivatives. Dimethicone copolyols have carbanol hydroxyl groups and silicone present. The introduction of fatty groups into the molecule using classical surfactant chemistry can result in a molecule that has a silicone portion, a polyoxyalkylene portion, and a fatty portion. This leads to a molecule that has a water-soluble part, an oil-soluble part, and a silicone-soluble part. Dimethicone copolyols conform to the following structure:

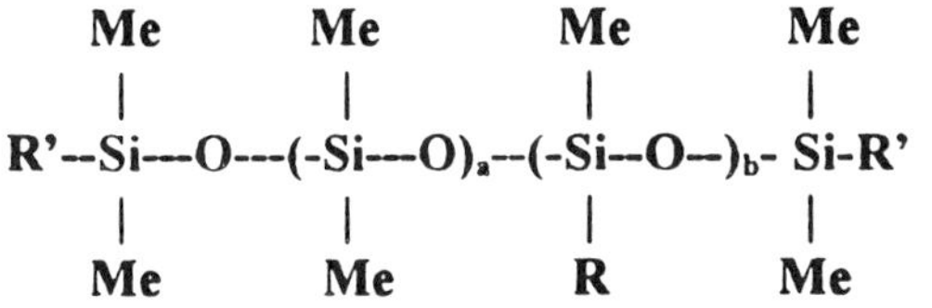

```
        Me       Me          Me       Me
        |        |           |        |
R'--Si---O---(-Si---O)a--(-Si--O--)b- Si-R'
        |        |           |        |
        Me       Me          R        Me
```

R' is $-(CH_2)_3-O-(CH_2CH_2O)_x-(CH_2CH(CH_3)O)_y-(CH_2CH_2O)_z-H$

Dimethicone copolyols are prepared using a process called hydrosilylation. In this process a polymer having a silanic hydrogen present (Si–H) is reacted with an allyl alcohol alkoxylate, for example, $CH_2=CH-CH_2-O-(CH_2CH_2-O-)_8-H$, in the presence of a suitable catalyst. Commonly the catalyst contains platinum.

Several types of organofunctional compounds can be made based on the location of the functional group. For example, the *comb* product conforms to the structure

```
   Me       Me          Me        Me
   |        |           |         |
Me-Si---O---(-Si---O)a---(--Si---)b---O---Si---Me
   |        |           |         |
   Me       Me        (CH2)3      Me
                        |
                        O--(CH2CH2O)x(CH2CH(CH3)O)y(CH2CH2O)zH
```

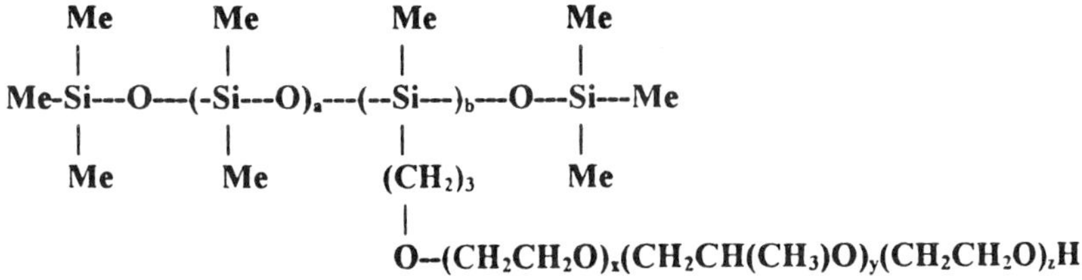

while the *terminal* product conforms to the structure

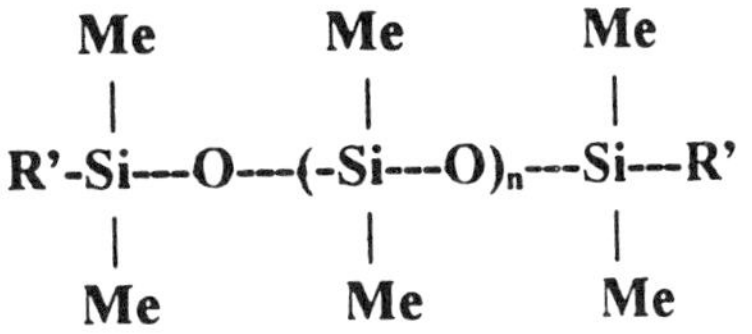

```
      Me        Me        Me
      |         |         |
R'-Si---O---(-Si---O)n---Si---R'
      |         |         |
      Me        Me        Me
```

R' is $-(CH_2)_3-O-(CH_2CH_2O)_x-(CH_2CH(CH_3)O)_y-(CH_2CH_2O)_z-H$

The term dimethicone copolyol, while descriptive of a class of compounds, does not describe a specific compound. This leads to confusion in the field when a formulator asks for dimethicone copolyol. The major variables that effect performance include (a) whether the compound is comb or terminal, (b) the molecular weight of the compound, (c) the ratio of silicone to polyoxyalkylene, and (d) the amount and location of polyoxyethylene or polyoxypropylene in the molecule. As a result, it is impossible to determine if a compound described as a dimethicone copolyol is water soluble or insoluble, or if it is to be used as a conditioner or an emulsifier.

2. Properties/Applications

In addition to being raw materials for preparation of other silicone surfactants, dimethicone copolyols have been used per se in many formulations. Dimethicone copolyol compounds have a wide range of applications in personal care applications.

a. Emulsification. One area is as an emulsifier for use in antiperspirants. In this application, the dimethicone copolyol generally is sold in cyclomethicone and is referred to as a formulation aid. The product emulsifies aqueous antiperspirants into cyclomethicone.

b. Emolliency. If the concentration of polyoxyethylene in a dimethicone copolyol is raised in the molecule, a solid waxy material results. Dimethicone copolyol compounds of this type are outstanding water-soluble emollients.

c. Conditioning. Because dimethicone copolyol compounds are nonionic, they are compatible with anionic surfactants, and consequently have found application in a variety of shampoos that contain conditioning agents.

d. Resin Modification. Dimethicone copolyol compounds are used to modify resins, for hairspray and mousse. The presence of silicone results in a plasticizer effect on the resin, making it less harsh on the hair.

C. Alkyl Dimethicone Copolyol

1. Structure/Background

Recently a series of compounds that have both alkyl groups and polyoxyalkylene groups have been promoted in the personal care market (9). These compounds conform to the following structure:

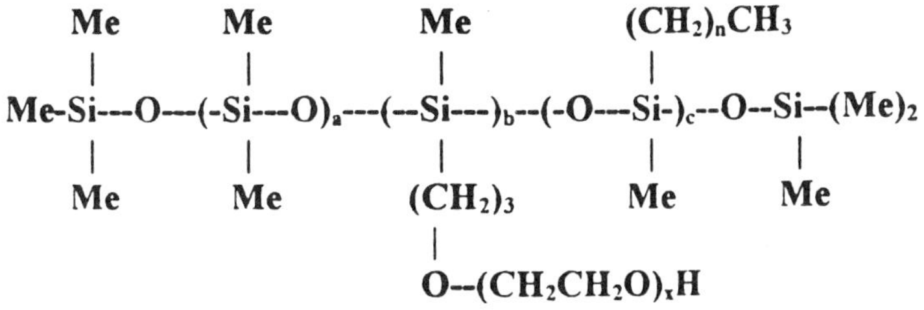

```
      Me        Me            Me            (CH2)nCH3
      |         |             |              |
Me-Si---O---(-Si---O)a---(--Si---)b--(-O---Si-)c--O--Si--(Me)2
      |         |             |              |        |
      Me        Me          (CH2)3           Me       Me
                              |
                            O--(CH2CH2O)xH
```

Alkyl dimethicone copolyols, like their non-alkyl-containing counterparts, are prepared using a process called hydrosilylation. In this process a polymer having a silanic hydrogen present (Si–H) is reacted with an allyl alcohol alkoxylate and an alpha-olefin. For example, a mixture of $CH_2=CH–CH_2–O–(CH_2CH_2–O)_8–H$ and $CH_3–(CH_2)_x–CH=CH_2$ can be co-hydrosilylated with the Si–H polymer in the presence of a suitable catalyst. Commonly the catalyst contains platinum.

2. *Properties/Applications*

These materials are used as emulsifiers in the preparation of both water-in-silicone and silicone-in-water systems. These products provide advantages over traditional hydrocarbon chemistries since they can be used in the preparation of emulsions without heat. These silicone polymers can be used to prepare products that contain little wax, contain a large concentration of water, and have a light spreadable feel on the skin.

D. Trimethylsilylamodimethicone

1. *Structure/Background*

Silicone compounds containing amino groups are widely used in personal care applications. They conform to the following structure (10):

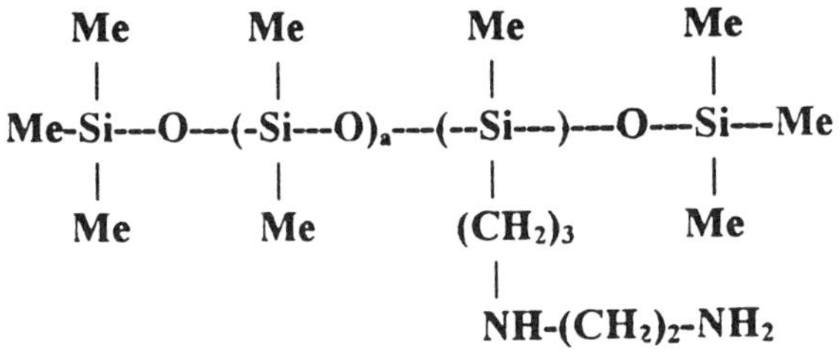

```
     Me        Me          Me          Me
     |         |           |           |
Me-Si---O---(-Si---O)a---(--Si---)---O---Si---Me
     |         |           |           |
     Me        Me        (CH2)3        Me
                           |
                         NH-(CH2)2-NH2
```

The CTFA name is trimethylsilylamodimethicone, or TSA, for short. Compounds conforming to this structure are affective conditioning agents.

2. Properties/Applications

Since the amino group has a pK_b of about 10.7, use of these materials at pH below this value results in the protonation of the amino group. Consequently, these compounds act as cationics at pH below their pK_b. The cationic nature of these compounds results in the adhesion to hair and skin using ionic interactions. Generally this class of product is used at around 1–3% concentration, and provides intensive conditioning.

E. Amodimethicone

1. Structure/Background

Amodimethicone is a 35% solids emulsion in water and is made by emulsion polymerization. The resulting emulsion contains water, the amodimethicone, surfactants, and fatty quaternium. The structure is

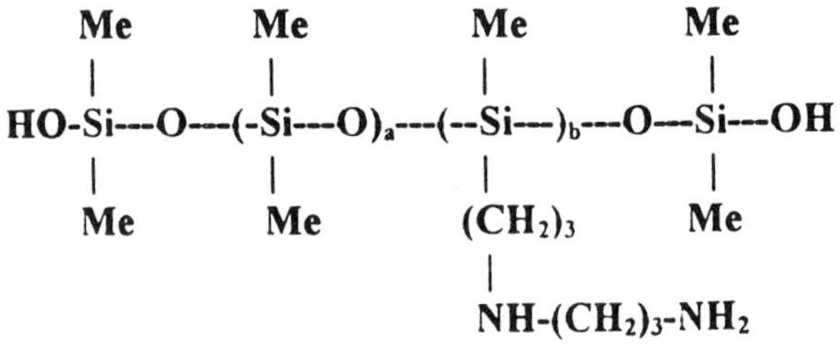

$$
\begin{array}{cccc}
 & \mathrm{Me} & \mathrm{Me} & \mathrm{Me} & \mathrm{Me} \\
 & | & | & | & | \\
\mathrm{HO}\text{-} & \mathrm{Si}\text{---O---(-} & \mathrm{Si}\text{---O)}_a\text{---(--} & \mathrm{Si}\text{---)}_b\text{---O---} & \mathrm{Si}\text{---OH} \\
 & | & | & | & | \\
 & \mathrm{Me} & \mathrm{Me} & (\mathrm{CH_2})_3 & \mathrm{Me} \\
 & & & | & \\
 & & & \mathrm{NH\text{-}(CH_2)_3\text{-}NH_2} &
\end{array}
$$

2. Properties/Applications

Amodimethicone differs from trimethylsilylamodimethicone in that the former has a reactive silanol group (Si–OH). This group, in the presence of the amino group, will polymerize to form a higher-molecular-weight polymer. This property together with the amino group, which below a pH of 10.7 is cationic, results in improved durability and substantivity in many formulations. These compounds are outstanding conditioners and have good skin feel. There are emulsions, however, and so are inherently less stable than solutions of the same product. Therefore, several considerations that must be made in formulating with this type of material. These include (a) freeze–thaw stability, (b) shear sensitivity, and (c) HLB incompatibility. These considerations may limit the amount of material that can be added to a given formulation, or may preclude the use of the material in certain formulations at all. Despite the fact that amodimethicone is available only as an emulsion, it has been used successfully for many years in many different applications both for hair and skin care.

F. Dimethicone Copolyol Amine

1. Structure/Background

More recently, a series of polymers related to amodimethicone has come onto the market (11). These products have the same structure, but in addition have a water-loving polyoxyalkylene glycol group present. Unlike amodimethicone, these products are not made by emulsion polymerization. They do not contain added surfactant or fatty quaternary, as can be seen in the following structure:

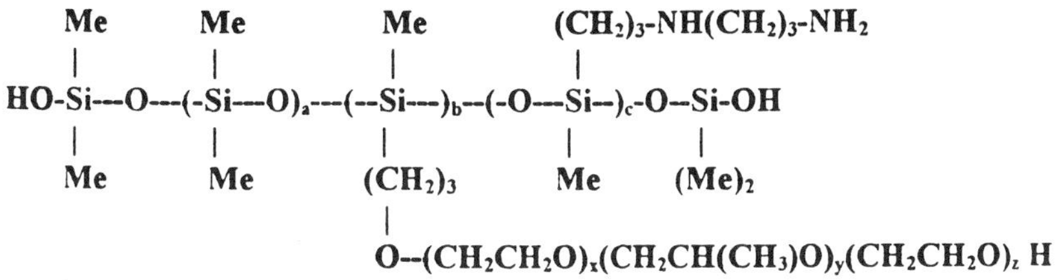

```
     Me        Me            Me            (CH2)3-NH(CH2)3-NH2
     |         |             |             |
HO-Si---O---(-Si---O)a---(--Si---)b--(-O---Si--)c-O--Si-OH
     |         |             |             |       |
     Me        Me           (CH2)3         Me     (Me)2
                             |
                             O--(CH2CH2O)x(CH2CH(CH3)O)y(CH2CH2O)z H
```

2. Properties/Applications

The introduction of the polyoxyalkylene glycol portion into the molecule allows for the regulation of water solubility. Products are available that are water soluble, self-emulsifying, or water insoluble. Conditioning comes from two groups: (a) those based on the value of n, and (b) the number of amino groups present. The a value results in a more greasy conditioning, like silicone fluid. The amino conditioning is a more slick feel. The durability of the polymer to a variety of substrates can be altered by the concentration of silanol (Si–OH) groups. Products run from mildly conditioning, for every day use, to extremely durable. Since these materials are not emulsions, they are freeze–thaw stable, shear stable, and can be used in formulations having a relatively wide range of HLB values. The presence of the polyoxyalkylene glycol prevents these materials from being gunky on the hair or skin. They impart a very luxurious feel to the skin.

G. Silicone Quaternium Compounds

1. Structure/Background

Fatty quaternary compounds, discussed elsewhere in this book, are well-known conditioners. There can be several undesirable attributes of fatty cationic products which limit their usefulness in certain formulations. For example, fatty cationic surfactants are incompatible with anionic surfactants, forming insoluble complexes when the two types of materials are combined (12). Cationics are considered somewhat toxic when ingested and they are eye irritants, but they tend not to be topical irritants (13). Fatty alkylamidopropyl dimethylammonium compounds are commonly found in conditioning treatments for hair, but are difficult to formulate in clear shampoo systems (14). Many of these limiting attributes can be mitigated by making silicone analogs of fatty amidopropyl dimethylamine quaternary compounds. A series of alkylamido silicone quaternary compounds based on dimethicone copolyol chemistry has been developed. These materials are compatible with anionic systems, over a limited range of concentrations; provide outstanding wet comb properties, antistatic properties, and nongreasy softening properties to hair, fiber, and skin; and are not based on glycidyl epoxide or alkanolamine chemistries. Compounds can be prepared that have varying amounts of polyoxyalkylene oxide in the polymer. The ability to regulate the type of alkylene oxide and the amount present in the silicone polymer results in a series of products ranging in water/oil solubility. Compounds of the comb type conform to the following structure:

```
    Me       Me          Me          Me
    |        |           |           |
Me-Si---O---(-Si---O)a---(--Si---)b---O---Si---Me
    |        |           |           |
    Me       Me          (CH2)3      Me
                         |
                         O--(CH2CH2O)x(CH2CH(CH3)O)y(CH2CH2O)zR'


                 Me
               + |
R' is   -C(O)-CH2-N-(CH2)3-NH-C(O)-R
                 |
                 Me
```

Compounds of this type are prepared in a two-step reaction sequence. First the dimethicone copolyol is esterified with chloroacetic acid. The resulting dimethicone copolyol chloro-ester is then used to quaternize a tertiary amine in aqueous solution. Typically, the tertiary amine is an alkylamidopropyl dimethyl amine.

2. *Properties/Applications*

Silicone quaternium compounds of the above type have been designated by the CTFA as silicone quaternium 1 through silicone quaternium 10. They can be easily formulated into a variety of personal care products, without emulsification, the use of elaborate thickening systems, or homogenization. These have outstanding compatibility with nonionic and other cationic materials and, more surprisingly, compatibility with anionic surfactants. For example, 3%

Table 1 Silicone Quaternary Compounds

CTFA name	Alkyl amido group
Silicone quaternium 1	Cocamidopropyldimethyl
Silicone quaternium 2	Myristamidopropyldimethyl
Silicone quaternium 8	Di-linoleylamidopropyldimethyl

silicone quaternium 8 can be incorporated into 15% sodium lauryl sulfate cold, with only mild agitation. This results in a clear simple base for the formulation of two-in-one shampoos that are nongreasy and do not build up. Wet combing studies also indicate significant conditioning benefits. Additionally, the silicone quaternium gives antistatic properties and gloss, while any attempt to incorporate silicone oil into this type of formula results in a hazy, low-foaming product that separates into two phases quite rapidly. The silicone quaternium compounds of this class that are available commercially are shown in Table 1.

Other noteworthy properties include the effect on foam properties and rheological characteristics. Silicone quaternium 8 was added to a conventional shampoo formula and evaluated for initial foam height and foam stability. Both factors were improved by the incorporation of the silicone quat.

When salt (NaCl) was added to a test formulation consisting of sodium lauryl sulfate, cocamide DEA, and silicone quaternium 8, the presence of the silicone quat lowered the viscosity from a peak viscosity of 15,000 cps to about 8,000 cps. The percentage added salt to reach peak viscosity was not altered. To obtain viscosities above those shown using the silicone quat, alternative thickeners can be used. These include polyacrylates, guar gums, and xanthanes.

H. Silicone Surfactants

1. Structure/Background

Formulators in the personal care field realize that there are a vast number of traditional surfactants from which to chose in the preparation of new products. Nonionic, cationic, amphoteric, and anionic products are available, and within each class there are numerous products. A neophyte formulator might ask, "Why are there so many types of surfactants?" The answer is clear: the structure of the surfactant determines the functionality. For example, if you are seeking a conditioner for hair, more than likely you would choose a quaternary compound, not a nonionic. It makes sense, then, that incorporation of functional groups into silicone-based surfactants should also have a profound effect

Table 2 Comparison of Hydrocarbon and Silicone Derivatives

Hydrocarbon products	Silicone products
Anionics	
Phosphate esters	Silicone phosphate esters (10)
Sulfates	Silicone sulfates (11)
Carboxylates	Silicone carboxylates (12)
Sulfosuccinates	Silicone sulfosuccinates (13)
Cationics	
Alkyl quats	Silicone alkyl quats (14)
Amido quats	Silicone amido quats (15)
Imidazoline quats	Silicone imidazoline quats (16)
Amphoterics	
Amino proprionates	Silicone amphoterics (17)
Betaines	Silicone betaines (18)
Phosphobetaines	Silicone phosphobetaines (19)
Nonionics	
Alcohol alkoxylates	Dimethicone copolyol
Alkanolamids	Silicone alkanolamids (20)
Esters	Silicone esters (21)
Taurine derivatives	Silicone taurine (22)
Isethionates	Silicone isethionates (23)
Alkyl glycosides	Silicone glycosides (24)
Free-radical polymers	
PVP/quats	Silicone free-radical quats (25)
Polyacrylates	Silicone polyacrylate copolymers (25)
Polyacrylamides	Silicone polyacrylamide copolymers (25)
Polysulfonic acids	Silicone polysulfonic acid copolymers (25)

on properties. Prior to 1990, dimethicone copolyols were the principal silicone surfactant available to the formulating chemist. Dimethicone copolyols are nonionic compounds analogous to alcohol ethoxylates. It is not surprising, then, that a series of surfactants based on silicone as a hydrophobe, which contain other functional groups similar to those seen in traditional surfactants, would be developed. In some instances, silicone is incorporated into a surface-active agent, with a polyoxyalkylene portion of the molecule and/or a hydrocarbon portion of the molecule. As will become clear, this results in several unique properties of the surfactant. It has been understood that simply using silicone surfactants and traditional hydrocarbon-based surfactants in mixtures results in unpredictable results. Hill (15) reports that the behavior varies from antagonistic (positive nonideal) to synergistic (negative nonideal). Incorporation of the hydrocarbon group into a silicone surfactant overcomes the interactions which provide nonideal performance in blended systems.

Table 2 shows the product classes available to the formulator both in silicone-based products and in the more traditional hydrocarbon products. The introduction of the functionalities onto the silicone backbone results in multifunctional products with a unique combination of properties.

In order to prepare silicone surfactants, one needs a raw material onto which the functional group can be placed. Two silicone polymers are used, dimethiconol (also known as silanols) and dimethicone copolyols, both of which contain reactive hydroxyl groups.

2. *Properties/Applications*

The following silicone surfactants illustrate the new properties that are attainable when one combines silicone and classical surfactant technology

I. Silicone Complexes

1. *Structure/Background*

Fatty quaternary compounds commonly are used in conditioning applications and are discussed elsewhere in this book. While they are well accepted in many applications, there are some areas in which improvement of properties would be of interest to the formulator. These include the following.

1. Fatty quaternary compounds are incompatible with anionic surfactants, since an insoluble complex is frequently formed when the two types of materials are combined (16).
2. The use concentration of many fatty quaternary compounds is limited by their irritation properties (17). At concentrations of 2.5%, most quats are minimally irritating to the eyes (18); but they are more irritating as the concentration increases.
3. Fatty quats are generally hydrophobic and when applied to substrate render the substrate hydrophobic. This can cause a loss of water absorbance of the substrate. It is not an uncommon situation for a traveler to a hotel to encounter a very soft towel that totally fails to absorb water. This is because the fatty quaternary gives softness but, being hydrophobic, also prevents rewetting. This situation also can be observed on hair: the conditioner becomes gunky on the hair and has a tendency to build up.

Many of these negative attributes have been mitigated by making silicone complexes with carboxy silicone compounds (19). The carboxy silicone complexes with quaternary compounds have altered properties that make them highly desirable in personal care applications. In order to study the complexes,

Table 3 Carboxy Silicone Fatty Quat Complexes

CTFA name	Fatty quat
Stearlkonium dimethicone copolyol phthalate (referred to as SDCP)	Stearlkonium chloride
Cetyl trimethylammonium dimethicone copolyol phthalate (referred to as CDCP)	Cetyltrimonium chloride (referred to as CETAC)

the following quaternary compounds were chosen as controls. Stearalkonium chloride is an excellent conditioning agent, having outstanding substantivity to hair. It has detangling properties and improves wet comb when applied after shampooing. The FDA formulation data for 1976 reports the use of this material in 78 hair conditioners, 8 at less than 0.1%, 18 at between 0.1% and 1.0%, and 52 at between 1% and 5%. Cetyltrimonium chloride, or CETAC, is a very substantive conditioner which in addition to having a nongreasy feel, improves wet comb and also provides a gloss to the hair. It is classified as a severe primary eye irritant (20). Therefore its use concentration is generally at or below 1%. These materials were complexed with a carboxy silicone to form clear, water-soluble complexes. These complexes provide outstanding wet comb properties, antistatic properties, and nongreasy softening properties to hair, and fiber and skin. They are minimally irritating to the eye and can be used to formulate clear conditioners. The carboxy silicone fatty quat complexes are shown in Table 3.

2. *Properties/Applications*

Fatty quat/carboxy silicone compounds are easily formulated into a variety of personal care products, without emulsification, the use of elaborate thickening systems, or homogenization. Carboxy/quat complexes are amenable to a variety of physical forms used in personal care products. They can be formulated into shampoos, gels, mousses, exothermic conditioners, and virtually any other form of product desired. They provide a number of beneficial properties, such as compatibility with anionic surfactants, lower irritation, and good combing properties.

Traditional fatty quats, like stearlkonium chloride and cetyltrimonium chloride, form water-insoluble complexes when combined with sodium lauryl sulfate in aqueous solution. This is due to the formation of an anionic/cationic complex that is insoluble. CDCP, like the traditional quats, forms an insoluble complex, but surprisingly, SDCP is compatible and clear, as shown by the following test. A 10% solids solution of sodium laruryl sulfate (28%) was

Table 4 Compatability of Quats with Anionic Systems

Quat	Endpoint	Observation
Stearalkonium chloride	0.3 mL	White solid
Cetyltrimonium chloride	0.2 mL	White solid
CDCP	0.5 mL	White solid
SDCP	35.8 mL	Haze develops

Table 5 Eye Irritation of Selected Quats

Material	Result	Interpretation
Cetyltrimonium chloride	106.0	Severely irritating
CDCP	8.3	Minimally irritating
Stearalkonium chloride	116.5	Severely irritating
SDCP	11.3	Minimally irritating

prepared. Separately, a 10% solids solution of the quat compound was prepared. The quat is titrated into 100 mL of the sulfate solution. The formation of a white insoluble complex, or the formation of a haze, is considered the endpoint of the titration. As can be seen from Table 4, SDCP is unique in that it is compatible with anionic systems.

Eye irritation is a major concern in the formulation of personal care products, particularly when working with quats. Primary eye irritation was tested using the protocol outlined in FHSLA 16 CFR 1500.42. The products were tested at 25% actives. The results are reported in Table 5.

It should be noted that at a concentration of 0.5%, stearalkonium chloride was minimally irritating. This rating of minimally irritating was the same as for CDCP and SDCP at 25% (or 50 times the concentration). As the data clearly show, the irritation potential of the complex is dramatically reduced, when compared to the starting quat.

The ability of hair to be rewetted is an important factor in selecting conditioning agents. This makes them different from the standard fatty quats, which make the substrate hydrophobic and gunky. When fatty quaternary compounds that have been complexed with carboxysilicone are used to treat hair, they make hair soft, but they do not make the hair hydrophobic.

Perhaps most important in the context of this book is the outstanding effect the incorporation of the silicone complexes has on both wet and dry combing. Compounds made using this complexation technology give outstanding wet and dry combing properties to hair.

J. Silicone Phosphate Esters

1. *Structure/Background*

A series of silicone surface-active agents containing an ionizable phosphate group has been developed. These silicone-based phosphate esters and their derivatives are the topic of numerous patents (21). The properties have been compared to and contrasted with traditional fatty phosphate esters (22). The structures of the silicone phosphate esters are as follows:

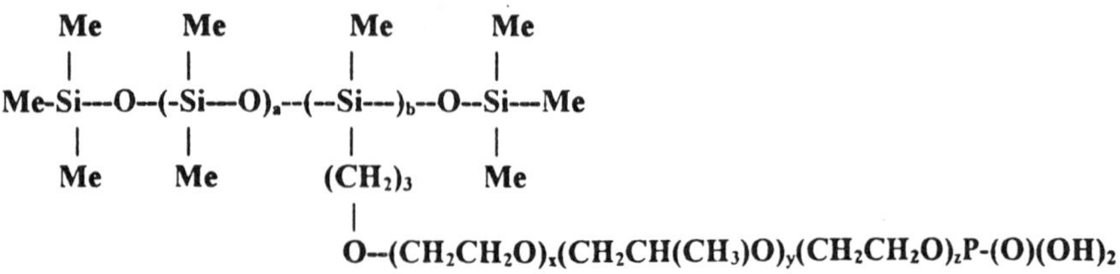

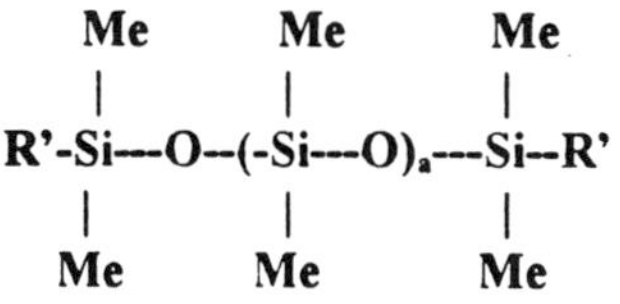

```
        Me       Me            Me        Me
        |        |             |         |
Me-Si---O--(-Si---O)a--(--Si---)b--O--Si---Me
        |        |             |         |
        Me       Me          (CH2)3      Me
                               |
                               O--(CH2CH2O)x(CH2CH(CH3)O)y(CH2CH2O)zP-(O)(OH)2

              Me        Me         Me
              |         |          |
          R'-Si---O--(-Si---O)a---Si--R'
              |         |          |
              Me        Me         Me
```

R' is $-(CH_2)_3O--(CH_2CH_2O)_x(CH_2CH(CH_3)O)_y(CH_2CH_2O)_zP-(O)(OH)_2$

2. *Properties/Applications*

Silicone-based phosphate esters are substantive to hair, skin, and fiber and provide antistatic properties. Since these compounds contain a pendant ionizable phosphate group, they provide antistatic and lubrication properties to the hair or fiber.

Silicone phosphate esters are acidic and can be neutralized to any desired pH with alkaline materials. The pH of the final formulation not only affects the solubility of the phosphate ester, but also has a profound affect on other surfactant properties such as wetting, foaming, and emulsification. The partially neutralized phosphate ester has solubility characteristics between that of the free acid and the completely neutralized phosphate ester. As emulsifiers, silicone-based phosphate esters are efficient, producing emulsions of the oil-in-water type. They are very useful as emulsifiers for personal care applications, such as moisturizing creams and lotions. They are particularly effective in producing creams and lotions containing sunscreens (both organic and inorganic), pigments, skin protectants, and medicaments. The emulsification properties allow for the use of these materials in one-step shampoos and other applications, including emulsion polymerization processes.

As water-soluble emollients, silicone phosphate esters can be used in aqueous systems. For example, they can be added to carbomer gels for emolliency without diminishing the clarity of the gel. The phosphating of dimethicone copolyol renders the molecule more water soluble, necessitating fewer moles of ethylene oxide for the same degree of water solubility. This allows for retention of much of the favorable esthetics associated with the increased content of the dimethylpolysiloxane portion of the molecule.

As foaming agents, silicone-based phosphate esters produce higher levels of copious foam than are produced by dimethicone copolyols. The sodium and potassium salts of the phosphate esters tend to be slightly better foaming agents than the phosphate esters in their free acid or amine salt form. Salts formed by the neutralization of dimethicone copolyol phosphate and myristamidopropyldimethylamine are excellent foam boosters and conditioners in shampoo systems. Salts formed by the neutralization of dimethicone copolyol phosphate and linoleylamidopropyldimethylamine are excellent emulsifiers for dimethicone in shampoo systems. As detergents, silicone-based phosphate esters show good detergency: surface tension reduction, wetting, emulsification, dispersing properties, and solubilization. Their detergent properties are generally considered to be equal to those of nonionic surfactants; however, the presence of the silicone in the molecule results in improved mildness, substantivity, and conditioning properties over conventional fat-based phosphate esters. These properties make the products excellent candidates for incorporation into shampoos and other detergent systems for personal care. Silicone-based phosphate esters can function as hair conditioners and have been found to have outstanding conditioning affects when applied to permed hair. A double-blind, half-head study was conducted on female subjects having permed and nonpermed hair. Subjects were instructed to apply conditioners after shampooing on wet hair, leave in for 2 min, and then rinse. After dry combing, no difference in conditioning was perceived by the subjects with unpermed hair. However, a dramatic difference was observed by all subjects having permed hair. The side where the dimethicone copolyol phosphate-containing conditioner was used showed greatly improved body, curl retention, and manageability compared to the side where the conditioner without the dimethicone copolyol phosphate was applied. It was concluded that dimethicone copolyol phosphate is a highly efficacious material for conditioning permed hair.

a. Toxicity. Silicone phosphate esters have been found to have a low order of toxicity. They are nonirritating to the skin when evaluated by FHSLS 16 CFR15 00.41 for primary skin irritation, giving values of zero. They are nonirritating to the eyes when evaluated by FHSLS 16 CFR 1500.41 for primary eye irritation, giving values of zero. They are nontoxic when evaluated by FHSLS 16 CFR 1500.41 for acute oral toxicity, having LD_{50} values above 5 g/kg. They are also noncomedogenic when evaluated using a standard comedogenicity assay. This profile suggests that these materials are outstanding candidates for personal care applications.

b. SPF Enhancement. Dimethicone copolyol phosphate has been demonstrated to boost the SPF of chemical sunscreens by as much as 47%, and nonchemical or pigmented sunscreen systems by as much as 17% (23). It has been postulated that the superior wetting properties of the dimethicone copolyol phosphate allows the sunscreening materials to cover the skin more efficiently, in a thinner, more uniform film.

c. Typical Formulations. Silicone phosphate esters find applications in a variety of personal care products. Their properties as outlined above make them excellent additives for hair care, skin care, and other personal care applications.

K. Other Silicone Esters

1. Structure/Background

Two new classes of silicone based esters have recently been prepared. The first class of materials is prepared by the esterification reaction of a dimethiconol and a fatty acid (24). These dimethiconol esters contain both a fatty group and a silicone polymer group. They conform to the following structure:

$$CH_3(CH_2)_x\text{-}C(O)\text{-}O\text{--}\overset{\displaystyle Me}{\underset{\displaystyle Me}{|}}\!\!\!\!Si\text{---}O\text{--}(\text{-}\overset{\displaystyle Me}{\underset{\displaystyle Me}{|}}\!\!\!\!Si\text{—}O)_a\text{---}\overset{\displaystyle Me}{\underset{\displaystyle Me}{|}}\!\!\!\!Si\text{--}O\text{-}C(O)\text{-}(CH_2)_xCH_3$$

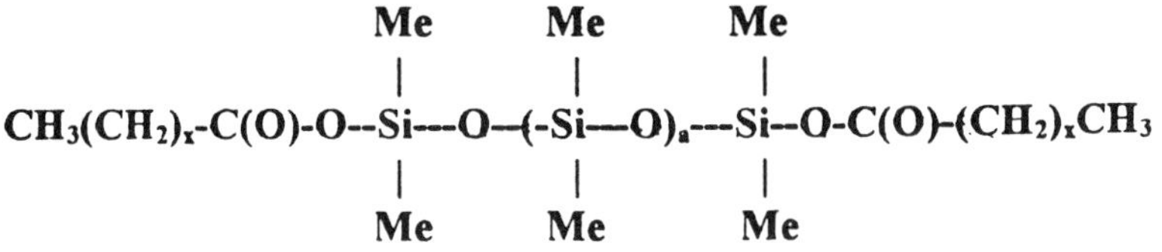

The second class results from the reaction of a dimethicone copolyol with a fatty acid (25). The dimethicone copolyol group initially contains a silicone group and a polyoxyethylene group. The incorporation of the fatty group by the esterification reaction results in a product that has a water-soluble, a silicone-soluble, and a fatty-soluble group present in one molecule. This class of compounds conforms to the following structure:

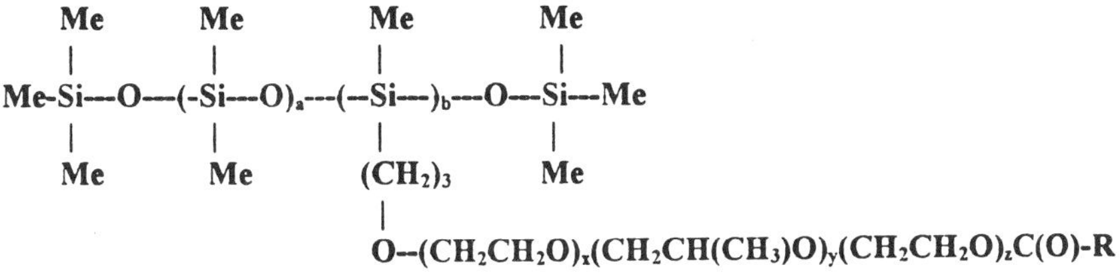

```
     Me         Me            Me          Me
     |          |             |           |
Me-Si---O---(-Si---O)a---(--Si---)b---O---Si---Me
     |          |             |           |
     Me         Me          (CH2)3        Me
                              |
                              O--(CH2CH2O)x(CH2CH(CH3)O)y(CH2CH2O)zC(O)-R
```

2. Properties/Applications

Typical properties of this class of compounds are as follows.

Dimethiconol stearate. A water-insoluble, nonocclusive, highly lubricious silicone was. The physical form is a white pastelike wax, which liquiefies under pressure. It has been used in personal care applications, including dispersing of pigments such as titanium dioxide and zinc oxide for sunscreen products, and in many other applications. This is a dimethiconol-based compound.

Dimethicone copolyol stearate. This material forms a microemulsion in water with no added surfactant. The product is a liquid. It is highly lubricious and tends to stay at the interface of water and glass. The product has been used as a hair conditioner, and in a variety of emulsification applications. This is a dimethicone copolyol-based compound. To illustrate the range of solubilities achievable using this technology, products were tested at 5% solids.

Table 6 illustrates that silicone can be chosen for inclusion in a particular formulation by considering the phase in which the formulation can use the benefit of silicone. It can be added as an emulsion to the phase in which the silicone functionality is desired. For example, if you want the silicone in the oil phase to effect pay off of an oil-in-water emulsion, pick a silicone soluble in the oil phase. If you want silicone in the aqueous phase to affect the spreadability of the emulsion, pick a water-soluble silicone. You may also pick a silicone that is insuluble in either phase. In addition to two distinct families of products which are prepared using fatty acids, two distinct classes of products are made from natural triglycerides. These materials provide conditioning, gloss, and softening properties when applied to the hair or skin.

Products derived from many other triglycerides are also available. Silicone derivatives from these function in formulations in the same manner as silicone products made from the analogous fatty acids, and are used predominantly for their name on the label. For example, dimethicone copolyol cocoabuterate might be a good additive for after-sun products. For deodorant sticks, dimethicone copolyol isostearate provides slip and negates the soapy feel of the

Table 6 Solubilities of Products

Solvent	5% product	
	Dimethiconol	Dimethicone copolyol
Water	Insoluble	Dispersible
Mineral spirits	Soluble	Insoluble
Silicone fluid (350 cs)	Dispersible	Insoluble
Polyethylene glycol (PEG 400)	Insoluble	Dispersible
Glycerol trioleate	Insoluble	Soluble
Oleic acid	Dispersible	Soluble
Mineral oil	Dispersible	Insoluble

sodium stearate. For day creams, dimethiconol stearate forms a hydrophobic, nonocclusive film on the skin. Dimethiconol fluoroalcohol dilinoleate is a unique fluorosilicone wax which provides outstanding barrier properties.

IV. CONCLUSION AND FUTURE DEVELOPMENTS

A recent publication (26) suggests that the balance between the oil-soluble portion, the silicone-soluble portion, and the water-soluble portion of a silicone surfactant is critical to the functionality. This relationship among the three different phases has been explored in a concept called three-dimensional HLB. The development of this concept promises to offer new assistance in the preparation of emulsions. In summary, new silicone polymers are used in many diverse application areas in the personal care market. There has been plethora of new silicone compounds that contain portions of the molecule that is not silicone based. These non-silicone groups in the molecule can be hydrocarbon based, polyoxyalkylene based, or based on other classes of raw materials. The new polymers have unique properties, partially contributed by the silicone and partially contributed by the other groups. Silicone compounds are not just oil phases any more. More work is expected on the application of these materials to different formulations. The future of these types of compounds in personal care applications will depend on the ability of companies to create the new compounds and the ability of creative formulators to utilize the compounds.

REFERENCES

1. U.S. Patent 3,964,500 (1976).
2. U.S. Patent 4,741,855 (1988).
3. U.S. Patent 5,100,657 (1992).

4. U.S. Patent 5,100,658 (1992).
5. U.S. Patent 5,100,646 (1992).
6. U.S. Patent 5,106,609 (1995).
7. Di Sapio A. Soap Cosmet Chem Spec 1994; 70(9). September.
8. O'Lenick AJ Jr, Parkinson JK. J Soc Cosmet Chem 1994; 45:247–256.
9. Dhams G, Zombeck A. Cosmetc Toilet 1995 (Mar.); 110:91–100.
10. U.S. Patent 3,661,964 (1972).
11. U.S. Patent 5,378,787 (1995).
12. Dhams G, Zombeck A. Cosmet Toilet 1995 (Mar.); 110:91–100.
13. Hunter A. In: Encyclopedia of Conditioning Rinse Agents. Micelle Press, 1983: 174–175.
14. Ibid., p. 99.
15. Hill R. ACS Symp Ser 1992; 501:278.
16. Dhams G, Zombeck A. Cosmet Toilet 1995 (Mar.); 110:91–100.
17. Ibid.
18. U.S. Patent 5,070,171 (1991).
19. O'Lenick A Jr, Sitbon Suserman C. Carboxy silicone quaternary complexes. Cosmet Toilet 1996 (Apr.); 111:67–72.
20. O'Lenick A Jr, Parkinson JK. J Soc Cosmet Chem 1994; 45:247–256.
21. Ibid.
22. O'Lenick A Jr, Parkinson JK. Phosphate esters: chemistry and properties. Textile Color Chem 1995; 27(11):17–20.
23. Imperante J, O'Lenick A Jr, Hannon J. Dimethicone copolyol phosphates to increase sun protection factor. Soap Cosmet Chem Spec 1996; 76(5):54–58.
24. U.S. Patent 5,051,489 (1991).
25. U.S. Patent 5,180,843 (1993).
26. O'Lenick A Jr, Parkinson JK. Cosmet Toilet 1996 (Oct.); 111:37–44.

10

Cationic Surfactants and Quaternary Derivatives for Hair and Skin Care

Matthew F. Jurczyk, David T. Floyd,
and Burghard H. Grüning
Goldschmidt Chemical Corporation, Hopewell, Virginia

I. INTRODUCTION

Quaternary ammonium compounds have traditionally been utilized for emulsified hair conditioners. Later advances led to the development of quaternaries, which were less irritating and are compatible with shampoos and liquid soap detergents. These innovations paved the way for commercial applications in skin care creams and lotions.

This chapter presents the definition, development, and application of quaternary derivatives for the personal care industry.

A. Quaternary Ammonium Salts

By general definition, quaternary ammonium salts are "a type of organic compound in which the molecular structure includes a central nitrogen atom joined to four organic groups as well as an acid radical" (1). At least one of these substitution groups is typically hydrophobic in nature. Quaternary derivatives based on fatty acids, proteins, sugars, and silicone polymers are all used in the cosmetic industry. Quaternary ammonium salts are cationic surface-active compounds and adsorb readily onto surfaces such as hair and skin.

It is the ability to adsorb onto substrates which makes the use of cationic surfactants so widespread. By 1996, one trade directory alone listed over 300 commercially available cationic surfactants and over 200 nitrogen-based amphoteric surfactants (2,3).

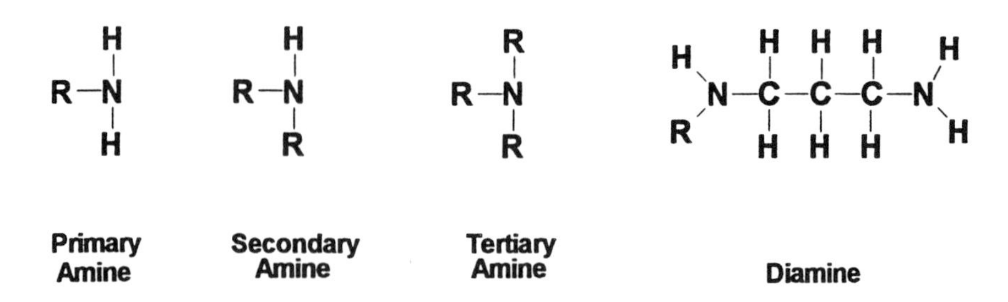

Figure 1 Representative amines. R = an organic radical based on one or more carbon atoms. (From Ref. 4.)

B. Amines

Fatty alkyl-substituted amines form the base intermediate for most cationic surfactants. Representative examples are shown in Figure 1.

The number of radicals combined with the nitrogen atom determines the order of substitution. These designations are known as primary (mono-substituted), secondary (di-substituted), and tertiary (tri-substituted). Primary amines are also used as intermediates in the preparation of diamines (4).

Quaternary ammonium salts are formed by the reaction of amines with alkylating agents such as methyl chloride, methyl sulfate, or benzyl chloride. Representative quaternaries are shown in Figure 2.

C. Amphoteric Surfactants

A related class of amine derivatives are nitrogen-based amphoteric and zwitterionic surfactants. Zwitterionic materials have both a positive and a negative charge at their isoelectric point. Amphoteric surfactants are characterized by a molecular structure containing two different functional groups, with anionic and cationic character, respectively (5). Most amphoteric surfactants behave in acidic mediums as cationic surfactants, and in alkaline medium like anionic surfactants. The alkyl betaines and alkylamido betaines are different in that they cannot be forced to assume anion-active behavior through an increase in pH value (6,7). While these compounds can function and are stable over a wide pH range, they exhibit cationic tendencies in acidic systems (8). Examples include alkyl betaines and alkylamido betaines, carboxylated imidazoline derivatives, and amphoacetates. These are shown in Figure 3.

One characteristic of cationic surfactants is that they tend to form insoluble salts (precipitates) in combination with anionic surfactants, such as fatty alcohol sulfates. Amphoteric surfactants, however, remain clear in anionic systems

$[CH_3-N(CH_3)_2-R]^+ X^-$

X = chloride; X could be methyl sulfate on some quaternaries

R = aliphatic saturated or unsaturated, C_{12}- C_{22} chain lengths

Monoalkyl Trimethyl Quaternary Ammonium Chloride

$[CH_3-N(R)_2-CH_3]^+ X^-$

X = chloride or methyl sulfate

R = aliphatic, saturated or unsaturated, normal or branched, C_8-C_{22}

Dialkyl Dimethyl Quaternary Ammonium Chloride

$[CH_3-N(R)_2-R]^+ X^-$

X = chloride

R = aliphatic, alkyl, normal or branched, C_8-C_{18}

Trialkyl Methyl Quaternary Ammonium Chloride

$[R-N^+(CH_3)_2-CH_2-CH_2-CH_2-N^+(CH_3)_2-CH_3] X^-$

X = chloride
R = aliphatic, saturated or unsaturated

Complex Diquaternary Ammonium Chloride

$[CH_3-N(CH_3)(R)-CH_2-C_6H_5]^+ X^-$

X = chloride
R = aliphatic, normal, C_{12}-C_{18}

Alkyl Benzyl Dimethyl Quaternary Ammonium Chloride

Figure 2 Representative quaternary types. (From Ref. 4.)

and are said to be "anionic compatible." This feature, along with their ability to enhance formula viscosity, mitigate irritation, and impart skin softness, has contributed to the increasing popularity of nitrogen-based amphoteric surfactants in the personal care industry.

D. Amine Oxides

Amine oxides are another group of quasi-cationic, nitrogen-based products which display anionic surfactant compatibility. Amine oxides are prepared by

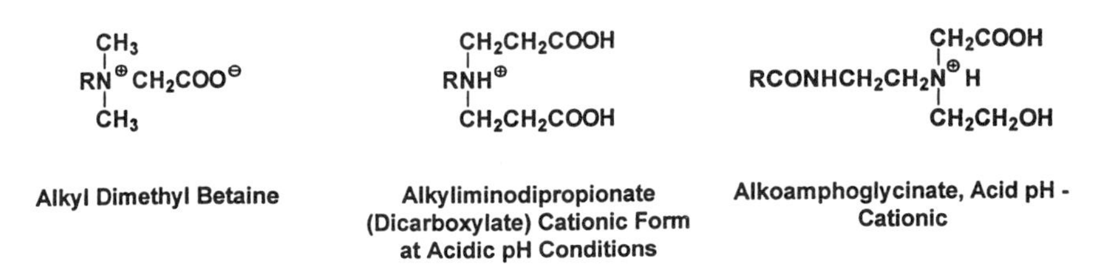

Figure 3 Representative amphoteric surfactants. (From Ref. 112.)

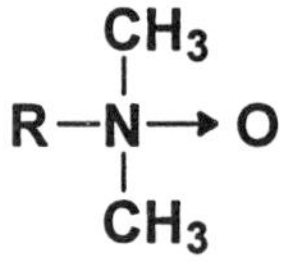

$$R-\overset{\displaystyle CH_3}{\underset{\displaystyle CH_3}{\overset{|}{\underset{|}{N}}}} \rightarrow O$$

Figure 4 Cocamine oxide. R represents the alkyl groups derived from coconut oil. (From Ref. 113.)

reacting tertiary amines with hydrogen peroxide. They are nonionic at alkaline pH conditions, but gain positive charges and exhibit cationic activity in acidic solutions. Although they have been incorporated into shampoos and other personal care products, amine oxides play their greatest commercial role in household, industrial, and institutional cleansers. See Figure 4.

E. Amidoamines

Amidoamines are an interesting class of substances characterized by the presence of one or more amide groups coupled with one or more amine functions. When compared to alkyl amines and their derivatives, amidoamines are generally more compatible with anionic surfactants, easier to formulate with, and function as better foaming aids (9). Representative amidoamine compounds and related derivatives are displayed in Figure 5.

II. HISTORICAL DEVELOPMENT OF THE CATIONIC SURFACTANT INDUSTRY

Groundwork for the commercial development of cationic surfactants was laid in the early years of the twentieth century, when Einhorn described the anti-

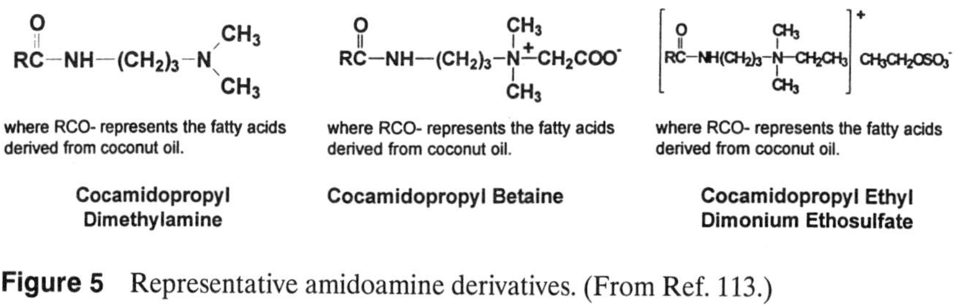

$RC(=O)-NH-(CH_2)_3-N(CH_3)_2$

where RCO- represents the fatty acids derived from coconut oil.

Cocamidopropyl Dimethylamine

$RC(=O)-NH-(CH_2)_3-N^{+}(CH_3)_2-CH_2COO^{-}$

where RCO- represents the fatty acids derived from coconut oil.

Cocamidopropyl Betaine

$[RC(=O)-NH(CH_2)_3-N(CH_3)_2-CH_2CH_3]^{+}\ CH_3CH_2OSO_3^{-}$

where RCO- represents the fatty acids derived from coconut oil.

Cocamidopropyl Ethyl Dimonium Ethosulfate

Figure 5 Representative amidoamine derivatives. (From Ref. 113.)

microbial properties of quaternary salts. This was followed by the first publication on the use of a quaternary as a surgical antiseptic in 1936 (10).

Significant toxicological data were also compiled during the early postwar period. Cutler and Drobeck noted that benzalkonium chloride appeared to be one of the most widely studied quaternaries between 1940 and 1950. Cutler further attributed this intense scrutiny to the proliferation of quaternary germicides in a wide range of industrial and household applications. These included farm disinfectants, industrial disinfectants, skin antiseptics, mouthwashes, and diaper sanitizers (11).

During the 1930s, other applications for cationic surfactants were being investigated. 1939 marks the first commercial production of long-chain fatty amines for an industrial application (potash flotation). Commercial procedures for the production of quaternary salts continued to be developed and refined in the 1940s. By the early 1950s, quaternaries such as dihydrogenated tallow dimethyl ammonium chloride were found in widespread industrial applications, ranging from organoclay additives for drilling muds to consumer fabric softeners, and eventually hair conditioners.

Institutional acceptance of quaternary ammonium compounds paved the way for their incorporation into the first personal care products. The timing of their commercial introduction (the late 1940s) coincided with a surge in market demand for hair conditioners used to mitigate damaging effects of popular permanent waves. While commercial hair conditioners had been available as early as the 1930s, development of a truly substantive cationic conditioner was not possible until the advent of stearalkonium chloride in 1946. Originally developed for the textile industry, this ingredient was readily adapted by cosmetic formulators and for many years was the workhorse of cream rinse formulations (12).

Further industry refinements have been directed at enhancing the compatibility of quaternary ammonium salts and other cationic derivatives with anionic surfactants. While true cationic surfactants are suitable for use in hair dyes, creme rinses, and some hair grooming preparations, their irritancy and incompatibility with anionic surfactants traditionally limited their potential for inclusion in high-foaming mass market shampoos, skin cleansers, and bath preparations.

In the early 1970s, a number of conditioning shampoos, based on polymeric quaternary ammonium resins such as polyquaternium 7 and polyquaternium 10, and polyethylene imines were introduced (13). Opportunities for incorporation of conventional quaternary ammonium salts in anionic shampoo systems, however, remained limited. In 1970 Jungerman commented that quaternary ammonium compounds with three or four long alkyl chains had not "attained any scientific or industrial significance" (14).

Within 15 years, this situation had changed dramatically. Procter & Gamble had patented the technology for a conditioning shampoo containing a "tri long chain alkyl mono short chain alkyl quaternary ammonium salt . . . together with a dispersed insoluble silicone phase" (5,16). Today, tricetylmonium chloride is prominent on the ingredient labels of brands which, when combined, account for over 30% of the U.S. shampoo market (17).

Other approaches aimed at improving quaternary compatibility with anionic surfactants resulted in the development and use of amphoteric imidazoline derivatives and cocamidopropyl betaine in a wide range of "mild" baby shampoos and skin cleansers. Betaines and imidazoline were found to be extremely effective in reducing irritancy while enhancing foam and viscosity performance. Their incorporation in cleansers and mild shampoos has greatly advanced the use and market acceptance of nitrogen-containing compounds in the personal care industry.

While the earliest quaternaries were originally developed for use in industrial applications and "adopted" by the personal care industry, many of the latest innovations have been synthesized specifically to satisfy unmet consumer needs. As the industry matures and becomes even more highly segmented, competition and the need to satisfy consumer expectations will continue to be a driving force behind the new technical innovations. A discussion of recent innovations is presented elsewhere in this chapter.

III. CATIONIC SURFACTANT CHEMISTRY AND FUNCTIONALITY

The main groups of cationic surfactants are alkylamines, alkyl imidazolines, ethoxylated amines, and quaternaries. These organo-modified polymers have been reviewed and will be further examined in this chapter (18–21).

A. Alkylamines

Alkylamines form a fairly large group of cosmetic surfactants. Alkylamines and their salts form a group in which the amino function is responsible for the surface-active properties and the water solubility of the compound. Primary, secondary, and tertiary alkylamines, with their salts and substances containing more than one basic nitrogen atom, are included in this group.

Alkylamines in this group possess long alkyl chains and are relatively hydrophobic. Their salts with inorganic and strong organic acids exhibit the type of solubility required for amphiphiles. These alkylamines are commonly prepared by reduction of the corresponding acylamide via the nitrile to the amine.

The synthesis of these alkylamines first requires the reaction of an acid chloride with 1 mole of a simple or substituted diamine to yield a series of substances sometimes referred to as "amido-amines." These substances are more polar than the straight-chain uninterrupted alkylamines and are more widely used in cosmetics as cationic emulsifiers. As a rule, these amines form precipitates with the more commonly used anionic surfactants.

The free amines in this group are waxy solids, while the salts and the interrupted amines exhibit higher melting points. The positive charge carried by the neutralizing amines is responsible for their substantivity to negatively charged biological surfaces, such as skin and hair.

These amines—usually in the form of salts from phosphoric, citric, or acetic acid—are used in cosmetics primarily in combination with fairly hydrophilic surfactants. The interrupted alkylamines are compatible with a variety of other surfactants. The interrupted alkylamines can be used, e.g., in combination with glyceryl monostearate to form emulsions with good tolerance to electrolytes. In lotion formulations, the interrupted amines prevent an increase in viscosity upon standing. Aqueous solutions formed by the acid-derived amine salts are effective hair conditiners and have antistatic properties.

Examples of these types of amines are presented in Figure 6.

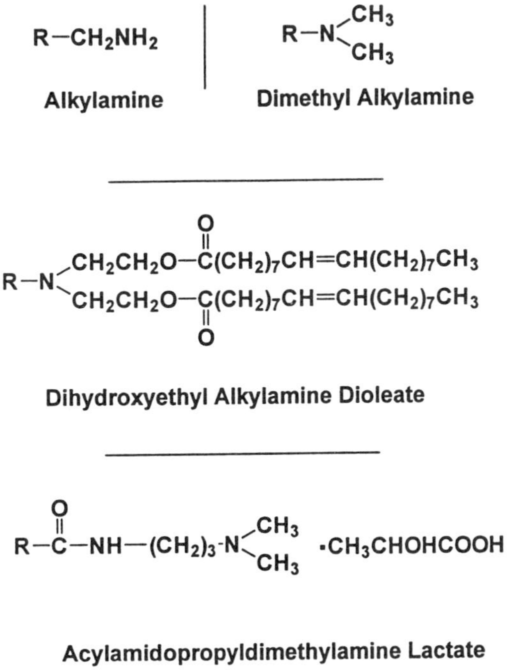

$R{-}CH_2NH_2$

Alkylamine

$R{-}N(CH_3)_2$

Dimethyl Alkylamine

$R{-}N[CH_2CH_2O{-}C(=O)(CH_2)_7CH{=}CH(CH_2)_7CH_3]_2$

Dihydroxyethyl Alkylamine Dioleate

$R{-}C(=O){-}NH{-}(CH_2)_3{-}N(CH_3)_2 \cdot CH_3CHOHCOOH$

Acylamidopropyldimethylamine Lactate

Figure 6 Examples of alkylamines.

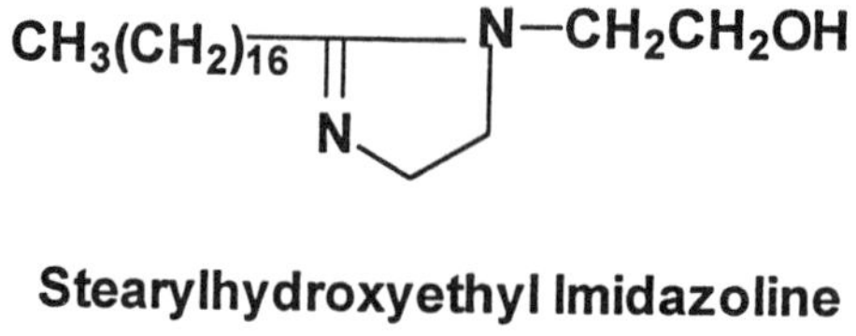

$CH_3(CH_2)_{16}$ — N—CH_2CH_2OH

Stearylhydroxyethyl Imidazoline

Figure 7 Typical alkyl imidazoline.

B. Alkyl Imidazolines

The alkyl imidazolines comprise a small group of basic heterocyclic substances. They are obtained by the reaction of aminoethylethanolamine with a suitable fatty acid. An example is given in Figure 7. During this synthesis, no linear amide is formed, but it is cyclized to a five-membered substituted ring. The fcarboxy-carbon of the fatty acid becomes part of this ring. For the sake of simplicity, the nomenclature retains the name of the originally employed fatty acid, although the pendant group has been shortened by one carbon atom. Reaction of the alkyl imidazoline with an alkylating agent opens the ring and yields the amphoteric derivatives of ethylenediamine. The compounds in this group are all 1-hydroxyethyl-2-alkyl imidazolines.

The alkyl imidazolines are liquids and normally distributed as aqueous solutions. Imidazolines themselves are not used widely in cosmetics. Amphoteric surfactants, however, can be derived from fatty imidazolines. They are employed in aqueous media. Imidazolines are subject to hydrolysis to the amide, and re-formation of the cyclic structure is probably pH dependent. It has not been established whether the cyclic imidazolines or their hydrolysis products constitute the active species.

A structure for a fatty imidazoline is given in Figure 8.

Four types of amphoterics are derived from a fatty imidazoline compound by a reaction from acetate or propionate, giving rise to monoacetate, diacetate, monopropionate, and dipropionate amphoterics (Figure 9) (22).

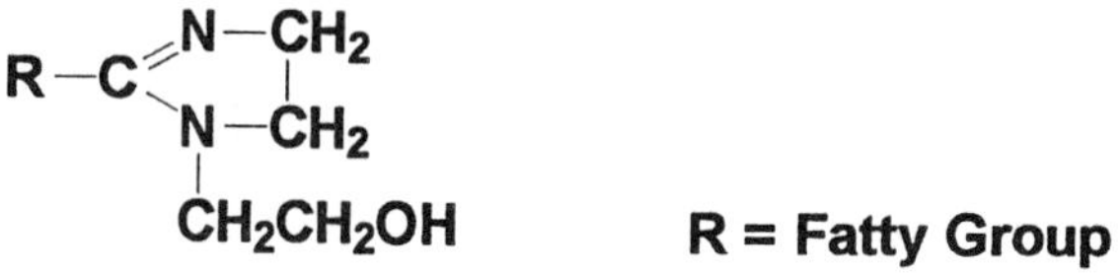

R—C(=N—CH_2)(—N—CH_2)
CH_2CH_2OH

R = Fatty Group

Figure 8 Structure for a fatty imidazoline.

Monoacetate

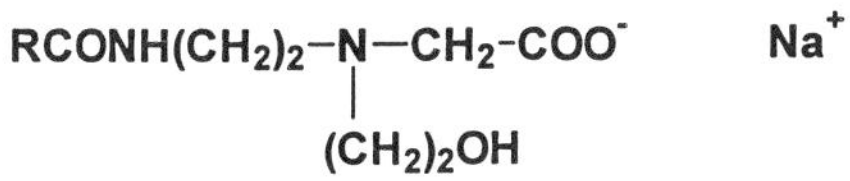

$$RCONH(CH_2)_2-N(-(CH_2)_2OH)-CH_2-COO^- \quad Na^+$$

Diacetate

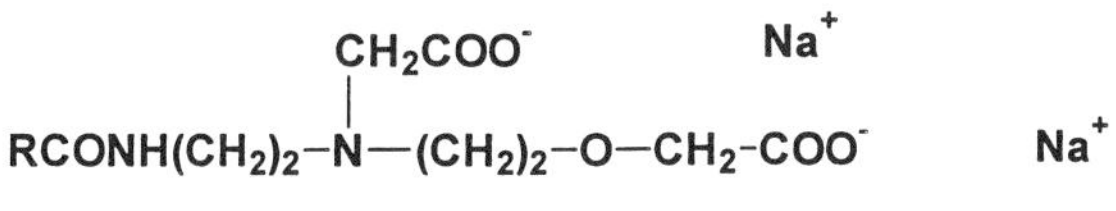

$$RCONH(CH_2)_2-N(-CH_2COO^-\ Na^+)-(CH_2)_2-O-CH_2-COO^- \quad Na^+$$

Monopropionate

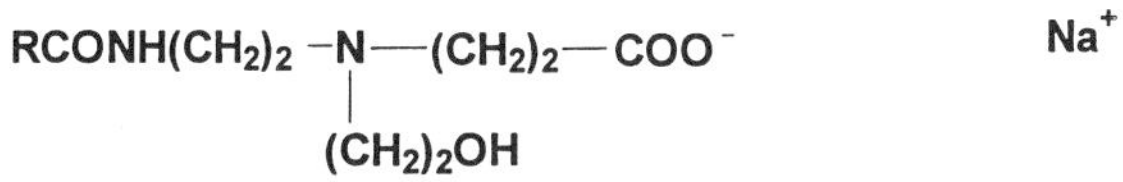

$$RCONH(CH_2)_2-N(-(CH_2)_2OH)-(CH_2)_2-COO^- \quad Na^+$$

Dipropionate

$$RCONH(CH_2)_2-N(-(CH_2)_2COO^-\ Na^+)-(CH_2)_2-O-(CH_2)_2-COO^- \quad Na^+$$

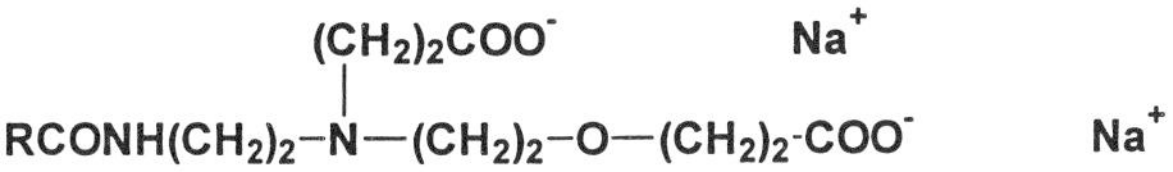

Figure 9 Four types of amphoterics.

In the U.S. market, three traditional amphoterics are used in most mild shampoos: disodium cocoamphodiacetate, sodium lauroamphoacetate, and disodium lauroamphodiacetate. While these are mild surfactants, their potential for conditioning is minor. They are selected for use mainly for irritation reduction and mitigation potential. They are most widely used in formulations for baby shampoo and personal cleansing.

C. Ethoxylated Amines

The ethoxylated amines constitute a group of nitrogen-containing surfactants in which the aqueous solubility is to a large extent dependent on the degree and type of alkoxylation. The basic nature of the amino group is not readily displayed in those substances, which carry long-chain polyoxyethylene (POE) groupings.

The simple POE amines are prepared from long-chain alkyl amines by ethoxylation. In some cases, the alkylamine is converted to a diamine before ethoxylation (Figure 10).

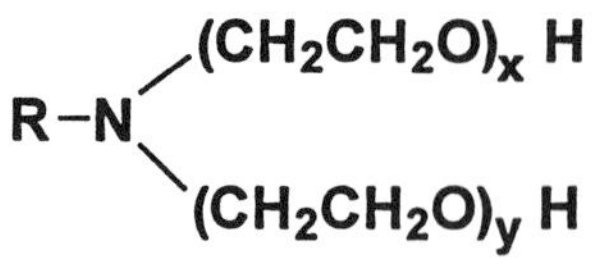

$$R{-}N\begin{cases}(CH_2CH_2O)_x\,H\\(CH_2CH_2O)_y\,H\end{cases}$$

PEG-n Alkylamine
(x + y has an average value of n)

Figure 10 Ethoxylated amine.

A more complex type of alkoxylated amine (the poloxamine) is synthesized by reaction of ethylene diamine with propylene oxide (Figure 11). This results in the formation of a hydrophobic tetra-substituted ethylene diamine. When this is then reacted with ethylene oxide, its hydrophobicity is decreased, depending on the ration of hydrophobe to hydrophile. Even more complicated,

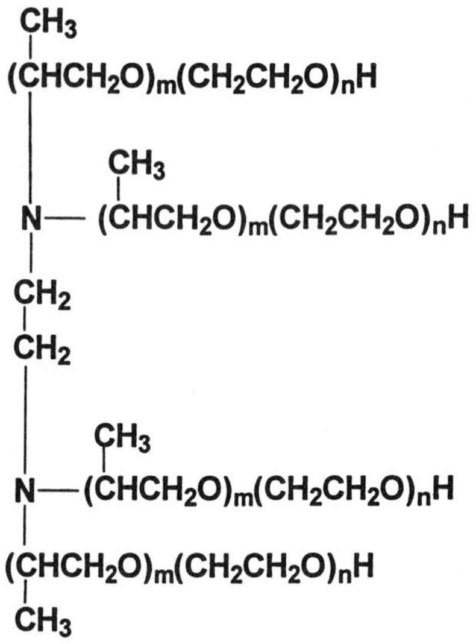

$$\begin{array}{l} CH_3 \\ | \\ (CHCH_2O)_m(CH_2CH_2O)_nH \\ | \\ N{-}\,(\overset{CH_3}{\overset{|}{C}}HCH_2O)_m(CH_2CH_2O)_nH \\ | \\ CH_2 \\ | \\ CH_2 \\ | \\ N{-}\,(\overset{CH_3}{\overset{|}{C}}HCH_2O)_m(CH_2CH_2O)_nH \\ | \\ (CHCH_2O)_m(CH_2CH_2O)_nH \\ | \\ CH_3 \end{array}$$

n = moles of ethylene oxide
m = moles of propylene oxide
m & n can have different ratios

Figure 11 Poloxamine.

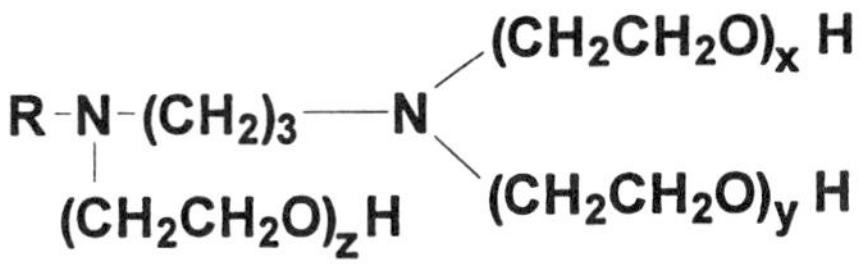

$$R-N(-(CH_2CH_2O)_zH)-(CH_2)_3-N\begin{cases}(CH_2CH_2O)_x H\\(CH_2CH_2O)_y H\end{cases}$$

PEG-n Alkylamine
(*x* + *y* + *z* has an average value of n)

Figure 12 Alkoxylate amine.

terminally fatty alkyl-substituted alkoxylated amines have been prepared, such as a PEG-*n* tallow polyamine (Figure 12).

Most ethoxylated amines are water soluble and are compatible with a wide variety of other surfactants, including anionics. They are relatively weak bases and do not require large amounts of acids to adjust their pH to the range normally used in cosmetics.

Ethoxylated amines are waxy solids which melt at relatively low temperatures. Ethoxylated amines are used as emulsifying and hair conditioning agents. They also can be used to aid in the dispersion of solids. Some members of this class increase product viscosity, while others are used to improve foaming. The substances making up this class are acid and alkali stable and are not subject to hydrolysis.

D. Quaternaries

While quaternary salts can be produced from the alkylation of primary or secondary amines, in commercial practice tertiary amines are typically employed (see Figure 13). This is done to reduce the need for excess alkylating reagent and limit the formation of free amine (5).

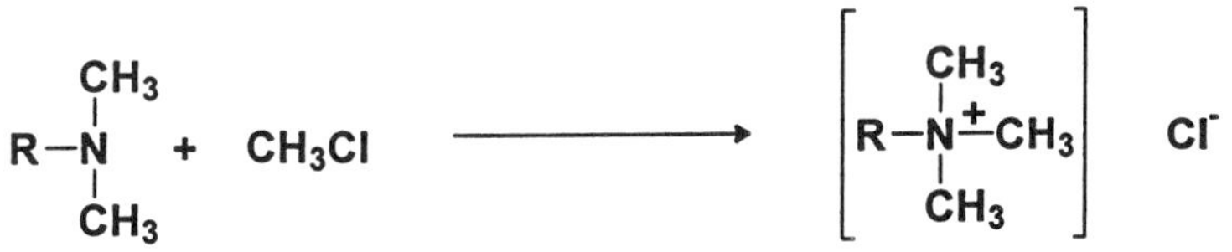

$$R-N(CH_3)_2 + CH_3Cl \longrightarrow \left[R-N^{+}(CH_3)_3\right] Cl^-$$

Tertiary Amine + Methyl Chloride ——→ Quaternary Ammonium Chloride

Figure 13 Quaternary synthesis.

The presence of an unreacted alkylating agent and a free amine are undesirable. To eliminate free amine, manufacturers acidify their product. This results in the formation of an odorless amine salt. It is important to recognize that since commercial quaternary ammonium compounds always contain amine salts, any formulation at alkaline pH conditions would result in odor, emulsion instability, and irritation problems stemming from reconversion of the amine salt to its free amine form. Development chemists must therefore maintain acidic pH conditions in their formulas (23).

E. Miscellaneous Cationics

No discussion of quaternary ammonium salts and cationic surfactants in the personal care industry would be complete without mention of the introduction of new derivatives which are based not on traditional animal or vegetable-based fatty substances, but on a wide range of natural and synthetic feedstocks.

During the 1960s, the substantivity and conditioning effects of collagen-derived polypeptides on hair were described and cosmetic proteins were included in shampoos, conditioners, permanent waves, bleaches, and other hair care preparations (24). The positive perception and consumer acceptance of cosmetic proteins led to the introduction of quaternized collagen hydrolysates by the early 1980s (25).

Similar advances were made using silicone technology.

> The acceptance of silicones by the cosmetic industry is substantiated by the increase in sales from only a few tons in the early 1970's to tens of thousands of tons in the 1990's. The development of amphoteric, amono functional and quaternized silicone derivatives offered a new range of mild products with improved aesthetic properties, including superior gloss, combing, and conditioning of hair. These compounds have also found a niche in the skin care industry where they can be used to enhance spreadability with reduced tackiness in creams and lotions (15).

The use of naturally derived glucose and hydroxyl substitutions has contributed to the relatively recent acceptance of quaternary ammonium derivatives in skin care applications. For instance, products such as lauryl methyl gluceth 10 hydroxypropyldimonium chloride are promoted for their ability to enhance skin moisturization (26). Cationic oil-in-water emulsifiers such as dipalmitoylethyl hydroxyethylmonium methosulfate can be used to enhance skin softening, lubricity, and emolliency (27).

F. Function of Quaternary Compounds on Hair

Daily hair care no longer encompasses just the cleansing of hair, but the expectation that attributes such as static control and conditioning will be realized (28). These expectations encompass a large number of attributes, most of which are closely related to Robbins et al.'s definition of the term "hair manageability" (29). An attribute that is closely associated with conditioning is ease of combing. An instrumental technique has been established to obtain a quantitative measure of this parameter (30).

The isoelectric point of untreated hair is pH 3.7 (31). This indicates that the hair surface attains a cationic charge at a pH under 3.7 while assuming an anionic character above that pH. Traditionally, quaternary ammonium salts (or quats) have been used as the main additive responsible for conditioning effects. Their structure can be represented as $[N^+(R)_a(CH_3)_{4-a}X^-$, where R is a long-chain alkyl group, X is the counteranion, and *a* is the number of alkyl groups attached to the quaternary nitrogen atom. There are, of course, many types of quaternized compounds that can be used in hair treatment, and their effects can be varied.

Quaternary ammonium compounds have been found to be efficient at depositing on hair as well as improving combability. Adsorption has been attributed to an electrostatic attraction between the anionic character of the hair fiber and the positively charged quat molecule (32–34). Hair has been described as being a "strong-acid ion exchanger," and one can expect some desorption of the quat upon rinsing with detergents. In addition to electrostatic forces, attraction owing to van der Waals forces has also been shown to play a part in the adsorption process (35). Increasing the number of alkyl groups, as well as increasing the length of the carbon chains on the quat, has been shown to be beneficial for the conditioning of hair (32,36). If the cationic charge character is responsible for bringing the quat to the hair surface, it is the long-chain alkyl groups that are believed to be responsible for lubricity and combing performance. Lunn and Evans (37) postulated that there are three principal factors which contribute to the "fly away" problem of hair or static charge effects. The first is the magnitude of charge which is generated by the contact and subsequent separation of hair and comb. The second factor is the mobility of charge and its rate of dissipation from the fibers. The third factor is the distribution of charge along the length of the combed fibers. In principle, the desired objective of reduced electrostatic effects can be approached by altering each of these factors. Either a reduction in the magnitude of charge generated or an increase in the mobility of that charge can be effective. Mutual repulsion of fibers can also be altered by changing the distribution of charge density along the length of the fiber.

The generation of static charges arises from an unequal transfer of charges across the interface between two bodies in contact when unlike objects are rubbed together. When the bodies are separated, they are each left with net charges of opposite sign and of magnitude equal to the differential charge transferred. Theoretical aspects of this process are discussed by Bick (38), Arthur (39), and Hersh and Montgomery (40). The charge generated by rubbing filaments together has been studied experimentally by Hersh and Montgomery (41). Henry et al. (42) measured both charge magnitude and the rate of its decay from rubbed textile fabrics. Barber and Posner (43) measured the charge generated by combing human hair. Mills et al. (44) also attempted to measure the charge generated by combing hair, but the method employed did not permit a distinction to be made between generation and dissipation mechanisms.

The distribution of charge along the length of a fiber, although noted by Ballou (45) as important, has received very little investigation. The only discussion of such phenomena is by Sprokel (46), who studied the variation of charge along a running textile yarn.

The charge generated on the hair by combing was found to be of positive sign for typical hair treatments and for all comb materials examined (37). This finding is consistent with two factors. First, keratin is at or near the positive end of the triboelectric series (47), and when it is rubbed against other materials which are lower than keratin in the series, a positive charge is developed on the keratin. [It is possible by certain treatments to alter the position of keratin in the triboelectric series (48).] Second, when two bodies are rubbed together under conditions where the bodies contribute unequal areas to the rubbing surface, the body which contributes the larger area tends to develop a positive charge (49). When hair is combed, it is the hair which contributes the larger area of contact.

When the experimental evidence was examined for these three charge properties of hair (37), it was found all of these need to be considered to fully understand the action of antistatic agents.

The charge generated by combing and the half-life of charge mobility both decrease with increasing relative humidity. The increased charge mobility is clearly a consequence of the greater water content at higher humidities, although the exact relationship is not well understood and other factors are also involved (50). Increased mobility of charges on the fiber leads to a decrease in the charge generated by combing, because of the charge conduction along the fibers as they are rubbed. This mechanism has been postulated to explain the decrease of generated charge with increasing relative humidity (40).

When a large concentration of a quaternary ammonium compound is present on the hair fiber, the surface conductivity is substantially increased. The negligible charge generated under such circumstances can be explained by this high conductivity and by the mechanism of charge conduction along the fibers. However, when the quaternary is rinsed off with water before drying, so that only small quantities remain on the fiber, the charge generated by combing remains relatively low, even though the charge half-life increases substantially and is comparable to that of untreated hair. The reduced charge generated cannot be explained by a mechanism of enhanced surface conductivity. An alternative mechanism must be sought.

It was hypothesized that the reduction of charge generated by combing, when hair is treated with quaternary ammonium compounds, is due primarily to the lubricating properties of these compounds on the dry hair, rather than to the enhanced conductivity. The quaternary acts as a lubricant and reduces tangling, so that the force required to pull a comb through the hair is substantially reduced, especially the end peak force. The reduced normal contact force between hair and comb leads to a reduced charge on the hair.

Medley (51) postulated that an antistatic agent need not be present as a continuous film in order to be effective. A discontinuous film would not give long-range conductivity and, therefore, the half-life of charge mobility would remain high. Medley proposed a mechanism requiring only localized conductivity at the contact site. This mechanism could be acting as a secondary effect. Another secondary effect could be a change in the chemical nature of the fiber surface, which would alter the magnitude of charge generated. Lunn and Evans (37) believed that the reduction of combing force by lubrication is the primary mechanism involved. We now know that quaternary ammonium antistatic agents do not normally achieve their effect by mechanisms of increased conductivity or of charge dissipation. Their primary effect is a lubricating action, which reduces substantially the force required to comb hair, especially the end peak force. The reduced normal contact force between hair fibers and comb leads to a reduction of static charge generated on the hair. The adsorption on hair of long-chain alkyl quaternary ammonium salts, cationic polymers, and complexes of cationic polymers with anionic polymers or anionic detergents can produce significant changes in the electrochemical surface potential of the fiber. This results in different charging characteristics in relation to polymers and metals. The effect of treatments such ad dyeing, bleaching, and permanent waving was also explored. Apart from altering the electrochemical potential, surface modification may also affect the conductivity of fibers (37,52). Modification of hair surface by reduction, bleaching, and oxidative dyeing results in very small changes of charging characteristics as compared to untreated fibers. They also have an insignificant effect on the fiber conductivities at low humidity.

During combing, such parameters as fiber elongation, stress, and magnitude of frictional forces between the comb and fiber undergo variations during the movement of a comb from the upper point of a tress toward the fiber tips. Consequently, nonuniform distribution of triboelectric charge density along the length of the fiber tress is usually observed (37). This might also effect the correlation between surface modification and triboelectric charging and lead to quantitatively different results.

G. Formulating with Quaternary Compounds

1. Hair Care

The chemistry of quaternary ammonium salts and allied nitrogen derivatives encompasses a broad range of structural variations. This versatility enables chemists to "custom tailor" formulas by selecting specific ingredients according to their performance characteristics.

When formulating cosmetic products with quaternary compounds, it is necessary to consider the effect of ingredient structure and molecular weight on parameters of solubility and compatibility (i.e., with anionic surfactants, salts, esters, bases, foam generation, and irritation). Factors such as ingredient substantivity (including buildup with prolonged use) and effect on moisture retention are also extremely important considerations for the formulation of both hair and skin care products.

One must also consider the effect functional ingredients have on hair tensile strength, combing performance, detangling performance, static (flyaway), and cuticle damage. Although individual cosmetic companies and testing laboratories may vary their procedures slightly, clinical and in vitro evaluation protocols for these parameters are well established and thoroughly described in the literature. Robbins offers a detailed treatment of this topic (53).

The number and chain length of fatty alkyl groups and the presence of hydroxyfunctional linkages have a direct bearing on the functional properties of quaternary salts. Short-chain (up to 16 carbon atoms) mono-alkyl quaternaries and ethoxylated quaternaries (i.e., cetrimonium chloride, PEG-2 cocomonium chloride, PEG-2 stearmonium chloride) are generally water soluble, whereas other mono-, di-, and tri-alkyl quaternaries (i.e., steartrimonium chloride, distearyldimonium chloride, tricetylmonium chloride) are not (54).

In one study, blond hair tresses were treated with test quaternary compounds, then rinsed, dried, and immersed in an anionic tracer dye. Tresses treated with dialkyl quaternary salts displayed greater color development, suggesting greater cationic substantivity, than tresses treated with monoalkyl or ethoxylated quaternaries. Studies involving solvent extraction of quaternary salts from treated hair tresses corroborated this finding. Detangling properties of quaternary ammonium salts also appear to be directly proportional to number and carbon chain length of the fatty moieties. See Figure 14 (55).

In a separate study, the ability of various conditioning agents to reduce static flyaway of hair was tested using a shadow silhouette hair tress method:

> For the measurement, every hair tress was combed with a coarse, sawed comb. With a punctual source of light (diameter 0.6 cm) the hair tress produced a silhouette on a screen in a distance of 15 cm. On this screen, concentric semi circles were marked [Figure 15]. The tip of the uncharged hair tress coincides with the centre of the semi-circle. In the shadow of the fixing point a ruler was attached to the screen so

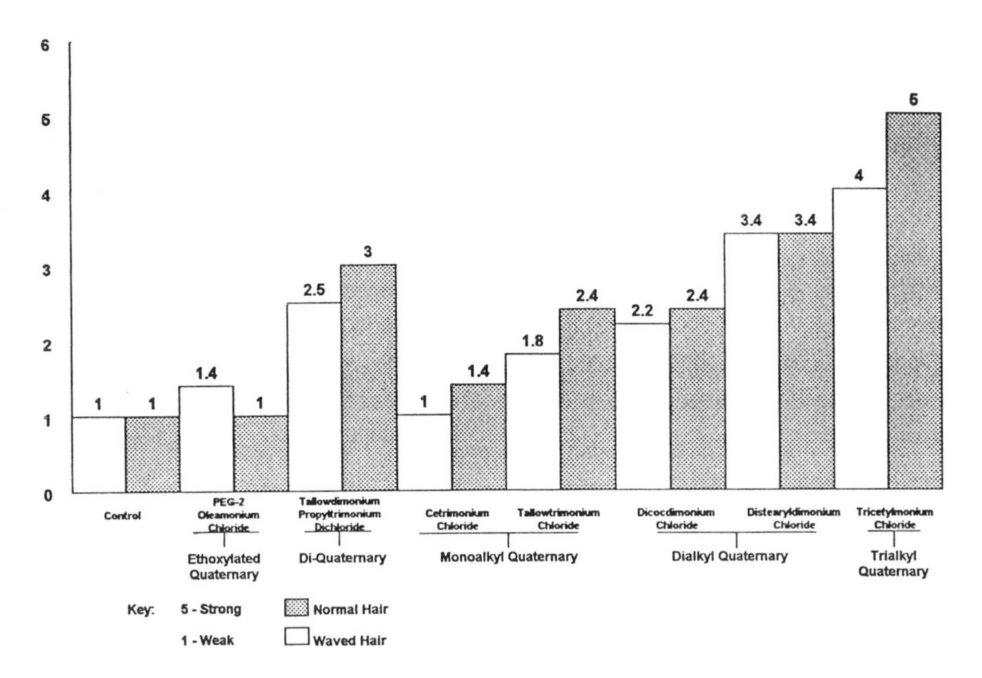

Figure 14 Detangling properties of quaternary salts. (From Ref. 32.)

that it could rotate. The ruler was turned to the silhouette so that the left and right border of the hair tress was marked. Then the radius of the semi-circle to which the ruler built a tangent, was read off. The sum of both radii results in the measured value, the stronger is the fly-away effect, caused by the electrostatic charge transferred to the hair by combing.

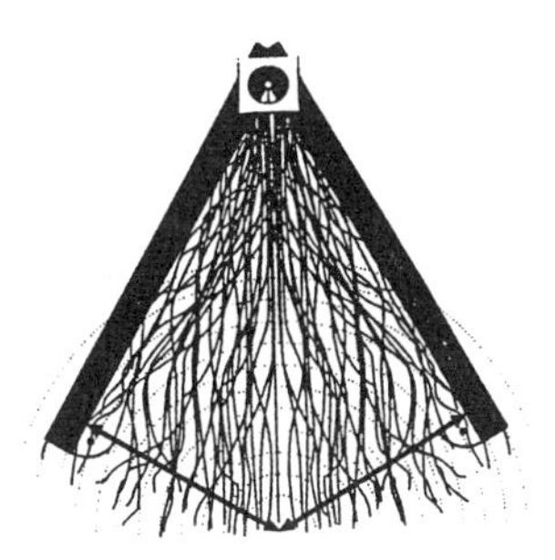

Figure 15 Flyaway effect measuring method. (From Ref. 56.)

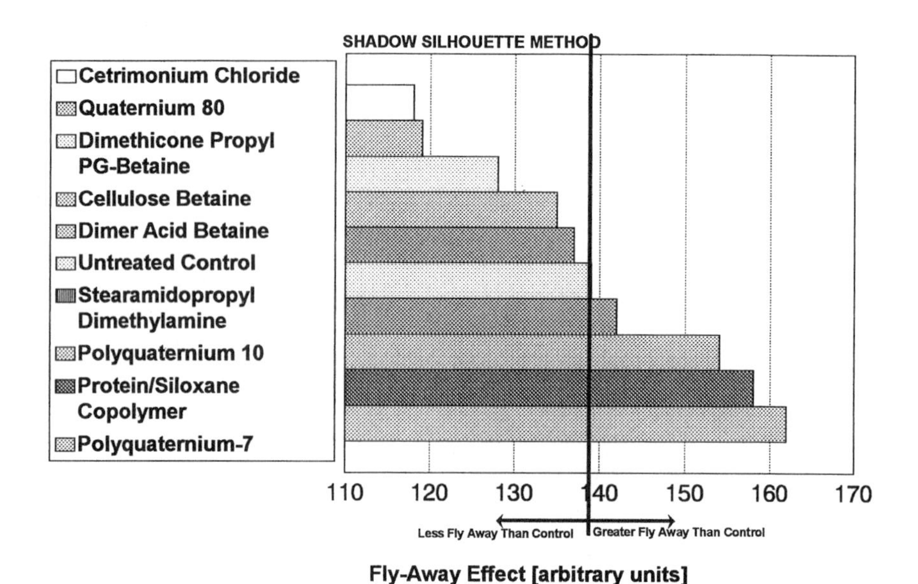

Figure 16 Flyaway effect by electrostatic charge. (From Ref. 56.)

Comparative hair static control data using selected quaternary, betaine, amidoamine salt, and polyquaternary species are shown in Figure 16. The shorter the bar, the less electrostatic charge is caused by combing (56).

2. *Skin Care*

Whenever creams and lotions are formulated, cosmetic chemists must pay particular attention to concerns pertaining to irritation and ingredient compatibility. Even though cationic emulsifiers can be balanced with anionic fatty acids to produce stable soap emulsions, cationic agents are normally precipitated by anionic emulsion stabilizers.

Cationic emulsifiers do offer certain advantages over anionic and nonionic components in the formulation of certain skin care products. Lanzet, in DeNavarre's book, states: "The pH of normal skin is 4.2–4.6. . . . Cationic emulsifiers are very compatible with the acid mantle and help to maintain it while anionic surfactant systems may promote keratin swelling and overtax or temporarily inactivate the buffering capacity of the skin, thus leading to irritation and lessened resistance to infection." Cationic and amphoteric surfactants are also compatible with quaternary germicides (57).

Many materials, such as polycationic resins, polymer complexes, cationic polysaccharides (including polycationic cellulose), cationic guar gum, etc., can be used by formulators to bridge problems associated with ingredient compatibility (58–68). Such anionic compatible cationics can now be found in many skin cleansers, which are themselves based on inexpensive, high-foaming, anionic surfactants. One example cites the use of a cationic guar gum to improve consumer skin feel of a bar soap (69). Related patents describe the use of other hydrated cationic polymers to improve mildness (67,70). In a completely different new personal wash application, cationic polymers are employed to enhance functional ingredient deposition on skin (71).

IV. COMMERCIAL APPLICATIONS

Quaternary ammonium salts and related surfactants continue to play a vital role in the development of new hair and skin care products. A search of U.S. patents between 1986 and 1996 yielded no fewer than 300 citations of cationic or quaternary hair- or skin care-related patents. These citations attest to the continuing evolution of quaternary ammonium salt technology. For instance, one example describes the preparation of quaternary ammonium substituted saponin esters derived from soy or ginseng feedstocks. These materials are claimed to offer excellent static control and improved conditioning benefits (72).

A. Hair Care

Long-chain mono- and dialkyl quaternary ammonium salts have been long recognized for their ability to improve hair texture and control static. Fatty alcohols are used together with such quaternaries in hair conditioners to improve smoothness and softness. One drawback to such systems is that they may display undesirable sticky or oily side effects. These problems can be mitigated, even on very dry hair, by using a branched quaternary ammonium salt in combination with a polyethylene glycol C_8–C_{10} alkyl ether (73).

Branched quaternary compounds have been demonstrated to offer other benefits to the cosmetic formulator. Stearyl octyl dimonium chloride (Figure 17) was found to offer comparable wet combing, dry combing, and static control compared to two unbranched quaternaries (dicetyldimonium chloride and stearlkonium chloride) in hair tress studies using bleached, waved hair. This ingredient was found to be water soluble and liquid at room temperature in concentrations suitable for use in clear cream rinses, conditioners, and pump sprays (74).

Similar benefits are cited in the use of isostearamidopropyl quaternary ammonium compounds. Isostearyl feedstocks are natural liquids at ambient con-

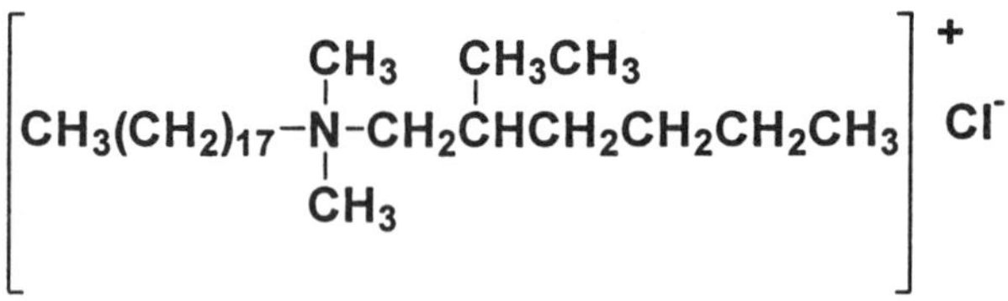

$$\left[CH_3(CH_2)_{17}-\overset{\displaystyle CH_3}{\underset{\displaystyle CH_3}{N}}-CH_2\overset{\displaystyle CH_3CH_3}{C}HCH_2CH_2CH_2CH_3 \right]^+ Cl^-$$

Figure 17 Stearyl octyldimonium chloride. (From Ref. 113.)

ditions, even though they are fully saturated. Ulike oleyl compounds, which are unsaturated, isostearyl derivatives are not subject to color degradation and rancidity concerns (75).

Alkylamidopropyl functions can be used to overcome disadvantages associated with long-chain fatty quaternaries. Typical problems associated with the use of quaternary compounds include anionic incompatibility, buildup on hair, negative efect on foam, and/or hazing in clear systems (76). Amidoamine salts can also be used to overcome these problems. Amidoamine salts also display a reduced tendency to build up on hair and can offer better "body" than their quaternary ammonium counterparts (77).

One of the more active areas of contemporary cosmetic technology is the use of silicones and silicone functional quaternary derivatives (78). Silicone polymers are used to improve lubricity, gloss, combing, and manageability in hair care products (79). They offer good surface tension modifications and remain fluid, even at high molecular weights. It is not surprising that a wide range of modifications can be made from them (18).

A recent search of the literature uncovered examples of silicone amphoterics (80,81), silicone betaines (82), silicone phosphobetaines (83,84), silicone alkyl quaternaries (85), silicone amido quaternaries (86), silicone acetate quaternaries (87,88), silicone imidazoline quaternaries (89), and silicone carboxy quaternaries (90).

Compared to organic quaternary salts, silicone quaternary derivatives are distinguished by the fact that they impart a "silklike feeling" to hair (91). The synthesis and preparation of polyquaternary polysiloxanes are described by Schaefer (92). One commercial example is shown in Figure 18. Janchitraponvej employs a silicone quaternary, together with an amidoamine salt, in the formulation of a clear hair conditioner (93,94).

Another interesting application for this type of polyquaternary siloxane is cited by Schueller et al. This patent describes the use of a diquaternary polydimethylsiloxane as the primary dispersant of water-insoluble silicone oils in aqueous hair care preparations, such as shampoos and conditioners. Silicone

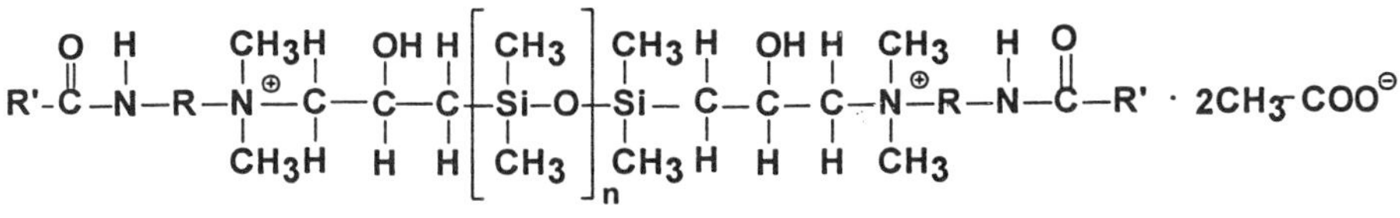

$$R'-C(=O)-NH-R-N^{\oplus}(CH_3)_2-CH_2-CH(OH)-CH_2-[Si(CH_3)_2-O]_n-Si(CH_3)_2-CH_2-CH(OH)-CH_2-N^{\oplus}(CH_3)_2-R-NH-C(=O)-R' \cdot 2CH_3COO^{\ominus}$$

R' = Cocos

Figure 18 Quaternium 80. (From Ref. 19.)

oils can "provide the hair with a silky, lubricious feel." While this silicone quaternary imparts substantivity and conditioning, its function as a silicone oil dispersing agent is unique and not typical of other commercially available quaternary hair conditioning agents (95).

O'Lenick describes the preparation of a quaternary containing both silicone and protein functional groups. The high-molecular-weight siloxane linkage contributes to oxidative stability and mildness, whereas the protein moiety contributes to film formation on hair and skin. Since such compounds are nonvolatile and display inverse cloud point characteristics, they are claimed to be ideally suited for use in personal care products (96).

Further discussion of the use of silicones in personal care products can be found in another chapter of this volume.

B. Skin Care

Sun care and tanning continue to be among the fastest-growing segments of the personal care industry. Several patents suggest the potential for use of quaternary salts of *para*-dialkyl amino benzamide as a new generation of non-irritating sunscreen actives (97,98). Another reference explores the possibility of binding cationic moieties to UV absorbers in order to enhance substantivity to skin (99).

In yet another application, quaternary ammonium salts are incorporated together with indole compounds to promote sunless tanning. "It is believed that melanin is an oligomer containing several indole segments formed by cyclization and polymerization of dihydroxy phenylalanine (DOPA) caused by exposure to the sun." Schultz et al. note, therefore, that indole compounds are being investigated for their potential applications in the self-tanning industry. The main problem associated with indoles as tanning enhancers is that they facilitate a color change slowly. It has been discovered that the formation of melanin to color the skin can be catalyzed by the use of quaternary ammonium salts. Indole/quaternary salt combinations are said to impart a deep, long-lasting tan which forms on the skin within 2 hr of application and exposure to sunlight (100).

Cationic emulsifiers do play an important role and can offer specialized benefits for cosmetic and pharmaceutical formulators. For instance, Papadakis describes the preparation of systems based on quaternized phosphate esters, such as linoleamidopropyl PG-dimonium chloride phosphate. These formulas display exceptional stability at acidic pH conditions and may be used in the preparation of cosmetics containing alpha-hydroxy acids (101). Zeigler notes that it is possible to prepare exceptionally mild, freeze–thaw-stable systems using combinations of quaternary ammonium functionalized phosphate esters together with cationic polysaccharides (102).

A European publication cites use of alkylamines and derivatives such as cholesteryl betainate in the formation of cationic oil-in-water emulsions for drug delivery systems. Cationic emulsions are said to be less likely to coalesce in the presence of sodium and calcium ions which are present in physiological fluids. Furthermore, these systems demonstrate enhanced affinity to biological membranes and can be used to facilitate the delivery of cosmetic (i.e., antioxidants, anti-free radicals, sunscreens) and pharmaceutical (i.e., steroid antiinflammatory drugs, beta-blocking agents for glaucoma treatment, antibiotics) agents (103).

Multiple water-in-oil-in-water (W/O/W) emulsions (Figure 19) with a ternary phase structure can be used to encapsulate active ingredients for controlled release in skin care and pharmaceutical preparations. Amphoteric and zwitterionic surfactants, such as lauramidopropyl betaine and dimethi-

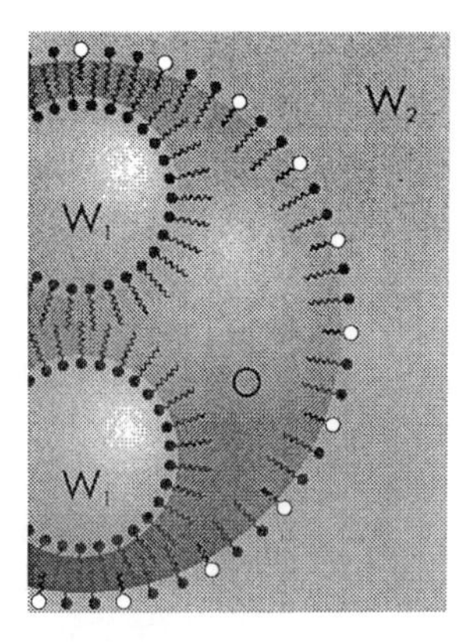

Figure 19 Structure of the interfaces in ternary emulsions. (From Ref. 104.)

cone propyl PG betaine, have been demonstrated to be effective hydrophilic stabilizers in such systems (104).

Wilmsmann's patent outlines the preparation of cationic oil-in-water emulsions based on a fatty acid salt, hydroxyethyl-ethylenediamine. This system is said to be highly substantive to skin, thereby forming a protective barrier against organic and alkaline irritants (105).

V. FUTURE DEVELOPMENTS AND CONCLUSIONS

While it is problematic at best to predict the course of future developments in any scientific field, an inspection of current patents and publications offers a glimpse into emerging technologies and market applications. Most recent work in this field seeks to understand the physiological effect of cationic surfactants in the context of pharmaceutical research. The following studies suggest that in the future we may see greater use of cationics in consumer oral care and medicated skin treatment preparations.

Cationic surfactants and nitrogen derivatives may play a role in facilitating healing. Examples include the use of amphiphatic peptides to stimulate fibroblast and keratinocyte growth (106) and use of "a skin protein complexing composition for the potentiation of the substantivity of aluminum acetate through the use of a cationic emulsifier" (107). Quaternized proteins in combination with anionic polymers have also been shown to be useful in treating keratinic substances (108). Ledzy describes the use of a cationic quaternary ammonium salt to increase the penetration of a pharmaceutical composition into skin (109).

Therapeutic studies involving the use of nitrogen-derived surfactants in oral care products are related to skin and pharmaceutical research. Polefka notes the importance of including a zwitterionic surfactant (in this case lauramidopropyl betaine) as a stabilizing agent in a mouth rinse containing a cationic antimicrobial aid (bis-quanide) together with anionic anticalculus agents (110).

Muhlemann (111) states that cationic diamine salts can be used to stabilize the hydrolysis of tin salts in aqueous and oral care compositions. The cationic additive preserves, and may even potentiate, the inhibiting action of the tin salts on plaque formation and dental caries.

An extensive volume of recent data also continues to document benefits for cationic surfactants in the field of sanitation. While this topic is beyond the scope of this chapter, it will nevertheless have a significant impact on the future marketability and consumption of cationics.

In conclusion, any survey of cationic surfactants, quaternary ammonium salts, and related nitrogen derivatives can include only a small sample of developments in the field. The authors hope that the reader will use the references cited to explore this topic in further detail.

REFERENCES

1. G. Hawley, ed. The Condensed Chemical Dictionary. 10th ed., New York: Van Nostrand Reinhold, 1981:877.
2. McCutcheon's Emulsifiers and Detergents. Vol. 1. Glen Rock, NJ: Manufacturing Confectioner Publlishing Co., 1996.
3. McCutcheon's Emulsifiers and Detergents. Vol. 2. Glen Rock, NJ: Manufacturing Confectioner Publishing Co., 1996.
4. Sherex fatty amines and diamines. Technical bulletin. Greenwich, CT: Witco Corp., undated.
5. Bluestein BR, Hilton CL. Amphoteric Surfactants. Vol. 12. New York and Basel: Marcel Dekker, 1982.
6. Moore CD. J Soc Cosmet Chem 1960; 11:13.
7. Ploog U. Seifen-Öle-Fette-Wachese 1982; 108:373.
8. J.M. Richmond, ed. Cationic Surfactants. Vol. 34. New York: Marcel Dekker, 1990:175.
9. Ibid., p. 51.
10. Jungerman E, ed. Cationic Surfactants. Vol. 4. New York: Marcel Dekker, 1970: 493.
11. Ibid., p. 528.
12. DeNavarre MG. The Chemistry and Manufacture of Cosmetics. Vol. 4. Carol Stream, IL: Allured Publishing Co., 1975:1102.
13. Gerstein T. Cosmet Perfum 1975 (Mar.); 90:38.
14. Jungermann E, ed. Cationic Surfactants. Vol. 4. New York: Marcel Dekker, 1970: 29.
15. Procter & Gamble. U.S. Patent 4,704,272.
16. Procter & Gamble. U.S. Patent 4,788,006.
17. Branna T. HAPPI 1995 (Dec.); 32:80.
18. Floyd D. Cosmetic and Pharmaceutical Applications of Polymers: Organo-Modified Silicone Copolymers for Cosmetic Use. New York: Plenum Press, 1991:49.
19. Floyd D, Jenni K. Polymeric Materials Encyclopedia. Vol. 10. New York: 1996: 7677.
20. O'Lenick A. Cosmet Toilet 1994; 109:85.
21. O'Lenick A. Cosmet Toilet 1996; 111:67.
22. Schoenberg T. Cosmet Toilet 1996; 111:99.
23. Gerstein T. Cosmet Toilet 1979 (Nov.); 94:35.
24. Balsum MS, Sagarin A, eds. Cosmetics, Science and Technology. Vol. 2. New York: Wiley Interscience,, 1972:348.
25. Stern ES, Johnsen VL. Cosmet Toilet 1983 (May); 98:76.
26. Technical Bulletin. Edison, NJ: Americhol Corporation, 1993.
27. Dunn CA. HAPPI 1996 (May); 33:126.
28. Nanavati. J Soc Cosmet Chem 1994; 45:135.

29. Robbins CR, Reich C, Clarke J. Hair manageability. J Soc Cosmet Chem 1986; 37:489–499.
30. Garcia ML, Diaz J. Combability measures on human hair. J Soc Cosmet Chem 1976; 27:379–398.
31. Jachowicz J, Berthiaume MD. Heterocoagulation of silicon emulsions on keratin fibers. J Colloid Interface Sci 1989; 133:118–134.
32. Jurczyk M. Cosmet Toilet 1991 (Nov.); 106:63.
33. Weatherburn AS, Bayley CH. The sorption of synthetic surface-active compounds by textile fibers. Textile Res J 1952; 22:797–804.
34. Loetzsch KR, Reng AK, Gantz D, Quack JM. The radiometric technique, explained by the example of adsorption and desorption of ^{14}C-labeled distearyl-dimethylammonium chloride on human hair. In Orfanos CE, Montagna W, Stuttgen G, eds. Hair Research: Proceedings of the International Congress, 1979. Berlin: Springer-Verlag, 1981:638–649.
35. Robbins CR. In Chemical and Physical Behaviour of Human Hair. New York: Van Nostrand Reinhold, 1979: chap 5.
36. Spiess E. The influence of chemical structure on performance in hair care preparations. Parfuem Kosmet 1991; 72:370–376.
37. Lunn O, Evans D. J Soc Cosmet Chem 1977; 28:549.
38. Vick FA. Theory of contact electrification. Br J Appl Phys 1953; suppl 2:S1–S5.
39. Arthur DF. A review of static electrification. J Textile Inst 1955; 46:T721–T734.
40. Hersh SP, Montgomery DJ. Static electrification of filaments: theoretical aspects. Textile Res J 1956; 26:903–913.
41. Hersh SP, Montgomery DJ. Static electrification of filaments: experimental techniques aned results. Textile Res J 1955; 25:279–295.
42. Henry PSH, Livesey RG, Wood AM. A test for lability to electrostatic charging. J Textile Inst 1967; 58:55–77.
43. Barber RG, Posner AM. A method for studying the static electricity produced on hair by combing. J Soc Cosmet Chem 1959; 10:236–246.
44. Mills CM, Ester VC, Henkin H. Measurement of static charge on hair. J Soc Cosmet Chem 1956; 7:466–475.
45. Ballou JW. Static electricity in textiles. Textile Res J 1954; 24:146–5.
46. Sprokel GJ. Electrostatic properties of finished cellulose acetate yarn. Textile Res J 1957; 27:501–515.
47. Shashoua VE. Static electricity in polymers: I. Theory and measurement. J Polymer Sci 1958; 33:65–85.
48. Landwehr RC. Electrostatic properties of corona-treated wool and mohair. Textile Res J 1969; 39:792–793.
49. Gruner H. An investigation of the mechanism of electrostatic charging of textile fibers. Faserforsch Textiltech 1953; 4:249–260.
50. Shashoua VE. Static electricity in polymers: II. Chemical structure and antistatic behavior. J Polymer Sci 1963; A1:169–187.
51. Medley JA. The discharge of electrified textiles. J Textile Inst 1954; 45:123–141.
52. Jachowicz J. J Soc Cosmet Chem 1985; 36:189.

53. Robbins CR. Chemical and Physical Behavior of Human Hair. 2d ed. New York: Springer-Verlag, 1988.
54. Jurczyk MF, Berger DR, Damaso AR. Cosmet Toilet 1991 (Apr.); 106:63.
55. Ibid., p. 67.
56. Leidreiter HI, Jenni K, Jorbandt C. SÖFW J 1994; 120:856.
57. DeNavarre MG, ed. The Chemistry and Manufacturing of Cosmetics, Vol. 3. Carol Stream, IL: Allured Publishing Corp., 1975:385.
58. Helene Curtis. U.S. Patent 5,417,965.
59. Calgon. U.S. Patent 5,338,541.
60. Conopco. U.S. Patent 5,338,540.
61. Helene Curtis. U.S. Patent 5,221,530.
62. Clairol. U.S. Patent 4,911,731.
63. Henkel. U.S. Patent 4,900,544.
64. Chesebrough Ponds. U.S. Patent 5,169,624.
65. L'Oreal. U.S. Patent 5,008,105.
66. ICI Americas, Inc. U.S. Patent 5,112,886.
67. Procter & Gamble. U.S. Patent 5,064,555.
68. Hi-Tek Polymers, Inc. U.S. Patent 5,009,969.
69. Procter & Gamble. U.S. Patent 4,946,618.
70. Procter & Gamble. U.S. Patent 5,296,159.
71. Cheseborough Ponds. U.S. Patent 5,543,074.
72. Pacific Chemical. U.S. Patent 5,182,373.
73. KAO. E.P. Patent 046887A1.
74. Jurczyk MF. Cosmet Toilet 1991 (Nov.); 106:91.
75. Gesslein BW. J Soc Cosmet Chem 1990 (May); 66:37.
76. Smith L, Gesslein BW. Alkylamidopropyl dihydroxypropyl dimonium chlorides. Technical bulletin. Inolex Chemical Corp.
77. Schoenberg T, Scafidi A. Cosmet Toilet 1979 (Mar.); 94:57.
78. Floyd D. Silicone Surfactants: Applications in the Personal Care Industry. New York: Marcel Dekker. In press.
79. Floyd D, Leidreiter H, Sarnecki B, Maczkiewitz U. Preprints 17th IFSCC Congress, Yokahama, 1992:297.
80. Siltech. U.S. Patent 5,073,619.
81. Goldschmidt. U.S. Patent 4,609,750.
82. Goldschmidt. U.S. Patent 4,654,161.
83. Siltech. U.S. Patent 5,091,493.
84. Siltech. U.S. Patent 5,237,035.
85. Siltech. U.S. Patent 5,098,979.
86. Siltech. U.S. Patent 5,153,294.
87. Goldschmidt. U.S. Patent 4,833,225.
88. Goldschmidt. U.S. Patent 5,891,166.
89. Siltech. U.S. Patent 5,196,499.
90. Siltech. U.S. Patent 5,296,625.
91. Meyer H. CTMS 1990 (Jan.); 15:5.
92. Goldschmidt. U.S. Patent 4,891,166.
93. Helene Curtis. U.S. Patent 5,556,615.

94. Helene Curtis. U.S. Patent 5,328,685.
95. Alberto Culver. U.S. Patent 5,306,434.
96. Siltech. U.S. Patent 5,243,028.
97. ISP Van Dyke. U.S. Patent 5,451,394.
98. ISP Van Dyke. U.S. Patent 5,427,774.
99. Pacific Chemical. U.S. Patent 4,987,183.
100. Clairol. U.S. Patent 5,049,381.
101. Helene Curtis. U.S. Patent 5,567,427.
102. Chesebrough-Ponds. U.S. Patent 5,135,748.
103. AM RCS. Dev. Co. W.O. Patent 93/8852 4155.
104. Hameyer P, Jenni K. Perf Kosmetic 1994; 75:844.
105. Donald Basiliere. U.S. Patent 5,229,105.
106. Demeter Biotechnologies. U.S. Patent 5,561,107.
107. C. Fox et al. U.S. Patent 4,879,116.
108. L'Oreal. U.S. Patent 4,796,646.
109. J. Ledzy et al. U.S. Patent 5,346,886.
110. Colgate. U.S. Patent 5,180,577.
111. GABA International. U.S. Patent 4,828,822.
112. Miranol technical and product development data. Technical bulletin, Rhone Poulenc, 1985.
113. Wenninger JA, McEwen AN, eds. International Cosmetic Ingredient Dictionary, 6th ed., Vol. 1 & 2, 1995.

11
Polymers as Conditioning Agents for Hair and Skin

Bernard Idson
University of Texas at Austin, Austin, Texas

I. INTRODUCTION

One class of conditioning agents is cationic (positively charged) quaternary ammonium salts, in which the cationic portion contains the surface active portion. Improvement of conditioning effects on hair and skin has been achieved by virtual replacement of monomeric quaternary ammonium compounds by cationic polymers. Polymers find their chief use as substantive conditioning agents in hair. However, a number of cationic resins find value in skin care because of their emollient smooth feel on the skin (1). Noncationic polymers are also finding a place as conditioning and substantive agents for skin and hair.

Included in this review are historical development, the physicochemical and structural bases for cationic polymer conditioning and substantivity, and discussion of the multiplicity of cationic polymers in primarily hair and secondarily in skin conditioning, including quaternized synthetics, cellulose derivatives, quaternized guar, lanolin, animal and vegetable proteins, substantive conditioning humectants, and aminosilicones.

A. Polymers (General)

The terms polymer, high polymer, and macromolecule are used to designate high-molecular-weight materials. Polymers can be natural, synthetic, or biosynthetic. A biosynthetic polymer is a natural polymer which has been modified with one or more synthetic functional groups.

Two major types of polymerization methods are used to convert small molecules (monomers) into polymers. These methods were originally referred to as addition and condensation polymerization. Addition polymerization is now called chain, chain-growth, or chain-reaction polymerization. Condensation polymerization is now referred to as step-growth or step-reaction polymerization.

The monomers normally employed in addition reactions contain a carbon–carbon double bond that can participate in a chain reaction. The mechanism of the polymerization consists of three distinct steps. In the initiation step, an initiator molecule(s) is thermally decomposed or allowed to undergo a chemical reaction to generate an "active species." This "active species," which can be a free radical, a cation, an anion, or a coordination complex, then initiates the polymerization by adding to the monomer's carbon–carbon double bond. The reaction occurs in such a manner that a new free radical cation, anion, or complex is generated. The initial monomer becomes the first repeat unit in the incipient polymer chain. In the propagation step, the newly generated "active species" adds to another monomer in the same manner as in the initiation step. This procedure is repeated over and over again until the final step of the process, termination, occurs. In this step, the growing chain terminates through reaction with another growing chain, by reaction with another species in the polymerization mixture, or by the spontaneous decomposition of the active site.

Two or more different monomers are often employed in a chain-reaction polymerization to yield a polymer containing the corresponding repeat units. Such a process is referred to as copolymerization, and the resulting product is called a copolymer. By varying the copolymerization technique and the amounts of each monomer, one can use as few as two monomers to prepare a series of copolymers with considerably different properties. The amount of different materials that can be prepared increases dramatically as the number of monomers employed increases. Thus, it is not too surprising that the majority of synthetic polymers used today are copolymers.

Condensation polymerization normally employs two difunctional monomers that are capable of undergoing typical organic reactions. For example, a diacid can be allowed to react with a diol in the presence of an acid catalyst to afford a polyester. In this case, chain growth is initiated by the reaction of one of the diacid's carboxyl groups with one of the diol's hydroxyl groups. The free carboxyl or hydroxyl group of the resulting dimer can then react with an appropriate functional group in another monomer or dimer. This process is repeated throughout the polymerization mixture until all of the monomers are converted to low-molecular-weight species, such as dimers, trimers, tetramers, etc. These molecules, which are called oligomers, can then react further with each other through their free functional groups. Polymer chains that have moderate molecular weights can be built in this manner. The high molecular weights common to chain-reaction polymerizations are usually not reached. This is due to the fact that as the molecular weight increases, the concentration of the flee functional groups decreases dramatically. In addition, the groups are attached to the ends of chains and, hence, are no longer capable of moving freely through the viscious reaction medium.

What makes polymers so "different"? A polymer's unusual physical behavior is due to the tremendous amount of interactions between its chains. These interactions consist of various types of intermolecular bonds and physical entanglements. The magnitude of these interactions is dependent on the nature of the intermolecular bonding forces, the molecular weight, the manner in which the chains are packed together, and the flexibility of the polymer chain. Thus, the amount of interaction is different in different polymers and quite often different in different samples of the same polymer.

B. Cationic Polymers for Conditioning

A great improvement in recent years has been the introduction into shampoos of cationic compounds, the hydrophilic functional groups of which are no longer at the end of the fatty chain but inserted within a polymeric structure, whereby potential for irritation is far reduced as compared with conventional products. Such resins, formerly used in setting lotions for their hair-holding properties, can be combined with amphoteric and nonionic surfactants, and some show remarkable compatibility with anionics, which is the most required property both for good foaming and cleaning power and for economic reasons.

Cationic polymers are a type of substantive raw material commonly used in hair-conditioning formulations. They are made by attaching quaternized fatty alkyl groups to modified natural or synthetic polymers. While structurally similar to quats, they have many more cationic sites per molecule and much higher molecular weights.

Once deposited, cationic polymers provide hair with slip, manageability, and good combability. They increase body in damaged hair, spread well and evenly, and can improve split ends. Their relatively high activity, which allows low use levels, and compatibility with anionic surfactants when properly formulated, make them ideal conditioning agents. A potential drawback of using cationic polymers is their tendency to build up with repeated usage. This can weigh the hair down and give it an unappealing look and feel.

Cationic polymers can be made from a variety of synthetic as well as natural polymers, such as guar gum and cellulosics. A polymer's physical characteristics vary according to the monomer and monomeric ratios used in its manufacture. Molecular weight is another factor affecting the polymer's physical characteristics and performance properties.

Proteins, polypeptides derived from various plant and animal sources, are common conditioning agents. Because of their similarity to the proteinaceous structure of hair and skin, they are naturally adsorbed. Once deposited, proteins are said to improve surfaces by attaching to the damaged sites. Also, their substantivity can be enhanced through reaction with quaternized materials.

II. HISTORICAL DEVELOPMENT

Early hair preparations contained shellac fixatives. The disadvantage, primarily sticky, hard-feeling hair, provoked formulation of shampoos with softer feels (conditioning) containing cationic surface-active agents. These surfactants are classified according to the charge on the hydrophobic (water-repellant) portion. Anionic surfactants possess a negative charge; $RCOO^-$ (R is usually a fatty acid from carbons 12–18). Cationic surfactants bear a positive charge. They are principally quaternary ammonium compounds, $R_4N^+X^-$ (monomeric) or $(R_4N+)_nX^-$ (polymeric). By virtue of their positive surface charge, they have a great affinity to negatively charged keratin in hair and skin, the basis of their conditioning and substantive properties. Most cationic polymers do not build up on the hair and are not as irritating. They do not interfere with foaming or cleaning as do fatty chain cationics (2). Cationic polymers are soluble in water with varying degrees of substantivity and are compatible with anionic, cationic, and amphoteric surfactants and electrolytes (see Section III).

Cationic polymers, proteins, and amino acids are very efficacious conditioning agents because of their substantivity to hair. Some studies suggest that amino acids will actually penetrate into the hair and increase the moisture content as well. Proteins and cationic polymers typically remain on the fiber surface, reducing combing forces and flyaway, and in some systems they provide enhancement of volume and body and an improvement in manageability. These ingredients may be found in shampoo formulations in concentrations ranging from 0.5% to 10.0%. Silicones may be incorporated into shampoos in this same concentration range to reduce eye irritation, combing forces, static charge, and drying times while increasing foam stability, body, shine, and manageability, depending on the particular silicone materials chosen.

In addition to hair conditioners, most of the cationic polymers impart smooth feel, emolliency, and substantivity to skin. Noncationic polymers often serve the same purpose as the charged macromolecule, but in lesser strength.

The *CTFA Cosmetic Ingredient Handbook Directory* defines monomeric cationic surfactants as "quaternium" compounds and polymeric quaternary compounds as "polyquaternium" compounds (3). Detailed descriptions of varied available cationic polymers follow in Section IV.

Shampoos have been formulated with certain cationic polymers such as polyquaternium 10 (Section IV.K), polyquaternium 7 (Section IV.C), and guar hydroxypropyltrimonium chloride (Section IV.M), which were compatible with anionic surfactants in the shampoo formula but became incompatible upon dilution with water and, as a consequence, were deposited on the hair during rinsing. The deposit conferred wet combing and styling benefits, but in many cases these polymers formed extremely stable complexes with hair keratin, and with frequent shampooing gave rise to a "buildup" that was difficult to remove (4). The buildup may arise from the fact that the polymer–surfactant complex has a lower stability constant than the polymer–keratin complex formed at the hair surface and, therefore, the anionic surfactant cannot compete with the hair surface and, consequently, cannot remove the polymer deposit.

Conditioning shampoos exemplified as "two-in-one" or "three-in-one" products came into vogue. What makes these multifunctional shampoos attractive to consumers is that they can get cleaning, conditioning (and/or dandruff control) from a single product in one step.

There is now a generation of conditioning shampoos in which the deposition of very hydrophobic materials such as silicones and long-chain alkyl compounds are being highlighted. Some products still contain cationic polymers such as guar hydroxypropyltrimonium chloride, and some include long-chain alkyl quaternary amines (4).

III. CONDITIONING AND SUBSTANTIVITY FROM CATIONIC POLYMERS

Cationic polymers are used chiefly in conditioning shampoos for their good wet combing properties. By choosing the proper polymer/surfactant combination, the complex of cationic polymers and anionics is deposited onto the hair surface during dilution (water rinsing) to provide lubrication and mending of damage (5).

The conditioning effect is based on the deposition onto the hair surface or into the hair fiber of certain functional components that have resistance to subsequent water rinsing. Since the isoelectric point of hair is approximately 3.67, its surface bears a net negative charge near the neutral pH where most shampoos are formulated. Thus, anionic surfactants bearing a negative charge are not very substantive to hair and leave the hair in an unmanageable condition (6,7).

The improvements in conditioning agents on hair and skin have resulted in major replacement of monomeric quaternary ammonium compounds ("quats") by cationic polymers, the hydrophilic functional groups of which are no longer at the end of a fatty chain but inserted within a polymeric structure, whereby the potential for irritation is reduced. While structurally similar to quats, they have many more cationic sites per molecule and much higher molecular weights. About one-third of the quaternary ammonium compounds used today for hair and skin care are polymeric in nature.

Cationic polymers are characterized by possession of a multiplicity of polar and/or ionic groups which confer water solubility. For adsorption onto keratin to occur there needs to be some attraction between the groups and the keratin surface. The actual hair conditioner is not the cationic polymer itself but rather a complex of the polymer and the anionic surfactant.

While the guidelines for adsorption of surfactants are similar to those of polymers, there are differences due mostly to the smaller relative size of the surfactant molecules. If there are both a cationic polymer and a cationic surfactant in the solution, there is a competitive adsorption. Due to its relative smallness and high mobility, the surfactant can reduce the uptake of the polymer.

Cationic polymers can be made from a variety of synthetic as well as natural polymers, such as guar gum and cellulosics. A polymer's physical characteristics vary according to the monomer and monomeric ratios used in its manufacture. Molecular weight is another factor affecting the polymer's physical characteristics and performance properties.

While all chapters of this volume deal with varied aspects of conditioning, it is worth redefining conditioning in the context of cationic polymers. Schueller and Romanowski (8) note that conditioning, such as with cationic polymers, helps hair and skin look and feel better by improving the condition of these surfaces. Hair conditioners are intended primarily to make wet hair easier to detangle and comb and to make dry hair smoother, shinier, and more manageable. Skin conditioners primarily moisturize while providing protection from the drying effects of the sun, wind, and harsh detergents.

An ingredient's ability to improve hair or skin condition depends on it being deposited onto surfaces and preferably remaining intact, even after rinsing. This resistance to rinse off is known as "substantivity" and can be achieved through the use of certain raw materials that, because of water insolubility or electrostatic attraction, stay on the hair and skin.

Cationic polymers must be substantive to condition effectively. As with "quats," a polymer's cationic nature allows substantivity via coulombic attraction to anionic surfaces. However, cationic polymers are also used in anionic surfacant-containing formulas, such as shampoos, where they are solubilized and expected to wash away during use. However, this does not happen because of a unique solubility mechanism. Anionic surfactant systems containing cationic polymers can be designed so that polymers are soluble in the product but become insoluble during rinsing and deposit on the hair. This occurs because of an association between the cationic polymer and the anionic surfactant in cosmetic formulations such as shampoos. With excess surfactant, the polymer is solubilized, creating a clear solution. However, during rinsing, the surfactant concentration falls below the critical level required for solubilization, and the polymer/surface complex deposits on the hair (8).

Once deposited, cationic polymers provide hair with slip, manageability, and good combability. They increase body in damaged hair, spread well and evenly, and can improve split ends. By choosing the proper polymer/surfactant combination, the complex of cationic polymers and anionics is deposited onto the hair surface during dilution (water rinsing) to provide lubrication and mending of damage (5).

It is known that replacing part of the anionic detergent(s) by amphoteric surfactants dramatically increases the substantivity of cationic cellulose derivatives to hair (9). However, too much substantivity can be undesirable because unwanted excessive buildup may occur after repetitive shampooing, leaving hair feeling stiff and heavy. Polyamine derivatives (pseudo-cationic secondary amines) are also used as conditioning agents. It is claimed that their weak cationic character does not allow for buildup on the hair shaft, even with repeated applications (10). Increasing use of silicones has helped avoid buildup. The adsorption of some cationic polymers on hair has been demonstrated to be reduced by addition of small amounts of electrolytes (such as sodium chloride), which also may partially desorb the adsorbed polymer (11).

The adsorption of quaternaries having long-chain fatty portions as part of the molecule is the basis for most conditioner formulas. The fatty portion, which is largely not attached to the substrate, acts as a lubricant. The lubricating action makes combing easier. It also allows for detangling of a snarled hair assembly. There is also a degree of hydrophobic bonding in most cases which is very common. This bonding is of the van der Waals type and has been shown to be very strong. "Polyquats" can have multiple-site bonding as well as hydrophobic bonding. This will bridge or bind along the hair shaft to form a coating which may be filmlike. This film-forming character of certain polyquat resins causes stiffening of the fiber. Stiffening can be a body-building property (12). The polymers act by adsorbing onto the keratin substrate (hair or skin). Thus, conditioning involves, at least in part, interfacial phenomena in which the surface properties of the keratin are altered (13).

The conditioning action of any agent may depend on its substantivity or adhesion to hair fibers or on its tendency to interfere with the degreasing action of pure or modified detergents. In the former case, the presence of the conditioner can be demonstrated with the aid of a labeled conditioner or with the aid of dyestuffs whose effect on hair is modified by the presence of the conditioning molecule (14,15).

Substantivity and film-forming properties make these polyquats excellent candidates for hair care products. Potential benefits include good set retention, high luster, improved body, and reduced conditioner buildup. Skin conditioning and fixative properties contribute to the elegance and permanence of aftershave or cologne formulas. The moisture barrier and fixative properties of these "polyquats" also lend themselves to formulation of antiperspirants or deodorants.

A. Skin Conditioning

Skin conditioning by cationic polymers has not had the degree of investigation as has had hair conditioning. However, a resin such as polyquaternium 19 finds value as a skin conditioner because of its emollient feel on the skin, as well as its substantivity. Substantivity to skin has been demonstrated by liquid scintillation counting. Faucher (11) used the same technique to show substantivity to calfskin for polyquaternium 10.

Quaterniums 4 and 6, in dry skin lotions, offer a combination of substantivity to the skin with a long-lasting smooth feel. Polyquaterniums 19 and 20 are claimed to provide polymeric moisture barriers or retard evaporation rates (16). Polymethylacrylaminopropyl trimonium chloride is a functional additive in a wide range of skin care products. Astringents, toners, colognes, and preshave products prepared with this cationic resin have a silky smooth afterfeel (17).

Further skin conditioning properties are discussed with individual cationic and noncationic polymers. The adsorption of a cationic polymer on skin shows the same general behavior as on hair: a sharp initial uptake followed by a slow approach to equilibrium. Like hair, the mechanism appears to involve slow penetration of the skin by the polymer, since the uptake by the polymer far exceeds that of a monolayer. It appears that skin keratin is more reactive than the harder, more highly crosslinked keratin found in hair.

IV. CATIONIC CONDITIONING POLYMERS

Cationic polymers are available in many forms initiated by PVP (poly-N-vinyl-2-pyrrolidone) and its copolymers (polyquaternium 11, 28, 16; see Section IV.B). One series of acrylates is (polyquaterniums 5, 6, 7, 18, 22, 28; Sections D–G, K). There are also quaternized polyvinyl polymers (polyquaterniums 19, 20; Sections G–H) and quaternized lonenes (polyquaterniums 2, 17, 18, 27; Section I). Cellulose cationic polymers include polyquaterniums 4, 10, 24 (Section K).

Polysaccharides such as guar hydroxypropyl trimonium chloride have gained in popularity (Section M). Natural polymers have been quaternized, such as lanolin and proteins (Section P). Aminosilicones are finding increasing use (Section R).

A. Polyvinylpyrrolidone (PVP)

The growth of cationic polymers as hair conditioners was initiated by the introduction of PVP (poly-N-vinyl-2-pyrrolidone) in 1950 by then-GAF [now International Specialty Products, ISP (18)] and then by BASF (19). A homopolymer is a molecule with repeating units of the same molecular structure.

Originally developed for incorporating into hairsprays, PVP has been found to have conditioning properties if applied via a shampoo. The transparent film on hair imparts smoothness and luster, as well as substantivity, to skin and hair. The films have good holding power and do not flake under moderate climate conditions. PVP has other properties which are used in shampoos and hair conditioners—it improves foam stabilization, increases viscosity and, in certain formulations, can reduce eye irritation (8). Today, PVP is occasionally used as a conditioning agent in shampoos. However, PVP and its copolymers are used primarily as styling resins.

As hair styles continued to progress, some of the deficiencies of PVP homopolymer began to be noticed. PVP is hygroscopic. It tends to become sticky, dull, and tacky as atmospheric moisture is adsorbed, and, of more significance, the adsorbed moisture plasticizes the film and causes ductile fracture of the bond. A further drawback of PVP homopolymer was its tendency to become brittle and flaky in dry weather (20).

B. Copolymers of PVP

Many of the objectionable properties of PVP homopolymer were overcome by the introduction of PVP random copolymers. A random copolymer tends to have properties which are intermediate between the properties of the homopolymers which would be formed by polymerizing the monomers separately. Thus, a polar hygroscopic homopolymer can be rendered more moisture resistant by introduction of a nonpolar comonomer (20).

1. PVP/Vinyl Acetate (VA) Copolymers

PVP/VA copolymer (Figure 1) was offered as an improved hairspray-conditioner (18). Four compositional ratios were offered: 70/30, 60/40, 50/50, and 30/70. The higher the ratio of VA, the less susceptible to atmospheric humidity and the harder the film, but consequently it is less easily shampooed off. The copolymers used have less moisture sensitivity than PVP, with good hold and good conditioning. The polymer will also coat the skin, leaving a conditioned, less oily feel (1).

2. PVP-α-olefin Copolymers

GANEX series (ISP 10) are linear copolymers of PVP and long-chain α-olefins. There are four varieties, differing in the olefin copolymer: P-904 (buty-

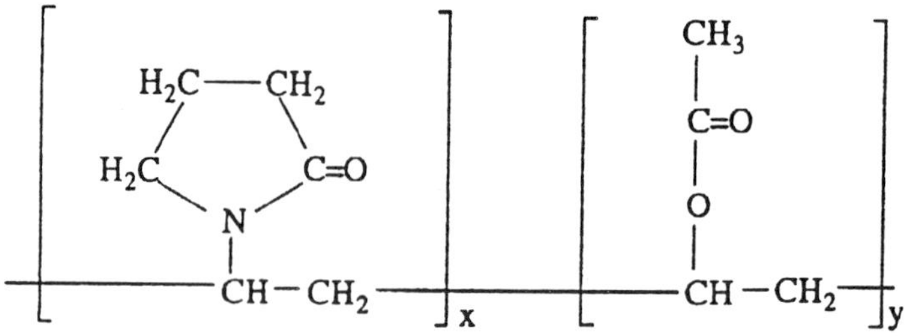

Figure 1 Vinylpyrrolidone/vinyl acetate copolymer.

lated PVP), V-216 (PVP/hexadecene), V-220 (PVP/eicosene), and WP-660 (tricontanyl PVP). They produce smooth substantive coatings on the hair. They also provide a unique "afterfeel" in skin care products (18).

3. PVP/Methacrylate Copolymers (18)

a. Copolymers 845, 937, 958 (ISP). Copolymers 845, 937, and 958 are PVP/dimethylaminoethyl methacrylate (DMAEM) copolymers with varied ratios of PVP/methacrylate (Figure 2). They find wide use in blow-dry conditioners as well as conditioning additives in cream rinses and shampoos. They give a smooth conditioning feel to the skin in creams, lotions, soft

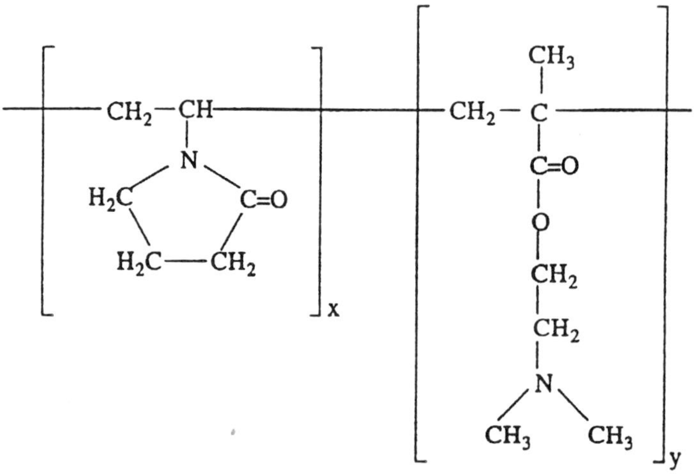

Figure 2 Vinylpyrrolidone/dimethylaminoethyl methacrylate.

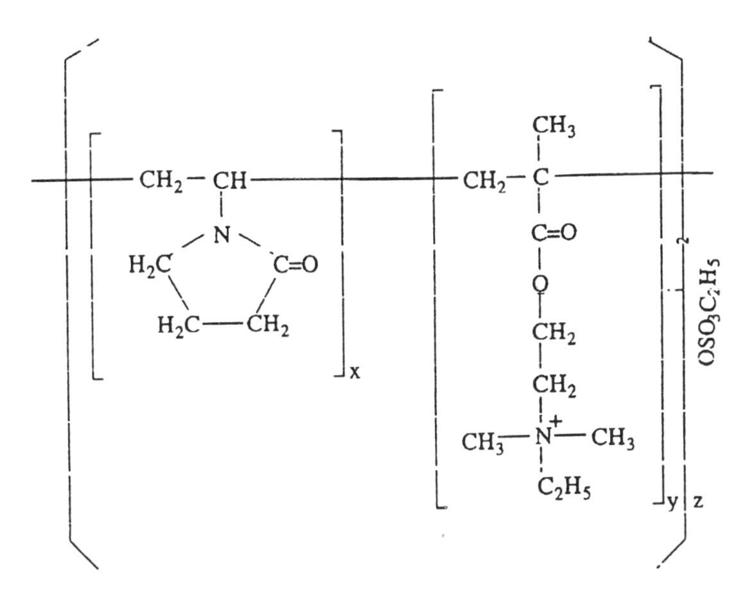

Figure 3 Quaternized copolymers of vinylpyrrolidone and dimethylaminoethyl methacrylate.

soaps, shaving products, deodorants, and antiperspirants. Copolymer 845 is slightly cationic and therefore mildly substantive to hair.

b. Quaternized PVP/DMAEMA (Polyquaternium-11). The cationic polymer, which is most widely used for its conditioning and palliative properties, is probably polyquaternium-11 [GAFQUATS, ISP (21)]. These compounds are quaternized copolymers of PVP and dimethylaminoethyl methacrylate (Figure 3). GAFQUAT 734 is a medium-molecular-weight product (ca. 100,000). GAFQUAT 755N has an average molecular weight of 1,000,000. Polyquaternium-11 adsorbs on the surface of hair from aqueous solution to give a more uniform and thicker layer than any other commercially offered cationic polyelectrolyte—a property which renders it especially useful as a conditioner or a palliative for damaged hair, as a preventive measure, and for simultaneous setting and conditioning benefits (20). Scanning electron microscopy (SEM) has been used to show substantivity to hair (21).

Polyquaternium-11 is an additive for improved skin feel in shaving products, skin creams and lotions, deodorants and antiperspirants, and liquid and bar soaps.

c. Polyquaternium 28 (GAFQUAT Hsl00). Polyquaternium 28 is the copolymer of PVP and methacrylamidopropyl trimethylammonium chloride (18).

Cationic nature gives substantivity to hair and skin, providing conditioning and manageability. Buildup with continued use is a drawback.

Polyquaternium 28 (and) dimethicone (PVP 5-10, GAFQUAT HSI, ISP) is a silicone encapsulate using a shell/core structure. These products combine the benefits of film-forming polymers and dimethicone, while minimizing the drawbacks associated with the greasy feel and buildup noted with silicones.

d. Vinylcaprolactam (VCL)/PVP/DMAEMA Terpolymer (GAFFIX VC-713-ISP). The VCL imparts film-fixative characteristics with reduced moisture sensitivity, and the DMAEMA confers cationicity, enhancing the substantivity to hair. The principal advantage of this terpolymer is the ability to combine both fixative and conditioning properties in one molecule. It is especially useful for gels, since it has good water solubility, excellent substantivity, and superior holding at high humidity and yet is easily removable by shampoos (20).

e. STEPANHOLD [Stepan (22)]. STEPANHOLD is a terpolymer of PVP/ethyl methacrylate/methacrylic acid and is said to be a good conditioner for hair and skin.

4. Polyquaternium 16

Polyquaternium 16 is a copolymer of PVP/methylvinylimidazoline (MVI) (LUVIQUAT, BASF) (19). Three different comonomer ratios are available: 70:30, LUVIQUAT F370; 50:50, FC 550; 5:95, FC 905. The higher the ratio of MVI, the less susceptible to atmospheric humidity and the harder the film is. With increasing type number the polymer becomes more substantive. On a comparative basis this means a moderately strong conditioning activity for the type 370, and an unusual intensive one for the type 905. The charge density of the three types has been found to have a linear correlation with the chemical structure. The charge density of the type 905 especially is extraordinarily high, indicating an unusual conditioning activity (23). Any combination of the three types of polyquaternium 16 may be applied in a single formula, which allows the composition of a "tailor-made" formulation by combining the respective benefits of the different polymer types.

C. Dimethyl Diallyl Ammonium Chloride (DDAC) Homopolymer (Polyquaternium 6)

MERQUATS (Calgon) (24) are a highly cationic series varying in molecular weight. Polyquaternium 6 (MERQUAT 100) is the homopolymer of dimethyl diallyl ammonium chloride. It has a weight-average molecular weight of approximately 100,000 and is excellently substantive to skin and hair to confer conditioning benefits. Unfortunately, it does not form good films and it has poor compatibility with anionic ingredients, owing to its very high charge distribution. For these reasons, it is not suitable for use in a dilution-depositing shampoo. This polymer is used, however, as a shampoo preconditioner. Polyquaternium 6, in dry-skin lotions, offers a combination of substantivity to the skin with a long-lasting, smooth feel. Polyquaternium 6 has been formulated into conditioners which contain amphoteric or nonionic surfactants rather than cationic surfactants and as palliatives for hair-straightening preparations (20). The incompatibility with anionic ingredients limits its use in shampoos, but in conditioners it contributes to detangling and improved wet and dry combability without a greasy feel (25).

1. DDAC/Acrylamide (Polyquaternium 7)

MERQUAT 550 and MERQUAT S (polyquaternium 7) are the copolymers of polyquaternium 6 with acrylamide (24). They have a lower charge density and higher molecular weight (weight average mol. wt. = 500,000) and are most suitable for inclusion as an active ingredient in body-building shampoos.

Polyquaternium 7 displays excellent substantivity to skin and hair, good film forming, and good compatibility with anionic surfactants, although a special grade is required for ethoxylated surfactants. This polymer also displays a sharp desolubilization of the polymer–surfactant complex at surfactant concentrations below the critical micelle concentration (see Section K) and it is therefore ideally suited for use in conditioning body-building shampoos.

These polymers also increase deposition of water-insoluble particles on hair during shampooing. This offers special benefits for enhancing deposition of zinc pyrithione in antidandruff shampoos (20).

The combination of high charge density and polymeric character improves hair manageability. As with other cationic conditioners or creme rinses, the conditioning effects are more apparent when the MERQUAT formulations are applied to hair damaged by chemical bleach or sunlight (26).

MERQUAT polymers impart soft, silky feel to the skin. When applied to dry skin, dilute solutions of MERQUAT 100 were found to dry rapidly, providing a smooth surface and velvety texture. Skin product applications include liquid and bar soaps/detergents, shaving creams, moisturizing or barrier creams and lotions, bath products, and deodorants.

2. DDAC/Acrylic Acid (Polyquaternium 22)

MERQUAT 280 (polyquaternium 22) is the copolymer of polyquaternium 6 with acrylic acid (24). It yields effective conditioning shampoos which are claimed to be easier to comb wet or dry, softer to the touch, have more body, and less tendency to build static.

D. Polymethacrylamideopropyl Trimonium Chloride

Polymethacrylamidopropyl trimonium mer chloride [Polycare 133—Rhone Poulenc (17)] is a highly charged cationic substantive homopolymer with good conditioning and hold without film buildup. Because of its stability to hydrolysis under alkaline conditions, Polycare 133 finds use as a conditioner for wave systems as well as a functional additive in skin care products (17).

E. Acrylamide/β-Methacryloxyethyltrimethyl Ammonium Methosulfate (Polyquaternium 5)

Polyquaternium 5 [RETEN—Hercules (27)] is a copolymer of acrylamide and β-methacrylyloxyethyl trimethyl ammonium methosulfate, recommended as a hair and skin conditioner. There are eight RETENS, varying in molecular weight. They enhance foaming and interact with keratin to give a conditioning effect. This polymer is claimed *not* to interact with anionic surfactants (20).

F. Adipic Acid/Dimethylaminohydroxypropyl/ Diethylenetriamine Copolymer [Cartaretin F-23, Sandoz (28)]

On hair, the polymer gives good lubricity and ease of wet combing without the feeling of oiliness sometimes associated with other cationic polymers. The performance of the polymer in conditioning shampoos, notably its substantivity to hair, is dependent on the types and concentrations of surfactants present in the formulation. In general, increased deposition is favored by use of mixed anionic/amphoteric or nonionic surfactant systems, and higher concentration of polymer solids.

G. Vinyl Alcohol Hydroxypropyl Amine (Polyquaternium 19)

Polyquaternium 19 [ARLATONE PQ-220, ICI (29)] is the quaternary ammonium salt prepared by the reaction of polyvinyl alcohol with 2,3-epoxypropylamine. Cationically modified resins find their chief use as substantive conditioning agents in hair preparations. However, a resin such as quaternium 19 finds value in skin care because of its smooth emollient feel on the skin, as well as its substantivity (30). Quaternium 19 also increases the protection afforded by sunscreen lotions.

H. Quaternized Polyvinyl Octadecyl Ether (Polyquaternium 20)

Polyquaternium 20 is a quaternized polyvinyl octadecyl ether [ARLATONE PQ-225, ICI (29)]. The superior moisture barrier properties result in resistance

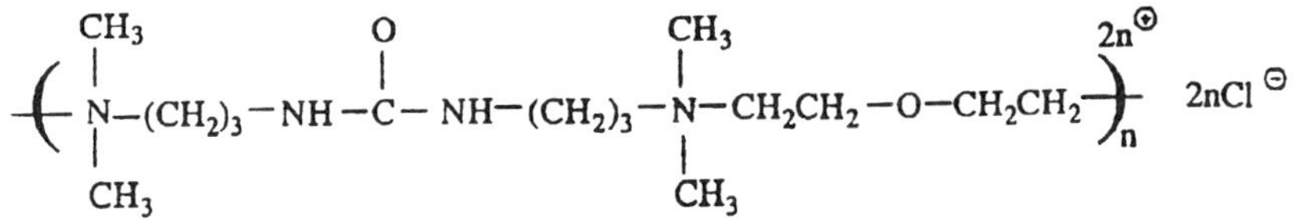

$$\left(\underset{CH_3}{\overset{CH_3}{N}} - (CH_2)_3 - NH - \overset{O}{\overset{\|}{C}} - NH - (CH_2)_3 - \underset{CH_3}{\overset{CH_3}{N}} - CH_2CH_2 - O - CH_2CH_2 \right)_n^{2n\oplus} \quad 2nCl^{\ominus}$$

Figure 4 Poly[N-[3-(dimethylammonio)propyl]-N′-[3-(ethyleneoxyethylene dimethylammonio) propyl/urea dichloride].

to water rinsing. Tests show that polyquaternium-20 augments the performance of a simple quaternium by improving hair setting properties and luster without excessive buildup (16).

I. Quaternized Ionenes (Polyquaterniums 2, 17, 18, 27)

The ionenes are cationic polyelectrolytes which are formed by condensation of di(tertiary amines) and dihalides via a quaternization reaction. Commercially available ionenes are polyquaternium 2, polyquaternium 17, polyquaternium 18, and polyquaternium 27.

Polyquaternium 2 [MIRAPOL A-I 5, Rhone Poulenc (31)] is defined as poly[N-[3-dimethylaminopropyl] N′-[3-(ethyleneoxyethylene dimethylamino) propyl] urea dichloride) (Figure 4). In common with all cationic polymers, polyquaternium 2 may be used in products primarily designed for hair conditioning, such as creme rinses, or in conditioning shampoos. In either case, long-chain anion-active surfactants should be present for complex formation.

Two chemically related polyquats, polyquaternium 17 and polyquaternium 18, have been introduced which are variations of the older polyquaternium 2. Polyquaternium 17 [MIRAPOL AD, Rhone Poulenc (31)] is the reaction product of adipic acid and dimethylaminopropylamine, reacted with dichloroethyl ether. Polyquaternium 18 [LUVIQUAT 500, BASF (19)] substitutes azelaic acid instead of adipic acid.

The products have a very similar function and are reported to improve wet and dry combability as well as flyaway. The main differences are found in a slightly better antistatic control by polyquaternium 17 and a better compatibility with anionics of polyquaternium 18. Like most cationic polyelectrolytes, the ionenes give good conditioning properties but poor setting properties. Acceptable setting in addition to conditioning can be achieved if a crosslinked ionene is used (20). It has been claimed that ionenes confer better wet and dry combing and shine on the hair than the traditionally used cetyltrimethyl ammonium bromide and do not build up on the hair with repeated application.

Clear shampoos require the use of amphoteric surfactants as coupling agents between polyquaterniums 17 and 18 and anionic surfactants. Appropriate amphoterics include imidazoline derivatives and amidobetaines. Polyquater nium 27 [MIRAPOLS 9, 95, 175, Rhone-Poulenc (31)] are block polymers formed by the reaction of polyquaterniums 2 and 17.

J. Polyquaternium 8

Polyquaterniuim 8 is the quaternary ammonium salt of methyl and stearyl dimethylaminoethylmethacrylate quaternized with dimethylsulfate. It is a useful conditioner for both hair and skin.

K. Cellulose Cationic Polymers

In order to replace hair body which has been lost by frequent shampooing and conditioning, formulators sought to introduce materials which would be deposited on the hair from the shampoo during the shampooing process. The bodying ingredient had to be dissolved or solubilized in the final product in such a way that it could be "triggered" to deposit on the hair at an appropriate point in the washing cycle—preferably during rinsing, when soil and sebum have already been removed. The breakthrough came with a quaternized hydroxyethylcellulose derivative, polyquaternium 10 (20).

1. *Hydroxypropyl Trimethyl Ammonium Chloride Ether of Hydroxyethyl Cellulose (Polyquaternium 10)*

Polyquaternium 10 [UCARE Polymer JR, Amerchol (32)] may be described as 1-hydroxypropyl trimethyl ammonium chloride ethers of hydroxyethyl cellulose. It is obtained by reaction of hydroxyethylcellulose with epichiorhydrin followed by quaternization by trimethylamine. It is a film-forming, water-soluble material which imparts good wet combing, curl retention, and manageability and has been shown to mend split ends. Studies employing radioisotopes have shown that this cationic polymer is substantive to hair and is removable by repetitive shampooing, the amount originally deposited depending on the surfactant present in the shampoo. In general, while it can be formulated with anionic surfactants, greater deposition occurs from shampoos formulated with a mixture of nonionic and amphoteric surfactants (33).

There are three different types of Polymer JR, JR 125, JR 400, and JR 30M, varying in molecular weight (32). The most commonly used product in conditioning shampoos, JR 125, has a number-average molecular weight of about 400,000 and about 1300 cationic sites (34). In the presence of an anionic surfactant, these cationic sites display coulombic attraction for the anionic head groups of the surfactant. At 1:1 charge neutralization of the cationic polymer, a hydrophobic complex precipitates from solution (20).

Surface-active molecules in aqueous solution undergo structural changes which indicate agglomeration of molecules to form aggregates. These aggregates are called micelles. The concentration at which the micelles begin to form is called the "critical micelle concentration" (CMC).

In the presence of excess polymer or excess anionic surfactant, clear solutions are obtained. Shampooing is typically done above the CMC in order to solubilize the sebum and oily soil. Polyquaternium 10 stays solubilized and out of the way of the cleaning during the shampooing process, but upon rinsing the composition on the head is diluted below the CMC and the polymer–surfactant complex is deposited on the hair (20). Presumably, this is due to the fact that above the CMC, hydrophobic interaction between the complex and free surfactant causes solubilization of the complex. Below the CMC, however, precipitation of the complex occurs at all ratios of surfactant in excess of the 1:1 molar equivalent ratio. The excess surfactant cannot solubilize the complex, because below the CMC, hydrophobic association of the surfactant molecules is not favored—the minimum free-energy state is one in which the surfactant molecules are individually dissolved.

The complex deposited from such cationic polymer/anionic surfactant systems appears to be liquid crystalline, with rheology that lubricates and improves the ease of wet combing. Further, the complex has antistatic properties (35). Thus, these systems have the dual advantages of conferring conditioning benefits on wet hair while enhancing the body and reducing flyaway of conditioned dry hair. They can help in mending split ends. In addition, Polymer 10 provides an emollient effect and a smooth conditioning feel on the skin (32). The substantivity of UCARE and CATREX polymers on skin translates into a perceptible silk smooth afterfeel, protecting the skin as well as conditioning it. Polyquaternium 10 has been included in formulations as a palliative in thioglycollate hair-waving compositions and as an additive to traditional conditioners and in oil-free conditioners.

Polymer JR is very substantive to hair and not easily removed. Use of high concentrations may lead to overconditioning and excessive buildup. Studies employing radioisotopes have shown that this cationic polymer is substantive to hair and is removable by repetitive shampooing, the amount originally deposited depending on the surfactants present in the shampoo (2).

Polyquaternium 10 is frequently used in shampoos, but polyquaternium 11 is the preferred ingredient for conditioners (25).

While structurally similar to UCARE Polymer JR, the analogous UCARE Polymer LR provides all the conditioning benefit of Polymer JR, but with less deposition on the hair. In general, while it can be formulated with anionic surfactants, greater deposition occurs from shampoos formulated with a mixture of nonionic and amphoteric surfactants.

Replacing part of the anionic detergent(s) by imidazolinium amphoteric surfactants dramatically increases the substantivity of cationic cellulose derivatives to hair (9). However, too much substantivity can be undesirable,

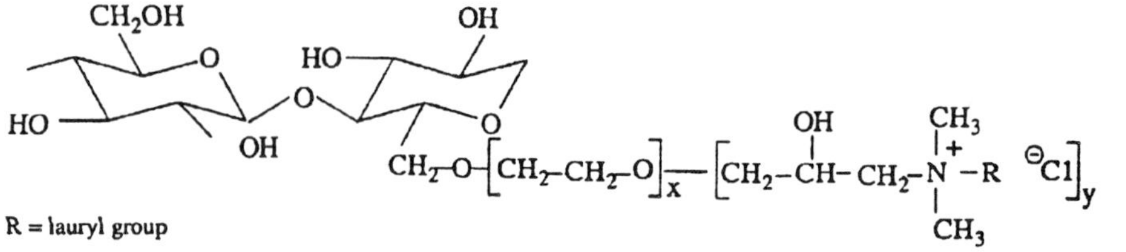

R = lauryl group

Figure 5 Polyquaternium 24.

because unwanted excessive buildup may occur after repetitive shampooing, leaving hair feeling stiff and heavy.

The CELQUAT-SC series of National Starch (36) (polyquaternium 10) is similar to the UCARE series of Amerchol. CELQUAT SC 240 is a lower-viscosity analog of CELQUAT SC 230M. BIOCARE SA [Amerchol (32)] is a combination of bioengineered hyaluronic acid and polyquaternium 10.

2. *Polyquaternium 24*

Polyquaternium 24 [QUATRISOFT POLYMER LM-200, Amerchol (37)] also has a cellulose backbone. It is a polymeric quaternary ammonium salt of hydroxyethylcellulose reacted with a laruyl dimethyl ammonium-substituted epoxide (Figure 5). It can be considered a hydrophobically modified polyquaternium 10. QUATRISOFT forms a nontacky film that is substantive to both hair and skin. It is substantive enough to withstand rinsings, but can create a buildup problem. This material acts not only as a conditioner but combines synergistically to increase the product viscosity.

ESCA (X-ray photoelectron spectroscopy) is a sensitive method used to detect the presence of adsorbed conditioning polymers on isolated, treated stratum corneum membranes, and to rank them in terms of the amount adsorbed (38). According to these studies, polyquaternium 24 adsorbs more strongly to skin membranes than polyquaternium 10 or chitosan. Polyquaternium 24 shows roughly equivalent deposition on both sides of the membrane, while polyquaternium 10 and chitosan deposit only on the inner surface of unwashed skin membranes. This is attributed to the presence of retained lipids on the outer surface during contact with the aqueous polymer solution.

3. *Hydroxyethylcellulose/Diallyldimethyl Ammonium Chloride (Polyquaternium 4)*

Polyquaternium 4 [CELQUATS H-100, L-200, National Starch (36)] is a copolymer of hydroxyethylcellulose and diallyldimethyl ammonium chloride. Cationic over the entire pH useful range, CELQUAT H-100 and L-200 are

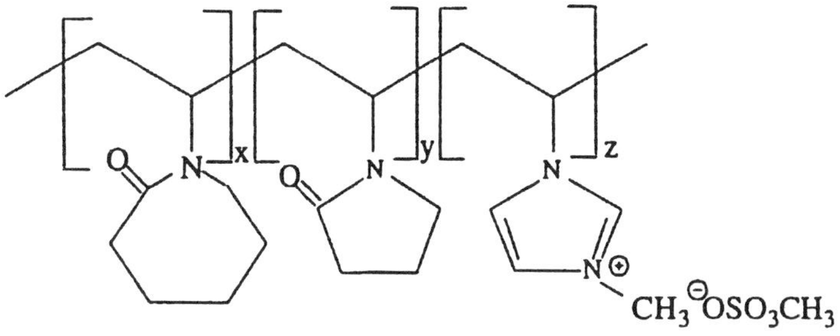

Figure 6 Polyquaternium 46.

substantive to anionic surfaces such as hair and skin. In conditioners, lotions, rinses, and shampoos, they significantly improve the appearance and manageability of hair. They give a lasting smooth, velvety feel to skin lotion and creams. They offer superior curl retention properties even in high-humidity conditions, excellent wet combing, and no buildup (25).

4. Alkyl Dimonium Hydroxypropyl Oxyethyl Cellulose

CRODACEL QL is the lauryl analog, QM the coco and QS the stearyl derivative. Not only do the CRODACELS have conditioning properties characteristic of their parent polymer, they also have high substantivity acquired as a result of quaternizing, while retaining compatibility with anionic surfactants. The CRODACELS can also be used in skin care products, where they impart a long-lasting and lubricious feel to the skin (26).

L. Polyquaternium 46

To keep in step with the market trend toward multifunctionality, BASF (19) developed new conditioning polymers with improved properties, especially for styling products: polyquaternium 46 (LUVIQUAT HOLD) (Figure 6). The conditioning effect of polyquaternium 46 lies within the optimum range for styling applications and is comparable with that of conventional polyquaternium 11 and PVP/VA copolymer blends.

M. Guar Hydroxypropyl Trimonium Chloride

Following the success of polyquaternium 10, it was not surprising that another cationic polysaccharide—a guar derivative—should emerge as a candidate for service in dilution-deposition shampoos (39).

Guar gum is a galactomannan with a β-1,3-linked mannan backbone to which are attached α-D-galactopyranosyl residues as 1,6-linked single-unit side chains. The cationic charge is placed on guar gum using the same chemistry as for polyquaternium 10, that is, by hydroxypropylation, addition of epichlorohydrin, followed by quaternization with a tertiary amine (20).

Guar hydroxypropyltrimonium chloride (GHPTC), not unexpectedly, functions in an identical way to polyquaternium 10, but it has been claimed to confer better wet combability. It has been used in conditioning and body-building shampoos. High cationic substitution confers substantivity to skin to provide conditioning. Because GHPTC is substantive, a protective polymer film remains on the skin.

N. Quaternized Lanolin (Quaternium 33)

Quaternium 33 [LANOQUATS, Henkel (40)] is quaternized lanolin. These products have many applications in cosmetics, particularly in the area of hair care products (17). Their compatibility with anionic surfactants makes them ideal shampoo systems, where their function is to deposit on the hair and combat the effects of static electricity, thus helping to control flyaway and aid in the ease of combing after shampooing (41).

The properties of these quaternaries also make them useful in hair conditioners and rinses applied after shampooing. The toxicological properties of lanolin acid quaternaries have been shown to be lower in magnitude than those of other commonly used "quats," and their solubility characteristics allow them to be incorporated in virtually any type of hair rinse/conditioner formulation (42).

O. Quaternized CHITOSAN (Polyquaternium 29)

CHITOSAN is the deacetylation product of chitin which is a major component of the exoskeletons of invertebrates. Combining CHITOSAN with pyrrolidone carboxylic acid (PCA) forms a water-soluble substantive humectant whose CTFA name is chitosan PCA and which forms polymeric films on the hair.

CHITOSAN that has been reacted with propylene glycol and quaternized with epichlorohydrin has the INCI (3) name of polyquaternium 29. Compatible with all types of surfactants and stable over the pH range from 2 through 12, polyquaternium 29 exhibits excellent hair fixative properties at low active levels.

P. Quaternized Proteins

Proteins and polypeptides derived from various plant and animal sources are common conditioning agents. Because of their similarity to the proteinaceous structure of hair and skin, they are naturally adsorbed. Once deposited, proteins are said to improve surfaces by attaching to the damaged sites. Also, their substantivity can be enhanced through reaction with quaternized materials (8).

Hydrolyzed homopolymer proteins are discussed in another chapter of this volume. This section is concerned only with quaternized cationic and noncationic proteins. Quaternization yields derivatives which exhibit even higher substantivity to hair, and are especially useful for selectively sorbing at damage sites. Steartrimonium, lauryl, cocoyl, and stearyl dimethylammonium hydrolyzed animal protein are all commercially available.

1. Quaternized Collagens

Brooks Industries (43) markets quaternized collagens: hydroxypropyltrimonium gelatin (QUAT-Coll IP10), cocodimoniumhydroxypropyl hydrolyzed collagen (CDMA), and steartrimonium hydroxyethyl hydrolyzed collagen (QS). IP-10 uses a low-molecular-weight trimethyl quat which shows excellent counterirritant properties in anionic surfactant systems. CDMA uses a fatty quat which has good wet combing, conditioning, and tangling effects on the hair. Croda (44) has the steardimonium analog (CROQUAT-S) and trimonium (CROQUAT Q), the coco dimonium (CROQUAT M), and the lauryl dimonium (CROQUAT L).

Collagen is the principal structural protein in the body. However, collagen is only one source of skin protein for cosmetics. Another useful source is keratin in the form of horn, hair, hoofs, or feathers. Milk, silk, vegetables, and yeast are other sources of protein for cosmetics applications.

2. Quaternized Keratin

Cocodimonium hydroxypropyl hydrolyzed keratin [CROQUAT, Croda (44); QUAT-KERATIN WKP, Brooks (43)] is derived from hydrolyzed keratin proteins obtained from wool. CROQUAT HH (Croda) is the same agent derived from the keratin of human hair. Containing cystine, it is recommended for conditioning perms and relaxer systems. With high substantivity, it can also be used in shampoos, conditioners, and creme rinses. Its affinity for keratin also renders it useful in nail care products.

3. Quaternized Vegetable Protein

With the “green movement” leaning toward nonanimal sources, companies have offered varied plant and vegetable-derived proteins. QUAT-VEG Q30 [Brooks (43), hydroxypropyltrimonium vegetable protein] is one such example. On the hair it possesses all the known conditioning and substantivity of the traditional animal protein quaternaries.

4. *Quaternized Wheat Protein*

HYDROTRITICUM QL, QM, and QS [Croda, alkyldimonium hydrolyzed whole wheat protein (45)] are cationic quaternized wheat proteins which are highly functional conditioning agents in both hair care and skin care products. HYDROTRITICUM QL is the lauryl analog, QM the coco, and QS the stearyl derivative. By virtue of their compatibility with and substantivity from anionic systems, they are suited for use in all types of hair care products—shampoos, conditioners, styling gels, mousses, and sprays—as well as skin care products—liquid soaps, facial scrubs, skin conditioning creams and lotions, skin cleansers, bath products, etc.

Hydroxypropyltrimonium hydrolyzed wheat protein (QUAT-WHEAT QTM, Brooks) shows improved substantivity, enhanced moisture binding, static reduction, and extra body-building effects on the hair. It may be used in hair and skin care products, e.g., shampoos, setting lotions, conditioners, and shower bath products. QUAT-WHEAT CDMA is the cocodimonium analog designed for softening, conditioning, and manageability of hair; SDMA is the soyadimonium derivative for cream rinse action.

5. *Quaternized Soy Protein*

Cocodimonium hydrolyzed soy protein (QUAT-SOY CDMA, Brooks) has an isoionic point over 9, indicating the cationic nature to show conditioning power and substantivity. QUAT-SOY LDMA is lauryldimoniumhydroxypropyl hydrolyzed soy protein, a softening and conditioning agent, insoluble on dilution with water.

6. *Hydrogenated Soyadimoniumhydroxypropyl Polyglucose*

Hydrogenated soyadimoniumhydroxypropyl polyglucose (BROCOSE Q, Brooks) is a hydrogenated soyabean fatty acid sugar quaternary based on a polysaccharide. BROCOSE Q can be considered to be a lipopolysaccharide with conditioning power, substantivity, softening, and moisturizing effects on hair.

Q. Substantive Conditioning Humectants

Lauryl methyl gluceth-10 hydroxypropyl dimonium chloride [GLUCQUAT 125, Amerchol (37)] and 6-(N-acetylamino)-4-oxahexyl-trimonium chloride (QUAMECTANT AM, Brooks) are substantive conditioning humectants. While not cationic polymers as such, they are worthy of inclusion. They achieve their substantive nature through quaternization which endows them with a positively charged nitrogen which adsorbs to negatively charged hair and skin. They are skin moisturizers, adjusting the skin's moisture content in line with the humidity.

R. Aminosilicones

Silicones offer a route to conditioning benefits without the buildup associated with dilution-deposition shampoos or the limpness associated with traditional rinse conditioners. They increase the hair's luster and ease wet combing. Applied to dry hair, they lubricate the fibers and ease dry combing.

Several types of silicone polymers have been used in shampoos and other hair care products. These include polydimethylsiloxane (PDMS) materials, silicone polyethers, and amine functional silicones.

Silicones are treated in detail in another chapter of this volume. This chapter deals with aminofunctional siloxane and silicone–protein copolymers. Two aminofunctional silicones have been assigned INCI names: amodimethicone and trimethylsilylamodimethicone (TSA). They are polydimethylsiloxane (PDMS) polymers in which some of the methyl groups attached to the polymer chain are replaced by organic groups of amine functionality. Polar amine groups along the siloxane chain have a profound effect on the silicone's conditioning properties, giving the polymer an affinity for proteinaceous surfaces such as hair. TSA polymers are useful as conditioning agents for clear shampoos based on a premix method of formulation (46,47). In effect, they provide conditioning while maintaining clarity, cleaning activity, and foam.

Amodimethicone is recognized for its extremely robust conditioning and for its ability to form clear products when used in high-surfactant shampoos (48,49). Amodimethicone is a useful ingredient in conditioners, gels, mousses, and permanents, but its use in shampoos has proved troublesome due to interactions between the cationic and the anionic surfactants, which can result in compatibility problems. However, the amodimethicone emulsion can be made compatible in high-surfactant-level shampoos (48).

The other class of amine-functional siloxanes is the trimethylaminodimethicones (TSA). Studies show that TSA does not retain its conditioning properties as long as amodimethicone does. TSA polymers are the silicones of choice for durable conditioning and silky feel in clear or opaque formulas. Fine TSA emulsions provide easy-to-prepare, clear, stable silicone-containing shampoos. General Electric markets Silicone Emulsion SM 2658, whose CTFA designation is amodimethicone (and) tridecth-12 (and) cetrimonium chloride. It dries to form a crosslinked polymer film which provides effective conditioning, particularly for damaged hair.

Although today's two-in-one conditioning shampoos are generally not clear, clarity can indeed be an important characteristic for some niche markets. For clarity with a more robust conditioning effect, a premixing step is typically required. This route to clarity depends on silicones that are "solubilized," that is, they are premixed with other ingredients to make them soluble. Use of the trimethylsilylamodimethicone (TSA) family of silicones is the preferred route

Table 1 Subjective Hair Tress Evaluations (Rated 1–5; 1 = Best)

	Wet comb	Wet feel	Dry comb	Dry feel
Fine TSA emulsion shampoo	2.5	2	1.5	1.5
Blank shampoo	4	4	2.5	2

for clarity, with strong conditioning and faster drying. A further approach to clarity depends on silicone-based (TSA) fine emulsions, which can be formulated into clear shampoos. The small, stable particles of silicone become dispersed and stabilized by simple "add-in" formulating that requires no premixing. An especially effective fine emulsion contains both amine-functional silicone and cationic surfactants. This route offers the formulator an additional method for achieving an easy-to-formulate, clear and stable silicone-containing conditioning shampoo. Table 1 shows hair-tress conditioning data for this formulation type and gives an indication of the combing and sensory improvements attainable. If clarity and good conditioning are requirements, but a premixing step is not desirable and an aging step is allowable, a high-surfactant shampoo that utilizes amodimethicone emulsion in "aged" shampoos may be the answer. High surfactant levels are useful in this conditioning shampoo to counteract the viscosity-reduction effect of the amodimethicone and to attain clarity. This method requires that the shampoo be aged from one to two weeks prior to use (49).

Compared to the TSAs used to create clear products, those used in forming pearlescent or opaque shampoos have higher molecular weights and lower amine functionality; they provide TSA-like conditioning properties, yet formulate more like dimethicones. The result is a pearlescent or opaque formulation that still requires a premix step, but that can have acceptable viscosity without the addition of water-soluble thickening polymers.

DC929 cationic emulsion [Dow-Corning (50)] is an emulsion of an amine-functional polymer. In hair care products it provides substantivity by acting as a conditioning agent that dries to a film by crosslinking. It leaves a soft feel on hair and durability without buildup. It remains on the hair through at least six shampoos (37). It has a very small particle size (~50 nm), which results in a translucent product that enhances conditioning performance. DC929 and SM 2658 enhance the performance characteristics of shampoos, rinses, and shampoo-in hair coloring. They improve wet combing ease, luster, and resistance to dry flyaway (50).

Quaternium 80 [ABIL QUATS 3270, 3272—Goldschmidt (51)] is a diquaternary polydimethylsiloxane used in conditioning shampoos and hair rinses. The conditioning effect of ABILQUAT 3270 and ABILQUAT 3272 improves combability of wet and dry hair; electrostatic charge is reduced and the hair becomes silky, shiny, and manageable. ABILQUAT 3272, which is of higher molecular weight than ABILQUAT 3270, is more compatible with anionic surfactants than the latter. Therefore, ABILQUAT 3272 is preferred for use in shampoos, shower and bath preparations, and liquid soaps, whereas the most important areas of application for ABILQUAT 3270 are conditioning hair rinses. In skin cleansing preparations the interaction of ABILQUAT 3270 or ABILQUAT 3272 with skin proteins provides a refatting effect that imparts a smooth and supple feel to cleansed skin.

1. Silicone Protein Copolymers

Hydrolyzed wheat protein polysiloxane copolymer [CRODASONE W, Croda (44)] retains some properties of both protein and silicone components, assuming in part the film-forming and substantivity of a protein and the lubricity, gloss, and spreadability of silicone. While the conditioning properties of proteins can help improve the integrity of hair and skin, the esthetic qualities of silicones help to enhance their appearance.

Phoenix Chemical (52) has developed a series of silicone–protein copolymers [PECOSIL (53)]. The compounds contain (a) a silicone portion of the molecule derived from a dimethicone copolyol, (b) an ionizable phosphate group on the dimethicone copolyol, and (c) a protein portion. All are linked in a covalent bond in one molecule. The presence of the phosphate group makes these materials silicone phospholipids. They have a tendency to form bipolar sheets rather than micelles. The products reduce irritation and provide a nonocclusive film on the hair and skin.

There are two classes of PECOSIL silicone proteins: one containing a phosphate group in the portion of the molecule that links the silicone and protein (PECOSIL SWP-83, SSP, SWPQ-40) and another that has no phosphate (PECOSIL SW-83, SWQ-40). All except SSP have a wheat protein souce. SSP is derived from soya. The phosphate-containing materials have a tendency to act as natural phospholipids, while the phosphate-free version acts more like quaternary compounds (54).

V. NONCATIONIC POLYMERS

The bulk of this chapter has been devoted to negatively charged (cationic) polymers. A number of high-molecular-weight neutral copolymers, as well as natural polymers, exist whose main property is the fixation of hair by forming clear films. They are also substantive to and condition hair. National Starch and Chemical has been foremost in marketing these copolymers (36,55).

A. Vinyl Acetate/Crotonic Acid Copolymer

These copolymers are known as the RESYN series. Resyn 28-2930 is a copolymer of crotonic acid and vinyl acetate. The unneutralized resin is hard and brittle. When neutralized by various aminohydroxy compounds, it becomes more flexible. The neutralizer acts as an internal plasticizer.

Resyn 2261 is an acrylic copolymer latex. It is a two-phase system consisting of submivroscopic particles of resin suspended in water. With the evaporation of the water, these resin particles flow together or coalesce to give a clear, water-resistant flexible film. It finds use in wave-set gels and conditioners.

B. Octyl Acrylamide–Acrylic Acid–Butylaminoethyl Methacrylate Copolymer (AMPHOMER)

AMPHOMER is an amphoteric acrylic resin developed specifically as a hair fixative. It is carboxylated at regular intervals along its molecular chain. The amphoteric character gives the polymer good curl retention, ease of combing, and conditioning effect.

C. Sodium Polystyrene Sulfonate (FLEXAN 130—National Starch)

FLEXAN 130 is a high-molecular-weight sodium polystyrene sulfonate. It is soluble in water, glycerine, and low-molecular-weight polyethylene glycols. FLEXAN can be plasticized by protein hydrolyzates, silicones, lanolins, and glycols. FLEXAN 130 is a primary active ingredient in conditioners and setting lotions with improved properties. It is a combination creme rinse conditioner with set which need not be rinsed out of the hair after comb-out.

D. Aminoethyl Acrylate Phosphate/Acrylic Acid (CATREX)

Aminoethyl acrylate phosphate/acrylic acid copolymer [CATREX—National Starch (55)], when suitability formulated, deposits onto hair from shampoo systems to yield improved wet combing as well as conditioning effect and body. It can be left on hair to impart gloss, body, antistatic properties, and set hold.

E. PEG Tallow Polyamine (Polyquart H—Henkel)

Polyamine derivatives (pseudo-cationic secondary amines) are also used as conditioning agents. It is claimed that their weak cationic character does not allow for buildup on the hair shaft, even with repeated applications. PEG 15 tallow polyamine [Polyquart H—Henkel (40)] is compatible with both anionic and cationic surfactants, allowing for simultaneous cleansing and conditioning of hair. Its antistatic property is derived from its pseudo-cationic character. It does not build up on the hair shaft even with repeated applications.

F. Hyaluronic Acid

Hyaluronic acid (HA) is a natural polyanionic polysaccharide (glycosaminoglycan) present in the intercellular matrix of most vertebrate connective tissues. Water-soluble complexes are formed by crosslinked HA (Hylan, Biomatrix) with polyquaternium 4, 11 and polyquaternium 24 (PQ 24). Water-insoluble complexes are formed with polyquaterniums 6, 10, and 7. In general, the polycation's charge density seems to be the most important factor in determining the solubility of its complex with Hylan.

An important property of the Hylan–PQ24 complex is that it significantly enhances the substantivity of Hylan to skin and hair. A PQ24:Hylan ratio of 10:1 resulted in greatest substantivity to hair. BIOCARE SA [Amerchol (32)] is a combination of bioengineered hyaluronic acid and polyquaternium 10.

VI. CONCLUSION

This review has attempted to trace the use of cationic and noncationic polymers in hair and skin conditioning. As styling trends required more chemical and thermal treatments, such as perming, coloring, and blow-drying, conditioning treatments became more important. Modern conditioners are designed to provide one or more of the following functions: provide ease of wet and dry combing; smooth, seal, and realign damaged areas of the hair shaft; minimize porosity; impart sheen and a silken feel to the hair; provide some protection against thermal and mechanical damage; moisturize; add volume and body; and eliminate static electricity. Today, polymers offer the consumer easier, more convenient, and more versatile styling and grooming than at any time in the past.

If the trend toward temporary, quick-change styles continues, we can expect polymers to continue to play a major role in hair care. This trend, however, is producing and will probably continue to produce sophisticated products which are targeted to perform just one function and to do that job very well (20). Advanced silicone technology can help reach these goals by offering a broad range of potential formulating techniques and product characteristics. Skin conditioning remains a largely unexplored area. Much work remains to prove cationic polymers as major softeners of dry skin and/or moisturizers.

REFERENCES

1. Idson B. Polymers in skin cosmetics. Cosmet Toilet 1988; 103:63.
2. Hunting LL. Can there be cleaning and conditioning in the same product? Cosmet Toilet 1988; 103:73.

3. CTFA International Cosmetic Ingredient Dictionary. 4th ed. Washington, DC: Cosmetic, Toiletry, Fragrance Association.
4. Lochhead RY. Conditioning shampoos. Soap/Cosmet/Chem Spec 1992 (Oct.):42.
5. Goddard ED, Philips TS, Hannan RB. Water-soluble polymer-surfactant interaction—Part I. J Soc Cosmet Chem 1975; 26:461.
6. Fox C. Introduction to the formulation of shampoos. Cosmet Toilet 1988; 103:25.
7. Robbins CR. Chemical and Physical Behavior of Human Hair. 3d Ed. New York: Springer-Verlag.
8. Schueller R, Romanowski P. Conditioning agents for hair and skin. Cosmet Toilet 1995; 110:43.
9. Bass D. Recent advances in imidazolinium amphoteric surfactants. Soap Perf Cosmet 1977; 50:229.
10. Idson B, Lee W. Update on hair conditioner ingredients. Cosmet Toilet 1983; 98: 41.
11. Faucher JA, et al. Textile Res J 1977; 47:616.
12. Smith L, Gesslein BW. Multifunctional cationics for hair and skin care applications. Cosmet Toilet 1989; 104:41.
13. Goddard ED, Schmitt RL. Atomic force microscopy investigation into the adsorption of cationic polymers. Cosmet Toilet 1994; 109:55.
14. Rieger M. Cosmetic use of natural fats and oils. Cosmet Toilet 1988; 103:59.
15. Scott GV, Robbins CR, Barnhurst JD. Sorption of quaternary ammonium surfactants by human hair. J Soc Cosmet Chem 1969; 20:135.
16. Davis RI. New polymers for cosmetic products. Cosmet Toilet 1987; 102:39.
17. Rhone Poulenc, Cranbury, NJ.
18. Specialty Products for Personal Care. Bulletin. International Specialty Products, Wayne NJ.
19. BASF Corp., Parisippany, NJ.
20. Lochhead R. History of polymers in hair care (1940–present). Cosmet Toilet 1988; 103:23.
21. Gafquat Quaternary Polymers for the Cosmetic Industry. ISP, Wayne, NJ.
22. Stepan, Northfield IL.
23. Goddard ED, Harris WC. An ESCA study of the substantivity of conditioning polymers on hair substances. J Soc Cosmet Chem 1987; 38:233.
24. MERQUATS. Calgon Corp., Pittsburgh, PA.
25. Hunting LL. The function of polymers in shampoos and conditioners. Cosmet Toilet 1984; 99:57.
26. Sykes AR, Hammes PA. Use of merquat polymers in cosmetics. Drug Cosmet Ind 1980 (Feb.):126.
27. Hercules, Wilmington, DE.
28. Cartaretin F-23. Sandoz, East Hanover, NJ.
29. ICI Americas, Wilmington, DE.
30. Good Housekeeping 1946; 122:119.
31. MIRAPOL. Rhone Poulenc, Cranbury, NJ.
32. Polymer JR for Hair Care. Booklet. Amerchol Corp., Edison, NJ.
33. Cannell DW. Cosmet Toilet 1979; 94:29.

34. Goddard ED, Hannan RB. Cationic polymer/anionic surface interactions. J Colloid Interface Sci 1976; 55:73.
35. Patel CV. Antistatic properties of some cationic polymers used in hair care products. Int J Cosmet Sci 1983; 5:181.
36. Cationic Cellulosic Polymers. National Starch and Chemical, Bridgewater, NJ.
37. Amerchol Corp., Edison, NJ.
38. Goddard ED, Harris WC. Adsorption of polymers and lipids on stratum corneum membranes as measured by ESCA. J Soc Cosmet Chem 1987; 38:295.
39. Wilkinson JB, et al., eds. Harry's Cosmetology. 7th ed. New York: Chemical Publishing Co., 1982.
40. Henkel, Ambler, PA.
41. Schlossman M. Quaternized lanolins in cosmetics. Soap Cosmet Chem Spec 1976; 52:10.
42. McCarthy JP, Laryea JM. Effects of the use of lanolin acid quaternary in hair conditioning preparations. Cosmet Toilet 1979; 94:90.
43. Brooks Industries, S. Plainfield, NJ.
44. Croda, Parsippany, NJ.
45. HYDROTRICUM. Croda, Parsippany, NJ.
46. Kohl G, Tassoff J, Chandra O. Conditioning Shampoos Containing Amine Functional Polydiorganosilicone. U.S. Patent 4,559,227.
47. DiSapio A, Fridd P. Silicone: use of substantive properties on skin and hair. Int J Cosmet Sci 1988; 10:75.
48. Halloran D. Silicones in shampoos. HAPPI 1991 (Nov.):60.
49. Halloran DJ. Silicone selection guide for conditioning shampoos. Soap Cosmet Chem Spec 1992 (Mar.):24.
50. DC 929 Cationic Emulsion. Dow Corning, Midland, MI.
51. Goldschmidt, Hopewell, VA.
52. Phoenix Chem., Somerville, NJ.
53. Imperante J. O'Lenick AJ, Hammon J. Silicone-protein copolymers. Soap Cosmet Chem Spec 1994 (Oct.):32.
54. O'Lenick AJ, Parkinson JK. Silicone quaternary compounds. Cosmet Toilet 1994; 109:85.
55. CATREX, National Starch, Bridgewater, NJ.

12
Formulating Conditioning Products for Hair and Skin

Mort Westman
Westman Associates, Inc., Oak Brook, Illinois

I. INTRODUCTION

This chapter is intended to provide the uninitiated chemist with a strategy for the formulation and development of products designed to condition hair or skin. This is done with the belief that a good strategic understanding of product development will continue to be applicable and of valuable service while specific formulation, or ingredient-based, approaches commonly become obsolete or outmoded.

It is not unusual for formulation technology to be replaced as the result of scientific innovation or other developments, such as regulatory requirements and commercial activities. An unfortunate but common example of the latter can be found when competitors seek to adopt or emulate the formulation approach employed by a commercially successful product. This too often results in the replication of a formula with limited technical and functional merit but one whose consumer acceptance is based on such unrelated considerations as clever market position and appealing advertising. With these considerations in mind, a diligent attempt has been made to augment the presentation of technical information with discussion of consumer, market, and corporate perceptions and requirements.

A. History/Background

Periodically, there have been important technical innovations in the formulation of hair and skin conditioners that have created new quantum levels of functionality. Predictably, they have spawned an increase of related patent activity and the adoption of similar formulation approaches by competitive manufacturers. An example related to hair product technology (where true innovation has been uncommon) is the development of the first anionic-based shampoo containing a cationic polymer during the late 1970s (1). The importance of this technology was made evident both through the numer of related patents granted to competitive manufacturers during the next few years and the number of shampoos employing similar technology introduced during that time period. The next innovation of similar importance to shampoo technology did not occur until the 1980s, when Procter and Gamble developed an effective means to reap the conditioning potential of silicone in a shampoo system (2). Once again, this development stimulated related competitive patent activity and formula emulation.

Skin care has produced many more true technical innovations. This is logical when one recognizes that skin is a highly functional, living organ capable of undergoing physiological changes (i.e., improvement). (This is to be contrasted with hair, which is not living and therefore is incapable of undergoing physiologically induced improvement.) During recent years the use of alpha-hydroxy acids has created a new technical "focal point" (3) for skin care formulations. Focal points of similar technical importance include the use of such ingredients as Retin A (4), liposomes (5), and biomimetic lipids such as ceramides (6). It is likely that future developments in skin biochemistry will spur further formula innovations.

B. Common Conditioning Ingredients

As discussed throughout this work, several classes of conditioning agents are commonly used for both hair and skin. Each class has its own unique benefits and formulating challenges. The reader should refer to specific chapters for a detailed discussion of these materials, but a few general comments are provided here.

1. Dimethicone and Silicone Derivatives

Dimethicone (polydimethylsiloxane) provides substantial conditioning effects to both hair and skin when properly delivered in appropriate quantities. Incorrectly formulated, dimethicone also has the potential to leave hair or skin feely tacky, greasy, and prone to attracting environmental dust and debris. The formulating challenge of delivering just the appropriate quantity of this category of materials is particularly acute for hair care products, since they are typically rinsed after application. If the dimethicone droplets are not sufficiently small, or adequately suspended, they will coalesce and settle out of the system. If they are emulsified by surfactants, they will, predominantly, be carried away with the rinse water and be of little or no value to the hair. Procter and Gamble researchers found a solution to this problem by suspending discrete dimethicone droplets with the aid of a natural gum.

In addition to dimethicone, the formulator should be familiar with the other silicones and silicone derivatives. Some may be selected for their excellent performance in specific areas of conditioning (e.g., phenyltrimethicone imparts excellent shine), while others may be selected for their compatibility with certain chemical systems (see Chapter 9 for a review of these materials).

2. *Esters and Oils*

The conditioning agents included in the broad category of esters and oils generally rely on their ability to smooth hair and skin with a lubricating film. One problem associated with the use of these materials is that, similar to silicones, they may leave behind an objectionable residue. Another is that they are not particularly substantive to hair or skin and therefore they are most effective when delivered from a leave-on product.

3. *Petrolatum*

Petrolatum is a highly effective skin moisturizing agent but is of limited utility due to its poor esthetic properties—primarily greasiness and tackiness. While it has traditionally been held to function as an occlusive agent, serving to retain the moisture in skin, this specific mechanism (not its functionality) has come under recent challenge. With the exception of its ability to impart shine, petrolatum's potential conditioning benefits to hair are limited primarily to those associated with grooming (i.e., imparting manageability and retaining a coiffure). Here too, the challenge is to reap these benefits while minimizing tackiness or heavy feeling.

4. *Proteins*

Proteinaceous materials are purported to have restorative properties to hair and skin. Readily discernable conditioning benefits can be achieved on hair by employing quaternized derivatives, while extremely pure forms of collagen have been employed medically for such purposes as reducing wrinkles and assisting wound healing. The protein-related claims made for cosmetic products have held great consumer appeal for a considerable period of time, and while it is likely that the preferred origin of these materials will change from animal to vegetable, there is no reason to conclude that this appeal will dwindle in the near future. When formulating with proteins, the chemist should be alert to potential stability issues such as discoloration, off-odor, and microbial contamination.

5. *Polymers*

Polymers can be very useful of imparting slip and smooth feel to both hair and skin. Cationic polymers are particularly effective for these applications because of their enhanced substantivity. This increased substantivity is essential to typical rinse-off hair conditioning applications, where cationic polymers provide dramatic improvement of wet comb and can minimize the "limpening" effect typical of cationic conditioning materials. Products containing these materials must, however, be carefully formulated to avoid causing "buildup" or creating the perception of dirtiness.

6. *Quaternary Ammonium Salts*

"Quats" are the workhorse of hair conditioning formulations. They impart excellent softness, manageability, excellent combing properties, and essentially eliminate flyaway, but (at injudicious concentrations) can be irritating to skin and the eyes. Overuse on hair can also result in limpening, greasy feeling, and the real—or perceived—need for more frequent cleansing.

7. *Humectants*

Humectants, usually polyols, are staples of skin conditioning formulations. Classically, these materials are perceived as attracting moisture from the environment and making it available to the substrate (presumably skin) with which they are in close proximity. Contrary to the lay concept of hair moisturization, this would not be of benefit to styled hair, since it would, in most cases, result in limpness, the emergence of undesirable curls, and frizziness (i.e., the protrusion of damaged areas of hair strands from the aggregate surface).

II. GENERAL FORMULATION STRATEGIES

Before focusing on considerations related directly to formulation, it is imperative that the chemist gain working knowledge of the structure, properties, and common problems associated with the intended substrate (i.e., skin or hair) and the ability to determine the impact of formulations on these substrates (i.e., performance testing). Discussions of these areas are considered beyond the scope of this chapter but may be readily found in technical texts and literature. Chapters 2 and 3 include excellent references on these topics.

When formulating conditioning products for hair or skin, the chemist is urged to impart long-term (or cumulative) improvement in addition to the normally expected immediate corrective benefits. Such long-term benefits may be achieved by providing a means of protection from future damage, or simply by repetitive application (i.e., cumulative therapy). Both approaches require excellent formulation skills and thorough understanding of the underlying problem(s).

Successful execution may, in some cases, require the ability to integrate a formula within a formula. A product example pertaining to both skin and hair care can be found in formulations intended to moisturize. Here current symptoms (e.g., rough, brittle/inflexible, dull, split/fissured skin or hair) should be corrected while striving for long-term elimination/improvement of the underlying problem. For hair this could mean the inclusion of lubricants and antistatic agents to minimize the damage incurred while styling. For skin this could translate into providing effective moisturization therapy combined with protection from harsh detergents and the environment (notably UV radiation). [The term "moisturize" is not employed literally in relation to the conditioning of hair, since the majority of moisture is of transient presence and of little benefit (and most often of detriment) to hair's appearance or condition. Instead, moisturization is treated as the correction of those maladies associated with dryness, including roughness, brittleness, dullness, and poor manageability.]

Before beginning the actual formulation process, the accomplished formulator weighs the impact of a myriad of direct and ancillary factors. It is important for the chemist to realize that certain product development "demons" will be encountered during the formulation process. These demons, some of which are unique to a given project, are technical issues which must be overcome to ensure project success. It is wise to anticipate as many of these issues as possible, because for every one problem that can be envisioned there are surely two or three lurking unseen, in the wings, waiting to spring at the most inopportune time. To this end, it may be helpful to draft a list of the ideal properties desired of the product to be formulated. The current state of the art will not permit development of the ideal product without significant compromise, but review of these attributes may help the chemist anticipate obstacles that will be encountered during the formulation process.

For example, take the case of a project intended to result in the development of the ideal hair conditioner. The properties of such a product are summarized in Exhibit 1. A similar list of ideal properties can be prepared for whatever product is being developed. It is important to consider these a starting point and to understand that there will be unavoidable limitations preventing many of these attributes from being achieved without compromise.

III. THE FORMULATING PROCESS

A. Product Profile

In most cases the formulator is a member of a multidisciplinary team within a structured organization. In order for the team to be successful, each member must deliver a specific, predictable benefit. If the overall plan is well conceived

Exhibit 1 Properties of the Ideal Hair Conditioner

One version for all types of hair (Remember, the *ideal* conditioner is being described.)

Visual appearance, texture, and fragrance that are supportive of product positioning and of functional benefits

Relative ease of dispensing from packaging, tempered by the need for flow properties to support perceived functionality (e.g., the perception of a formula being "rich in conditioners" would not be supported by water-like dispensing)

Ease of distribution throughout the hair, with "hand feel" consistent with product positioning and anticipated end benefits (e.g., it probably would be in error for a therapeutic conditioner to have a thin, runny hand feel)

No eye/skin irritation and will not induce allergic response

Ease of water rinsing

Wet feel varying from smooth to slippery, consistent with degree of conditioning required

Wet and dry conditioning attributes achieved without the slightest feeling of weighing down hair, without negatively impacting desired properties (e.g., body and appearance of volume, but will vary with product positioning)

Absolute ease of wet comb, dry comb, and manageability during styling

Instant drying

No flyaway

Dry feel consistent with positioning

Flexibility, resiliency, and freedom of movement (consistent with positioning)

SHINE; SHINE; SHINE

and the various team members are competent, when combined these benefits will result in a commercially successful product.

In order for the formulator to deliver a specific and predictable benefit, he or she must have a clear understanding of what is expected of the formulation from the very onset of the program. These expectations are most effectively established and ratified in a standardized format, which can be routinely employed. One approach involves creating a *product profile*. At minimum, this product profile must include details of desired physical attributes, esthetic properties, functional benefits, and cost parameters. Where possible, individual competitive (or internal) products should be cited as reference standards (or controls). It is also imperative that other important but less obvious information and requirements be included. At a minimum, the profile should include the following considerations:

Composition and design of packaging (both container and closure).
Desired advertising/marketing claims. These may be formula or ingredient related (e.g., "pH balanced," "98% biodegradable," or "contains vitamins, . . . protein, . . . sea kelp") or performance oriented (e.g., "leaves hair cleaner longer," "moisturizes better than the leading brand").
"Performance demonstrations," required for advertising or sales materials.

Perhaps most important, the method and measure by which successful formulation is to be determined (i.e., criteria for successful completion) should also be clearly described in the product profile. This is of paramount importance to efficient project execution. An example of a product profile designed for hair care products is provided in Exhibit 2.

The product profile is typically provided as a formal document prepared with multidisciplinary input (certainly including that of R&D) and ratified at appropriate levels of management. (Ideally, ratification should be documented at all levels of the organization that will eventually be given the opportunity to evaluate, or comment upon, formula performance or esthetic properties.) While the product profile is intended to establish a complete and definitive plan, at times it is prudent to adopt major change during its execution. Such occasions should be prompted by response to market and/or technological changes, not individual fancy or lack of forethought. Clearly, the capable and enlightened formulator will demonstrate both the ability to execute a well-thought-out plan and the flexibility to rapidly accept and respond to its modification.

At times this process may require the chemist to set aside the burning desire to demonstrate technical virtuosity in favor of compliance with what may appear to be a mundane product description. Consolation for such magnanimous compromise can easily be found in the recognition that extremely difficult technical challenges are frequently cloaked in what appear to be extremely pedestrian assignments. In other words, it may be more difficult to formulate an easily manufacturable, inexpensive shampoo for daily use than its luxuriously high-lathering "department store" counterpart.

B. Technical Evaluation and Review

Armed with a well-crafted product profile, the formulator can truly begin the formulation process. Presuming that the targeted formulation can be achieved

Exhibit 2 New Product Description—Hair Products

Project Identification: ______________________________

General Product Category: ______________________________

Form Prepared by: ____________________ **Date:** __________

I. AESTHETICS:

A. Appearance:

1. Comparative Standard(s): ______________________________

____

2. Color/Hue: ______________________________
3. Intensity: Pale ____ Light ____ Moderate ____ Dark ____
4. Clarity: Clear ____ Translucent ____ Opaque ____
5. Visual Effects: Pearlized ____ Other ____
6. Other ______________________________

B. Consistency:

1. Comparative Standard(s): ______________________________

____

2. Description of Flow Characteristics (choose one):

a. ____ Water-like	f. ____ *Nonflowable* Gel
b. ____ Very flowable, but not water thin	g. ____ *Nonflowable* Creme
c. ____ Flowable, slow pouring	h. ____ *Nonflowable* Paste
d. ____ Barely flowable, thick	i. ____ *Nonflowable* Wax
e. ____ Flowable only after squeezing, shaking package	j. ____ Other (describe) ____________

II. PERFORMANCE/FUNCTIONALITY

A. Foam Properties:

1. Comparative Standard: ______________________________
2. Volume: None ___ Slight ___ Moderate ___ Copious ___

(continued)

Exhibit 2 Continued

3. Feel: Thick/Rich with Lubricious Film ___ Thick/Rich ___
   Moderate ___ Thin ___ or Describe: ______________
4. Other (describe): ________________________________

B. Cleansing Properties:

1. Comparative Standard: ________________________________
2. Degree of Cleansing: Deep ____ Moderate ____
   Everyday ____ Mild ____
3. Feel of Hair After Rinsing: **Scale**: 0 (squeaky clean) to 10 (conditioned film)
   Enter number based on above Scale: ☐
4. Other (describe): ________________________________

C. Conditioning Properties:

1. Comparative Standard: ________________________________
2. Degree of Conditioning: Deep ____ Normal ____
   Everyday ____ Light ____
3. Other (describe): ________________________________

D. Styling Properties:

1. Comparative Standard: ________________________________
2. Drying Rate: Immediate ____ Rapid ____ Slow ____
3. Hold Level: Ultra ___ Extra ___ Normal ___ Light ___
4. Other (describe): ________________________________

E. Application Characteristics (feel, visual clues):

Describe, refer to existing products if applicable: ________________

____

F. Other Performance Characteristics:

Describe, refer to existing products if applicable: ________________

____

(continued)

Exhibit 2 Continued

III. TOTAL COST OF ALL FORMULA INGREDIENTS (excludes compounding, filling, packaging, etc.):

A. Maximum cost (e.g., $0.475/16 fl.oz. product): ____________

B. Line or individual SKU limitations: ____________

C. Other: ____________

IV. CLAIMS:

A. Ingredient Based Claims:

Note Claims for: a. The presence of specific ingredients or type of ingredients (i.e., Formula contains INGREDIENT XYZ).
b. The performance benefit that is to result (only when applicable).

	a. Ingredient/Ingredient Type	b. Resultant Performance Benefit
1.	____________	____________
2.	____________	____________
3.	____________	____________
4.	____________	____________

B. Performance Based Claims:

1. ____________
2. ____________
3. ____________
4. ____________

C. Competitive Product Based Claims:
(Describe claim and specific competitive product(s)/product version(s).)

1. ____________
2. ____________
3. ____________
4. ____________

(continued)

Exhibit 2 Continued

D. Other Claims:

1. ______________________________

2. ______________________________

E. Demonstrations:

1. ______________________________

2. ______________________________

V. CRITERIA FOR SUCCESSFUL COMPLETION/TEST REQUIREMENTS:

This section should describe, in as definitive terms as possible, the type of testing that will be required and the results that must be achieved in order for the formulation to successfully achieve the requirements of this project.

	Required Testing & Test Design	Results for Successful Completion
1.	______________	______________
2.	______________	______________
3.	______________	______________
4.	______________________________	

VI. PACKAGING

A. Primary Container:

1. Form: Non-Aerosol ____ Aerosol ____ Comment: __________

2. Size/Capacity (based on net content): ______________

fl.oz. _____ or wt.oz. _____

3. Type (non aerosol): Bottle _____ Jar _____ Pouch _____

Other ______________

4. Material of Construction (incl. type of plastic): ______________

5. Supplier & Model, if selected: ______________

6. Pigmented? Yes _____ (Describe ______________)

No _____

7. Other (Specify) ______________

(*continued*)

Exhibit 2 Continued

B. Closure:

1. Material of Construction/Resin type: ____________________
2. Dispensing Closure? Yes ____ (Describe ______________)
   No ____
3. Pigmented? Yes ____ (Describe ____________________)
   No ____
4. Supplier & Model, if selected: ____________________
5. Other (Specify) ____________________

VII. ADDITIONAL COMMENTS/CONSIDERATIONS:

____

____

____

____

____

within the current realm of technology, the first step toward this end is to collect and evaluate information regarding the functionality and safety of available raw materials. This will serve to help determine an appropriate starting point for formulation efforts. Such information is available from a variety of sources.

The veteran chemist has the distinct advantage of having years of experience on which to rely. However, in certain instances, even the veteran may be working in an unfamiliar area. In these cases chemists of all experience levels must turn to other sources for their direction. Seeking the advice of colleagues within a company, or sphere of acquaintances, can be helpful. One can, and should, draw upon the experience of peers whenever possible—perhaps a similar project was explored in the past and a co-worker's efforts may be beneficial to your project. In addition, it is critical that formulators know where to look for additional technical guidance. Many sources of such technical information are available.

1. *Scientific Journals*

Of key value is the *Journal of the Society of Cosmetic Chemists* and its international counterpart, the *IFSCC Journal*. These publications provide cutting-edge discourse by some of the leading researchers in the field. Unfortunately, they are available on a relatively infrequent basis—however, back issues can be very helpful if they are accessible. Other journals that may be of interest to cosmetic formulators, such as *The Lancet*, are available from the American Chemical Society.

2. *Technical Literature*

Numerous reference works are available to the formulator; two of the best known are the *Chemistry and Manufacture of Cosmetics*, by Maison G. de Navarre (Continental Press), and *Cosmetics Science and Technology*, by Balsam and Sagarin (Wiley-Interscience). The advantage of such texts is that they are very thorough and often combine a discussion of cosmetic theory with practical formula information. The disadvantage is that they are published infrequently and may not contain the most current information.

3. *Trade Publications*

Several monthly trade periodicals are available; these are extremely up to date and often focus on issues of specific interest to the formulator. They also provide information on new product releases and trends in the marketplace.

4. *Vendor Publications*

Most major suppliers of cosmetic raw materials make available lists of suggested formulations. Many vendors also provide technical support through their applications laboratories. These resources are not to be overlooked.

5. *Patent Literature*

Patent searches can reveal a wealth of information related to specific raw materials and formulation approaches. In addition to providing information on how others have formulated similar products, they may also help ensure that your work does not infringe upon existing patents.

6. *Internet Sources*

Recent advances in on-line information has brought powerful research techniques into the hands of every chemist with a computer and a modem. Many vendors maintain World Wide Web sites where the formulator can learn more about the raw materials they are using. Several excellent articles have been written about cosmetic science resources on the Web (7).

Regardless of the sources employed to gather information, this initial research phase is helpful in two ways. First, fundamental research provides information about chemical mechanisms and processes, essential to the development of products based on novel or innovative technology. Second, such research can help establish a "master list" of interesting raw materials and potential starting formulations. This list must be pared down to eliminate raw materials which may be unacceptable due to a variety of factors. For example, certain raw materials may be overly expensive, or they may be too difficult to handle to allow reasonable manufacturing, or their use may be prohibited by regulatory statutes.

Ideally, one would choose the most effective ingredients available without consideration of cost. In reality, however, cost considerations may be key to the product's successful commercialization and hence affect formula development. If one were to use only the most expensive raw materials in unrestricted concentrations, it is likely that project budget constraints would be exceeded. This is to be contrasted with the prospect of being unable to meet anticipated performance or esthetic goals due to adherence to unrealistic cost parameters. (Fortunately in this regard, expensive ingredients are often functional at very low levels. For example, fragrance is often the most expensive ingredient employed in a cosmetic formulation, but it is generally used at a relatively low level.) Here, it is critically important for the formulator to advise R&D management and Marketing if significant formula improvements can be achieved with the relaxation of cost constraints.

The manufacturing requirements of a given prototypical formula must be considered during the early steps of the formulation process, even though actual commercialization may be months (perhaps years) away. To this end, it is helpful for the cosmetic chemist to be familiar with typical compounding, processing, and filling equipment and the specific capabilities of the intended manufacturing facility. Ideally, prototypes should be designed for ready manufacture in this facility; however should this not be the case, special requirements must be well (i.e., "loudly" and promptly) communicated and planned for as early as possible. This will allow time for the purchase of equipment, or the evaluation of alternative manufacturing facilities or, of course, the consideration of an alternative formulation approach.

Regulatory factors may also affect the formulation process—even when considering products intended to impart conditioning. Such considerations may be of great importance to such over-the-counter products as intensive skin moisturizers or, perhaps, an antidandruff conditioner. The knowledgeable formulator should strive for an awareness of current and impending regulations which might affect cosmetic products, particularly those related to the product category in which he or she is involved.

C. Prototype Preparation and Testing

The next step in the formulating process is to use the accumulated information to prepare candidate prototypes. The primary functional objectives desired of the final product and the options for achieving these should be carefully considered before initiating actual formulation. Furthermore, it is recommended the formulation be built on careful determination of the ability of candidate primary active ingredient(s) to deliver the necessary functionality. Whenever feasible, this should be determined on a primary basis before beginning actual formulation. For example, before formulating a hair conditioner which focuses on detangling, it is recommended that the ability to ease wet comb of 2% solutions of candidate active materials (presumably quaternary ammonium compounds or amine oxides, with solubilizer if needed) be evaluated on laboratory tresses.

At this point, preliminary ingredient selection should also have been guided by careful examination of basic toxicological test data (generally, primary skin irritation, eye safety, and LD_{50} or some other measure of ingestive toxicity). For formulations built on conventional ingredients, the formulator should also consider prior usage of the candidate ingredients and ingredient combinations. Some degree of confidence can be gained through knowledge that the given ingredients have previously been employed in similar products/chemical systems at similar concentrations, but this should not be taken as a guarantee of safety. Exceptions exist, and additional medical safety testing of the finished products may be required (8–10). In any event, become very familiar with your company's policies regarding such safety testing, and gain expert input during the early stages of the project.

Once a suitable starting point for prototype development has been established, the chemist can prepare small-scale batches. Several excellent references discuss the issues associated with laboratory batching of cosmetic products (11,12). These developmental batches should be subjected to preliminary evaluations to ensure that they are reasonably functional and stable. (Presumably, considerations related to the "safety for use" of the prototypes were well investigated prior to the addition of chemicals to a beaker.) Depending on the conditioning requirements of the product, such preliminary performance evaluation may, for a hair product, be limited to laboratory tress testing and cursory salon evaluation, and to confirmation of esthetic properties for a skin treatment product. Further evaluation (e.g., consumer and/or clinical testing) can take place later in the program.

Similar to conducting preliminary performance evaluations, representative samples should be stored at elevated temperatures (typically at temperatures in the range of 37°C and/or 40°C and/or 45°C) to ensure that primary prototypes are free from gross stability defects. Based on feedback from these preliminary tests, the prototype can be modified until the formulator is reasonably confident the formula is adequately robust to merit further consideration. Since the inherent design of this type of testing is time dependent, it is recommended that promising prototypes be routinely placed in the oven in order to minimize the possibility of late disappointments. Should formulations fail in this early round of testing, the chemist may have to pursue alternative strategies. Accordingly, it is advantageous to pursue multiple formulation strategies concurrently.

Once the "best" formula/formulas has/have been selected, more exhaustive testing can be conducted. At this point, thorough performance testing should be done to confirm that the product meets the requirements established in the product profile. These evaluations may include instrumental evaluations, tress testing, and salon studies for hair conditioning products. Skin conditioning products may require more extensive (and costly) clinical studies. Chapters 13 and 14 focus specifically on test methodologies for these respective areas.

Given the availability of adequate resources and methodology, it is advisable for performance-related parameters to be evaluated initially under the most controlled conditions and with the procedure/instrumentation that will provide the most precise results. For hair products, this could mean testing first with a Diastron, Instron, or Goniometer (or, lacking their availability, conduct laboratory tress testing) under controlled environmental conditions, with carefully controlled application amounts and procedures, and employing relatively uniform samples of hair. Such testing will reap the most precise and reproducible results and reveal the presence or absence of even the slightest degree of functional improvement. *If you don't have it here, you don't have it.* It is pointless, wasteful, and potentially misleading to proceed to a less exact level of testing, such as that conducted in a salon or clinical setting or through consumer testing, without first determining if the targeted benefit can be detected under precise laboratory conditions.

On the other hand, it would be naive to limit testing to that conducted under controlled conditions, because that does not necessarily reflect reality. For example, the degree of improvement observed in the laboratory may not be perceivable under conditions of consumer use—and abuse. Also, controlled testing isolates measurement to one phenomenon at a time, while the consumer does not. The consumer, and to some degree the salon evaluator, integrates and considers simultaneously clusters of performance (and esthetic) characteristics. It is therefore critical to proceed to less controlled forms of testing such as salon testing, clinical testing, or consumer testing. Here, it is absolutely essential that studies are properly designed and adequate replications are conducted. Without such measures (which, in the case of salon testing, is too often the case) there is significant likelihood that results will be erroneous and misleading. *Poor formulations may be adopted and excellent formulations overlooked.* In short, when conducted properly, such testing is relatively slow, laborious, and expensive, but very valuable. Given these considerations, it makes great sense to first determine, via more rapid and accurate laboratory testing, which prototypes are worthy of such a drain of resources.

Formal stability testing should be conducted under a variety of conditions. In contrast with the preliminary testing mentioned above, a well-designed formal stability test scheme exposes the product to a wider range of temperatures, takes into consideration the designated packaging, even challenges the antimicrobial preservation system of the product after heat storage. Ultimately, stability testing should be performed on samples manufactured under actual production conditions. A thorough discussion of such testing is beyond the scope of this chapter, but several references are provided for the interested reader (13,14).

Once the formulator is reasonably satisfied that the formula meets the basic benchmark requirements of the program, final evaluations can be conducted. Consumer testing, for example, is usually conducted late in the program, because of its expense and relatively long time requirements. As noted above, such testing is very important because the consumer's integrated response to fragrance, color, consistency, and performance characteristics cannot be predicted soley through expert judgment. Successful completion of consumer testing is one of the most significant hurdles a potential product must overcome.

D. Formula Finalization

Eventually, if the product development "demons" have been held at bay, the formulator will arrive at a formula which meets all the criteria established at the onset of the project. Hopefully the corporation will then proceed to commercialization as scheduled, but it is not uncommon for factors unrelated to the formula to prevent this from occurring. Regardless of whether the formula is to be employed immediately or held in reserve for later use, it is helpful for the formulator to capture key information related to the development and testing of the product for later reference. Experience in the corporate world has also proven the value of maintaining this information in a concise and standardized format, and in a single, central location that facilitates ease of retrieval. The chemist or manager can easily refer to this important information should questions arise regarding some aspect of product development (e.g., why medical testing was or was not conducted, or why some portion of stability testing was waived).

There are specific areas of product development and testing which favor such documentation. These include the following: key internal tracking numbers (e.g., formula numbers, batch numbers), critical information related to the testing (safety, stability, and performance) which determined the subject formula(s) to be acceptable for commercialization, critical laboratory notebook entries, ingredient listings, relevant memos, and so forth. A sample form

Exhibit 3 Important Information Related to Formula Commercialization

Formula: ______________________; New Product ☐ Formula Revision ☐

Product Name: __________________ **Project Number:** __________

Related Formula Numbers: __________; __________; __________;

A. **Ingredient Cost:** ______________; **Within guidelines? Yes/No** ____

Comments: ____

____

____

B. **Stability Testing:** (Comment/explain, if no testing was required. Attach relevant memos/reports.)

Formula No.	Batch No.	Results

C. **Microbiological Testing:** (Comment, if no testing was required. Attach relevant memos/reports.)

Formula No.	Batch No.	Study	Results

D. **Process Development:** (Comment, if no testing was required. Attach relevant memos/reports.)

Formula No.	Batch No.	Study	Results

Exhibit 3 Continued

E. Clinical or Salon Testing: (Comment, if no testing was required. Attach relevant memos/reports.)

Formula No.	Batch No.	Study	Results

F. Medical Safety Testing: (Comment, if no testing was required. Attach relevant memos/reports.)

Formula No.	Batch No.	Study	Results

G. (Other) ______________________ **Testing:** (Attach relevant memos and reports.)

Formula No.	Batch No.	Study	Results

H. Ingredient Listing (attached): Prepared by: ______________________

Date ______________________

I. General Comments: ______________________

____

____

(continued)

Exhibit 3 Continued

J. Approvals:

Approved by:	____________________	**Date:**	____________
Approved by:	____________________	**Date:**	____________
Approved by:	____________________	**Date:**	____________
Prepared by:	____________________	**Date:**	____________

---

accommodating this important information related to formula commercialization is provied in Exhibit 3.

REFERENCES

1. Gerstein T. Shampoo Conditioner Formulations. U.S. Patent 3,990,991, Nov 9, 1976.
2. Bolich Jr, et al. Shampoo Composition Containing Non-Volatile Silicone and Xanthan Gum. U.S. Patent 4,788,006, Nov 29, 1988.
   Gtete, et al. Shampoo Composition. U.S. Patent 4,741,855, May 3, 1988.
   Oh, et al. Shampoo Composition. U.S. Patent 4,704,272, Nov 3, 1987.
3. Van Scott E, Yu RJ. Hyperkeratinization, corneocyte cohesion and alphahydroxyacids. J Am Dermatol 1984; 11:867–879.
4. Kligman AM, Grove GL, Hirose R, Leyden JJ. Topical tretinoin for photoaged skin. J Am Acad Dermatol 1986; 115:836.
5. Burmeister F, Bennet SE, Brooks G. Liposomes in cosmetic formulations. Cosmet Toilet 1996; 111(9):49.
6. Pauly M, Pauly G. Glycoceramides, their role in epidermal physiology and their potential efficacy in treating dry skin. Cosmet Toilet 1995; 111(8):49.
7. Klein K. The chemist's Internet. Cosmet Toilet 1996; 111(9):31.
8. Zeffren E. Preparing a cosmetic product for marketing: integrating safety testing into product development. Cosmet Toilet 1983; 98(11):48.
9. Waggoner WC. Clinical Safety and Efficacy Testing of Cosmetics, Cosmetic Science and Technology Series—Vol 8. New York: Marcel Dekker, 1990.
10. Whittam JH. Cosmetic Safety, A Primer for Cosmetic Scientists, Cosmetic Science and Technology Series—Vol 5. New York: Marcel Dekker, 1987.
11. Schueller R, Romanowski P. Laboratory batching of cosmetic products. Cosmet Toilet 1994; 109(11):33.
12. Oldshue J. Fluid Mixing Technology. New York: McGraw-Hill, 1983.
13. The Fundamentals of Stability Testing, IFSCC Monograph. Dorset, England: Micelle Press, 1992.
14. Idson B. Stability testing of emulsions. Drug Cosmet Ind 1988; 103:35–38.

13

Evaluating Effects of Conditioning Formulations on Hair

Janusz Jachowicz
International Specialty Products, Wayne, New Jersey

I. INTRODUCTION

A. General Background

Hair conditioners are an important hair care product category with an estimated retail market of about $940 million in the United States alone and probably a few times higher number in sales worldwide (Chain Drug Review, 1996). These products are designed to serve several functions, with the most important being to improve hair appearance, combability (detangling), and manageability. Some conditioning formulations can improve shine, change color, or impart styling if appropriate ingredients are included in the formula. Also, an important product category which has emerged during the last 20 years is that of conditioning shampoos.

One of the principal reasons conditioners are used is to prevent damage to hair by grooming and weathering, or to reduce the perceived damaging effects of chemical treatments. Everyday grooming procedures such as washing, rubbing, and combing primarily affect the fiber surface. This results in breakage and gradual erosion of the cuticle cells (1,2). Weathering, a combined action of solar radiation, temperature, and humidity, results in surface damage, decomposition of protein structural components (i.e., tryptophan, cysteine), and formation of oxidation products (3). Exposure to high temperatures by the use of hot irons or hair dryers can also produce similar modifications to hair. By far the most severe damage is caused by chemical treatments such as bleaching, dyeing, waving, and relaxing (1). These treatments result in increased surface roughness, oxidation of disulfide or thiol groups, and removal of hair lipids. The immediate and most obvious adverse effect of chemical treatments is an increase in hair porosity, which provides an altered feel that is often described as dry and raspy. Recent results of X-ray analysis suggest that chemical treatments can also affect crystalline, alpha-helical components of hair structure (4). Repeated use of chemical treatments may lead to even more profound changes, readily perceived by consumers, such as wrinkling of cuticle cells, complete stripping of cuticles, fibrillation of the cortex, and the appearance of split ends.

B. Chemistry of Hair Conditioners

The selection of an appropriate method for conditioner analysis, and the interpretation of the experimental data, is based on the understanding of the mechanism of interaction for various classes of conditioning agents with hair. These include oils (pure or in the form of emulsions), cationic surfactants, and cationic polymers. The compositions may also contain a number of ancillary agents such as fatty alcohols, nonionic or anionic polymers, surfactants, sunscreens, antioxidants, preservatives, etc. Frequently, these formulation aids can interact with actives by forming liquid crystals, gels, lamellar phases, or complexes and participate in the precipitation of conditioning layers on the surface of hair. The function of some additives may extend well beyond the typical functions of conditioning agents, as in the case of sunscreens or antioxidants. In such cases, special methods have to be used to evaluate the effectiveness of the conditioning product. This area will not be covered in this chapter.

Early conditioners used by ancient Jews, Egyptians, Assyrians, Babylonians, Persians, Greeks, and Romans through the Middle Ages and until the end of the nineteenth century consisted primarily of vegetable and animal oils (5). Olive, almond, and behen oils as well as softer fats of the ox, clarified beef marrow, veal suet, and hog's lard were the materials of choice when formulating products such as creams, pomades, generants, pomatums, bear greases, etc. Fragrancing was usually accomplished by the addition of essential oils including oil of bergamot, lemon, roses, rosemary, lavender, etc. A recommended method for testing an oil, when determining its effectiveness as a treatment for hair, was to smear it on a glazed tile and expose it to air at a temperature comparable to scalp temperature for a period of two or three days. If the oil did not become too thick or sticky (as a result of polymerization or crosslinking), or too rancid (hydrolysis), it would pass the test and would be deemed acceptable for hair conditioning purposes.

The selection of oils for contemporary formulations is very wide and includes animal-based oils such as beef tallow, lanolin, and mink oil. In addition, vegetable-based oils have gained widespread use, such as those derived from soy beans, wheat germ, olive, mineral oils (i.e., paraffin oil), petrolatum, silicone oils, and synthetic oils (polymeric hydrocarbons, synthetic esters, or perfluorinated ethers). The ability of various oils to perform conditioning functions is related to their molecular characteristics, such as surface energy, cohesive forces in surface layers, and surface shear viscosity (6,7). It also depends on the format of the entire hair care formulation and the presence of other ingredients. Typically, shampoos, daily conditioners, deep conditioners, revitalizing treatments, and conditioning shampoos are formulated as oil-in-water emulsions. On the other hand, one-phase nonaqueous oil systems are used for shine-enhancing or frizz-eliminating treatments.

Although oils are still in widespread use, they have been largely replaced by other specialized chemicals such as cationic surfactants and cationic polymers. Cationic surfactants, the most widely used ingredients of conditioning formulations, are represented by a wide variety of structures with differing performance characteristics. The mechanism of hair conditioning by cationic surfactants has been investigated by several researchers, with the earliest work completed by Robbins et al. (8) and Finkelstein et al. (9). This work helped to elucidate the surfactant structure–property relationships, the effect of hair structure on the sorption of cationic surfactants, and the mechanism of sorption from multicomponent systems containing anionic surfactants and/or polymers. Solution depletion methods in conjunction with kientic data analysis as well as staining techniques employing orange II, Rubine dye, and methylene blue were employed in these studies. At the same time, a considerable amount of experience was gained by formulation and sensory analysis of finished products. All of this information led to the formulation of some general rules regarding the use of cationic surfactants in conditioners. For example, a class of saturated single-chain quaternary surfactants with shorter chain lengths (C_{14}–C_{16}) and characterized by faster desorption rates and thinner deposition layers, are used primarily in mild formulations for thin or normal hair. On the other hand, surfactants with longer chain lengths (C_{18}–C_{22}), characterized by their high substantivity to hair and thicker deposition layers, are employed in systems designed for coarse and/or damaged hair. Other classes of cationic surfactants such as saturated multiple-chain quaternary compounds, alkylamidoamines, perfluorinated cationic surfactants, etc., may also be added to a formula to perform some special function. These functions may include triboelectric charge control, perceived faster drying of hair, enhanced softness, "sealing" of damaged hair, increased moisture content, better color retention after dyeing, increased mechanical strength, split-end mending, increased luster, and improved rinsability or washability. These properties are normally a function of the composition of the entire formulation and can be quantified by the selection of an appropriate measurement method.

Another class of raw materials for hair conditioners is cationic polymers. This class of compounds tends to impart a different feel to hair than quaternary surfactants, and their substantivity can be controlled by a selection of monomers, degree of quaternization, etc. They may also be deposited onto hair effectively from formulations containing a large excess of anionic or amphoteric surfactants, such as shampoos, oxidative hair dye lotions, etc. Early work on the interaction of hair with cationic polymers was reported by Chow (10), Woodward (11), and Goddard (12). These authors have demonstrated that strong interactions exist between hair keratin and cationic polymers such as polyethyleneimine, cationic cellulose, poly(dimethyldiallylammonium chloride), and co(vinylpyrrolidone-methacryloxyethyltrimethylammonium methosulfate). A variety of different techniques were used to perform fundamental studies on a polymer's affinity to hair. Radio-tracer techniques with ^{14}C-labeled compounds (11–13) and colloid polymer titration methods (14) were employed for quantitative sorption studies, while X-ray photoelectron spectroscopy (15), electron spectroscopy for chemical analysis ESCA (16), wettability (17), and electrokinetic measurements (14,15,18) were used to characterize the surface of hair after modification with polymers.

Finally, proteins constitute an important class of conditioner raw materials. They are claimed to impart softness to the hair, increase tensile strength, add body and gloss, enhance springiness, and improve the overall look and feel of hair. While there is little peer-review literature to support the effect of proteins on the physicochemical properties of hair, the substantivity of polypeptides to hair keratin is well established. Quantitative sorption work has been reported by Karjalla et al. (19), Turowski et al. (20), Mintz et al. (21), and other authors.

II. REVIEW OF TESTING METHODOLOGIES

A. Instrumental Techniques

1. Combing Analysis

Quantitative combing measurements are frequently used to evaluate the effectiveness of hair care products. The development of this technique can be attributed to Newman et al. (22), Tolgyesi et al. (23), Garcia et al. (24), and Kamath et al. (25). A variation of the method, termed spatially resolved combing analysis, has been recently presented by Jachowicz et al. (26). Figure 1 presents a photograph of an experimental setup based on a Diastron miniature tensile tester. In the usual procedure, a comb is passed through a hair tress and the force is measured as a function of distance. The experimental variables include the dimensions of a tress, the density of combing teeth, and the type of comb material. Most of the work reported in scientific and patent literature relates to hair tresses with a length in the range of 6–7 in., a weight of 1–3 g,

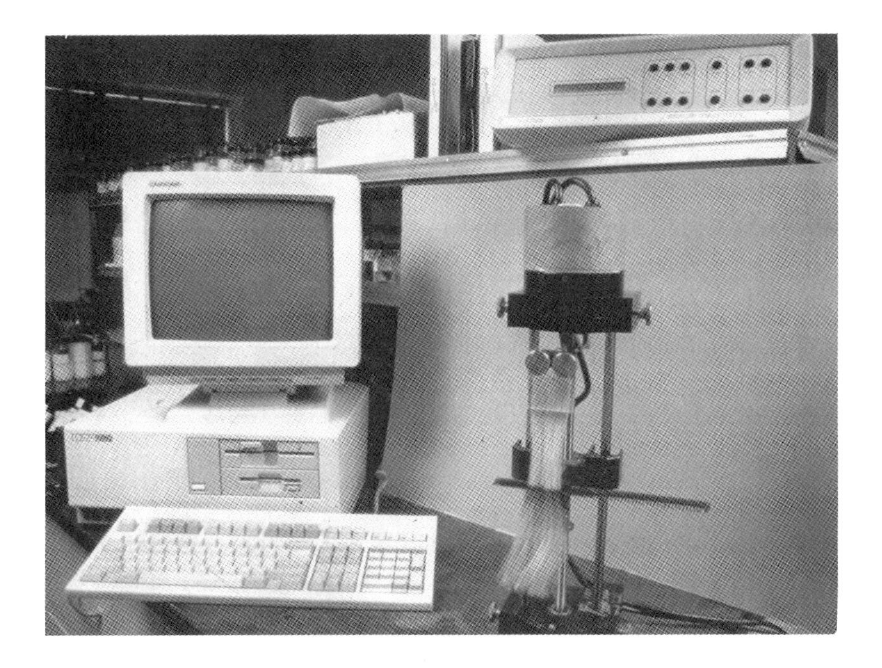

Figure 1 Experimental setup for performing combing measurements by employing a Diastron tensile meter.

and a tress width of about 1 in. (24). The magnitude of combing forces is higher for tresses with a higher density of hair. The combing force may also be increased as the distance between the teeth in the comb are decreased. Combing forces are, however, independent of the comb material, indicating that the forces are primarily due to fiber–fiber friction rather than fiber–comb friction (27). The measurements can be performed on both dry and wet fibers. Examples of dry and wet combing curves are included in Figure 2. The wet combing curve suggests relatively high combing forces throughout the whole tress, with a small disentanglement peak corresponding to the fiber tips. On the other hand, dry combing traces show small forces for most of the hair tress, with the exception of a large peak at the fiber ends arising from the disentanglement of fiber tips. To illustrate the magnitude of combing forces and associated errors, Table 1 shows typical results obtained for different types of intact and damaged hair. The data are presented in terms of wet and dry combing works, obtained by integrating combing forces values over the tress length. Note that the standard deviations for wet combing range from about 15% for easy-to-comb Piedmont (unpigmented) hair to 50% for damaged hair. Similar trends, with lower errors for undamaged hair, are evident from dry combing data.

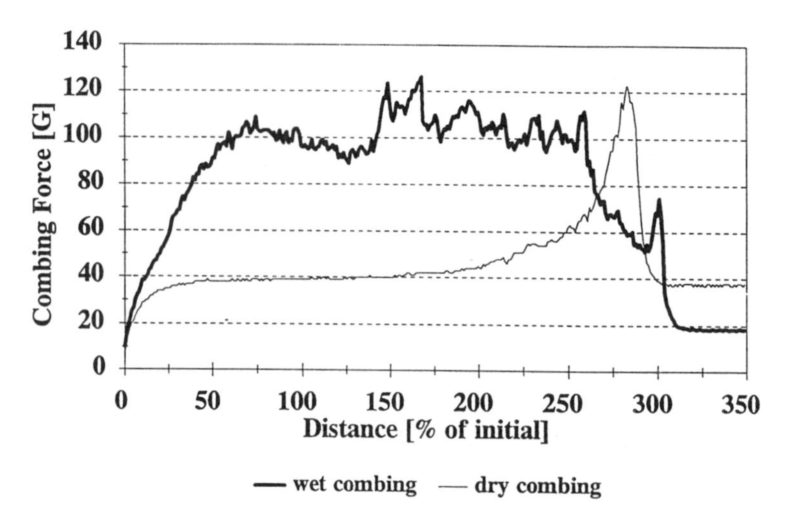

Figure 2 Wet and dry combing curves obtained by using a setup shown in Fig. 1.

A basic procedure for the evaluation of the effect of a conditioning treatment consists of (1) measurement of the dry and wet combing characteristics of untreated hair, (2) treatment with a conditioner followed by rinsing, (3) measurement of the dry and wet combing characteristics of treated hair, (4) shampooing, and (5) remeasurement of combing characteristics in order to ascertain the removability of a treatment. This sequence of treatment, shampooing, and measurement can be modified to include multiple treatments for the study of formulations such as conditioning shampoos, conditioning hair coloring products, or conditioning perms, which typically produce relatively small lubrication effects after a single application. The data obtained in such

Table 1 Wet and Dry Combing Works on Various Types of Hair

Type of hair	Wet combing work (g-cm)	Dry combing work (g-cm)
Piedmont	860 ± 130	166 ± 50
Dark brown Caucasian	2179 ± 675	678 ± 171
Oriental	2510 ± 890	—
Fine brown Caucasian	3257 ± 712	308 ± 28
Permed	3353 ± 897	—
Bleached	4878 ± 1054	309 ± 243
Dyed	5448 ± 2447	—

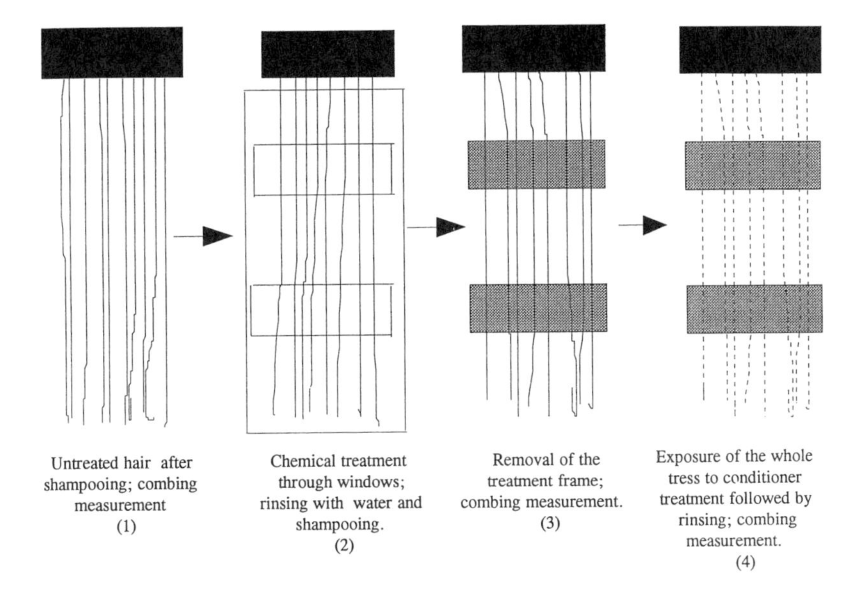

Figure 3 A scheme of the experimental protocol for using spatially resolved combing analysis of hair.

experiments allow for an accurate assessment of the effectiveness of a conditioning treatment and its substantivity to hair. To obtain statistically significant data, the measurements must be performed in replicate on several tresses.

A modification of the basic procedure, termed spatially resolved combing analysis, can be particularly useful to detect and quantify conditioning effects on hair damaged by chemical treatments and by heat or photo exposure. A detailed description of the technique is provided in Ref. 26 with one of the experimental protocols presented schematically in Figure 3. In this method, special frames are employed which allow the application of a treatment to selected areas of the fibers while shielding the remaining portions of the tress, thereby providing internal reference sections. The combing traces of hair treated in such a way show positive or negative peaks depending on whether the treatment results in an increase or a decrease in friction against the hair surface. The damaging or conditioning effects can be further quantified by calculating the differences in combing force or work values corresponding to the treated areas. Figure 4 presents an example of the application of this method to study the effect of bleaching followed by conditioning with a cationic polymer. Bleaching results in a three- to fourfold increase in combing forces as compared to untreated hair, while subsequent application of a polymer solution decreases the combing forces. The shape of the combing curve

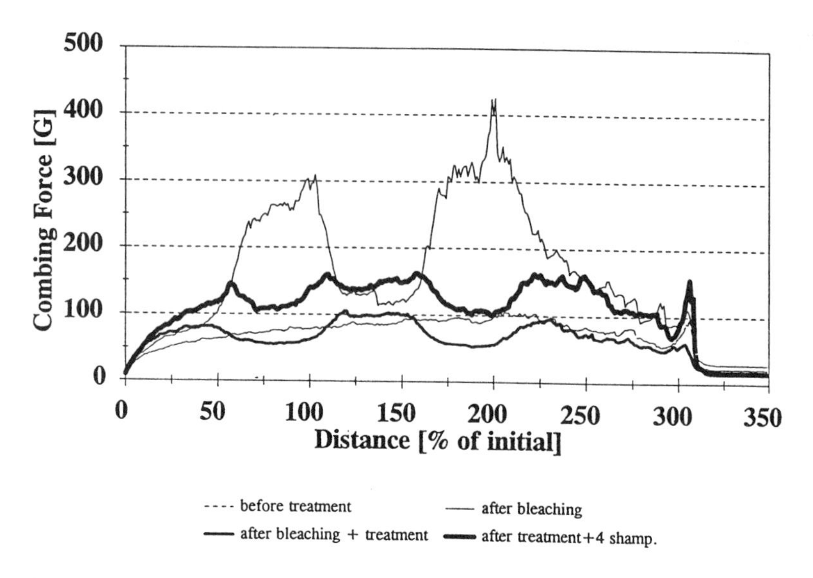

Figure 4 Wet combing traces for untreated and bleached hair obtained by following a procedure of spatially resolved combing analysis presented in Fig. 3.

after the polymer treatment, with two minima corresponding to the bleached sections of hair, indicate that polymer-modified damaged hair is actually easier to comb than untreated hair. Moreover, adsorbed polymer cannot be removed by four shampooings, as evidenced by low combing forces and the shape of combing trace characterized by the minima in the bleached sections. It is noteworthy that the data presented do not point to a buildup of the analyzed polymer on hair. Buildup is usually analyzed by a consecutive application of the same formulation on hair and can be quantified by the measurements of the amount or the thickness of the surface deposits. The data presented in Figure 4, or results obtained in similar experiments, can be further analyzed by curve subtraction, integration, averaging, or minima/maxima selection to provide indices characterizing the extent of conditioning, removability of conditioning agents by shampooing, and the buildup of conditioning actives.

2. *Tensile Strength*

Mechanical measurements are not frequently used to study the effect of conditioners on hair because it is generally assumed that the conditioner actives primarily produce a surface modification of hair and do not affect its bulk properties. Since mechanical modulae and viscoelastic properties of keratin are determined by the internal elements of hair structure, it is expected that the deposition of few milligrams per gram of a conditioning agent could not significantly alter the mechanical properties of hair. However, this view is contradicted by a recent report suggesting that the adsorption of polyquaternium-10 or guar hydroxypropyltrimonium chloride can affect the mechanical strength of hair (28). Measurements of hair strength can be performed by taking standard stress–strain curves for single fibers employing a tensile tester (Instron or Diastron) (2,29). The parameters calculated from the data are the tensile strength (stress at break) or the Young's modulus, which is calculated from the initial slope of the stress–strain curve. An automated tensile testing device for fibers, devised to shorten the processing time of a large number of single fibers and to obtain statistically valid information, has also become available from Diastron.

3. *Hair Body*

Hair body is an important hair attribute associated with the volume and mechanical properties of hair fiber assemblies. The importance of this hair property has been discussed by several authors, who stress that it is a psychophysical attribute which can be assessed by visual or tactile perception. The variables affecting hair body include fiber density (hair/cm^2), bending stiffness, hair diameter, hair configuration or shape, and fiber–fiber interactions (30–32). Several instrumental methods have been proposed to quantify hair body, such as the ring compressibility method (33,34), radial compressibility method (35), interfiber adhesion measurements (35), and tress volume measurements (36). In the ring compressibility method, a hair tress attached to a load cell of an Instron tensile meter is passed through compressing rings with various diameters and the force is measured as a function of distance. Increased body or springiness of hair is indicated by higher compression forces. In the radial compressibility method a tress is radially compressed twice by a compression ring attached to a load cell. From this experiment several parameters, including the fiber adhesion index (FAI) and ΔE_2, can be calculated as follows:

$$\Delta E_2 = \frac{100(E_{2,\text{ treated}} - E_{2,\text{ untreated}})}{E_{2,\text{ untreated}}}$$

$$\text{FAI} = \frac{100(E_{1,\text{ treated}} - E_{2,\text{ treated}})}{E_{1,\text{ treated}}}$$

where E_1 and E_2 are the energies of the first and second compressions, respectively, and treated or untreated refers to treated or untreated hair. The FAI gives information about the recovery of treated fibers from compression, while ΔE_2 quantifies the effect of treatment on the energy of the second compression. ΔE_2 was found to be less dependent on tress preparation which consists of combing and fluffing than the analogously defined ΔE_1 parameter.

Measurements of tress compressibility may be complemented by measurements of the interfiber adhesion, which can be performed by pulling a single fiber out of a bundle and recording the forces with a microbalance. The parameters characterizing hair body, ΔE_2, interfiber adhesion, FAI, and E_1 were found to correlate with the body rating of tresses by expert panels.

4. *Measurements of Luster*

The gloss or luster of hair is an important property which can be visually assessed by a consumer and is also considered to be a desirable hair attribute. Several instrumental methods have been employed for quantitative characterization of hair luster and for the evaluation of the effect of various cosmetic treatments on hair gloss. These include goniophotometric techniques and image analysis.

a. Goniophotometric Techniques. The use of goniophotometers is the most precise way of quantifying optical properties of hair. The foundation of hair gloss research was laid by a series of papers by Stamm et al. (37) which explored the effect of hair morphology on hair shine and gave a quantitative and qualitative description of light-scattering curves. Goniophotometers can perform measurements of light scattered by fiber tresses (37,40) or by single fibers (38,39) as a function of an incidence or receptor angle. In the most frequently employed procedure, a light source illuminates a hair sample at an incident angle and the light intensity is recorded for different receptor angles resulting in a light-scattering curve. An example of a typical light-scattering curve, with light intensity as a function of the scattering angle, is given in Figure 5. It consists of a primary peak due to specular reflection (for the incident angle of 45° the specular reflection should appear with the maximum at about 90°) and a secondary peak arising from light which penetrated hair and is reflected or scattered on the internal elements of the fiber structure. In highly pigmented fibers the intensity of the secondary peak is small. While the light contributing to the specular reflection is largely co-polarized (i.e., polarized in the same direction as the incident light), the light corresponding to the secondary peak is significantly cross-polarized. In addition to this, the intensity (or sharpness) of the light-scattering curve can be affected by the polarization of the incident and reflected light. The reason for this is that the reflected s-polarized light (light with its electronic vector perpendicular to the plane of incidence) is approximately five times more intense than the reflected p-polarized light, which, in turn, is due to the difference in reflection coefficients for s- and p-polarized incident light.

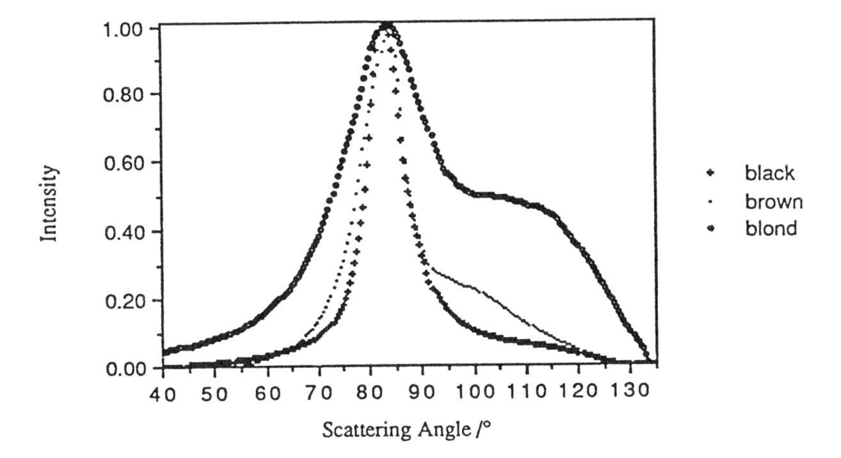

Figure 5 Goniophotometric curves of fibers illuminated (at an incident angle of 45°) using unpolarized white light. Data have been normalized so peak intensity corresponds to unity. (From Ref. 38.)

A quantitative measure of luster can be calculated by the following formula:

$$L = \frac{S}{DW(\frac{1}{2})}$$

where L is luster or shine, D is integrated diffuse reflectance, S is integrated specular reflectance, and W is the width of a specular peak at half-height.

The shape of the light-scattering curve for untreated hair, and thus the intrinsic luster of hair, depends on the state of the hair surface and the degree of hair pigmentation. An increase in melanin content in dark-colored hair reduces the intensity of light scattered on the internal cuticle surfaces and on other internal elements of hair structure. This results in lowering the diffuse reflectance and sharpening the specular peak, consequently providing an increase in hair luster. For damaged hair, in which the surface layers of cuticles are jagged and uneven, the intensity of scattered light (diffuse reflectance) is high to the point that the incident reflection peak can appear as a broad plateau rather than a sharp peak. It has been demonstrated that some types of hair treatments, including conditioning hair rinses, can increase the "sharpness" of the specular reflection and thus improve the luster of damaged hair. On the other hand, it was found that the luster of intact, undamaged hair, characterized by a well-defined and "sharp" specular reflection, cannot be increased by application of conditioning rinses or conditioning shampoos, which usually leave a light-scattering deposit on the hair surface. An increase in shine for this kind of hair can be produced by the use of oil-based shiners or clear hairspray resins which are capable of forming a continuous film on the surface of the fibers.

b. Image Analysis. Luster measurements can also be performed by using a color image processor which can analyze a pattern of light reflected by a natural-hair wig on a model head (41). In this study, hair was illuminated with white light at an angle of 30°. The data were obtained by scanning across highlighted and dark areas of the obtained image, and could be presented in a plot similar to a photogoniometric scattering curve. In addition to this, the reflected light could be resolved into three color signals, R, G, B (red, green, and blue) or into L^*, a^*, b^* parameters. Good agreement was found between calculated luster, defined as the ratio of the lightness in the highlight area to the lightness in the dark area, and the rating of luster obtained by a visual inspection of hair tresses. An even better correlation was found by employing a contrast parameter defined as

$$L_{ph} = \frac{B_h}{R_d}$$

where L_{ph} is luster, B_h is blue light intensity, and R_d is red light intensity in the dark area. The values of this parameter were found to be consistent with the visual ratings of luster for hair treated with hair dyes and a variety of hair grooming products. According to this work, oil-based luster sprays, styling gels, and pomades were effective in increasing the luster of hair. In contrast to other reports, the use of a conditioner after shampooing also resulted in an increase in specular reflection and a decrease in scattered reflection. In addition to this, the technique was found useful in substantiating the luster variations in colored hair, such as in blue-dyed Oriental black hair.

5. Visual Evaluation of Luster

The method of subjective shine evaluation is commonly used and can provide useful guidance for formulators. The results can be in good agreement with instrumental methods provided the experimental setup assures uniform orientation of hair samples and reproducible (from sample to sample) illumination conditions. The method usually employs special mounting frames which expose the tested tresses to artificial light illumination in such a way as to produce highlight (specular reflection) and dark areas. For example, Reich et al. (39) reported using hair tresses mounted at the root end and stretched over the cylinder while secured at the tip. Six tresses were mounted in a row on the testing rack and were illuminated by four bulbs in a single row approximately 10 in. above the tresses. The tresses were evaluated by 16–20 panelists, who ranked the tresses in order of relative shine. After each evaluation, the tress positions were interchanged to minimize any positional bias. An alternative method of visual evaluation was described by Maeda et al. (41), who used hair bundles attached to a model wig which were illuminated at an angle and evaluated by five experienced analysts.

It should also be mentioned that visual evaluations and tress rankings, such as those described above, can be complemented by simple light-intensity measurements using a photographic spot meter (for example, selected models manufactured by Minolta) to obtain a quantitative measure of gloss according to the formula

$$L = \frac{S - D}{S}$$

where S is the intensity of the specular reflection and D is the intensity measured in the background, diffuse scattering area.

6. Dynamic Electrokinetic and Permeability Analysis

Although streaming potential analysis of fibers has been used in the textile chemistry for some time, the simultaneous measurement of electrokinetic parameters (streaming potential and conductivity) and permeability of fiber plugs is a relatively new approach to study both model conditioning actives and complete commercial formulations (18,42). A simplified scheme of a DEPA experiment is presented in Figure 6. A typical experimental protocol includes the measurements of untreated hair, treatment with a conditioning agent, and measurements of the kinetics of sorption/desorption of ions during rinsing with a test solution (5×10^{-5} M KCl). The streaming potential data, converted into zeta potentials by means of the Smoluchowski equation, give information about the state of the fiber surface and the presence of adsorbed anionic or cationic groups. Conductivity, on the other hand, is related to the presence of free ions in the test solution (although surface conductivity of hair may also be a contributing factor), and its variation in the course of experiment is due to the desorption of ions into the test solution. In addition to this, changes in the flow rates (permeability) indicate variations in the volume of the fibers (i.e., swelling or shrinking) or deposition of surfactant or polymer on the fiber surface.

The method was employed to study single- and multicomponent surfactant or polymer solutions and permitted a comparison of their ability to modify hair surface. Quaternary ammonium surfactants, fatty amines, cationic polymers, and silicone emulsions were investigated as model systems and provided background information about the behavior of various colloids (18,42). Prototype and commercial formulations of various types were studied as 1% aqueous solutions (42). The main criteria for performance were (1) changes in zeta

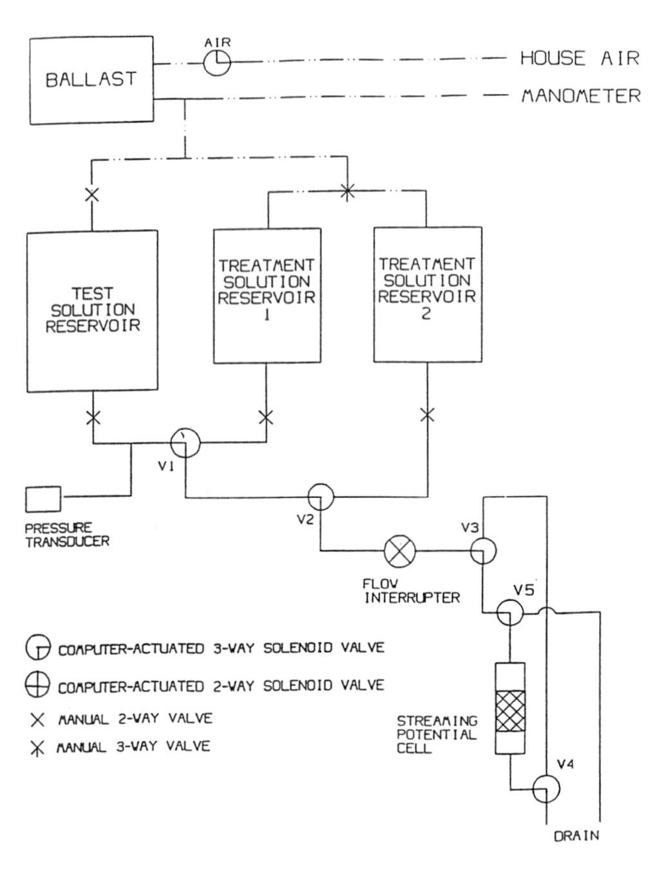

Figure 6 A scheme of an apparatus for performing dynamic electrokinetic and permeability analysis of hair plugs.

potential as a result of the application of a conditioning treatment including those occurring during the rinsing period, (2) changes in permeability of the plug and thickness of a deposited conditioning layer, and (3) changes in plug conductivity. Typically, hair treated with conditioners containing water-soluble and less substantive ingredients is characterized by thinner deposits of conditioning agents and by lower zeta potentials. Low-substantivity conditioners may also desorb during rinsing with the test solutions, as evidenced by a gradual decrease in zeta potentials and an increase in permeability. On the other hand, conditioners based on water-insoluble quats such as polymers and amino-functional silicones result in higher positive zeta potentials, thicker deposited layers, and less variation of these parameters as a function of rinsing time. Figure 7 presents the DEPA analysis of three commercial conditioners,

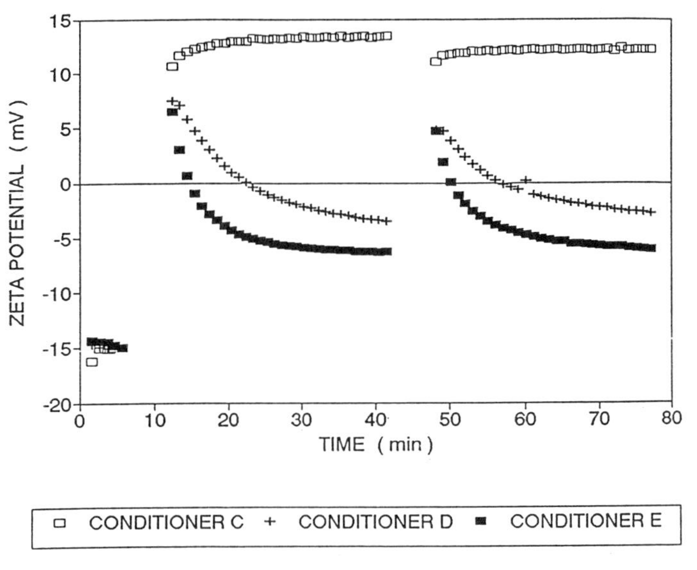

(a)

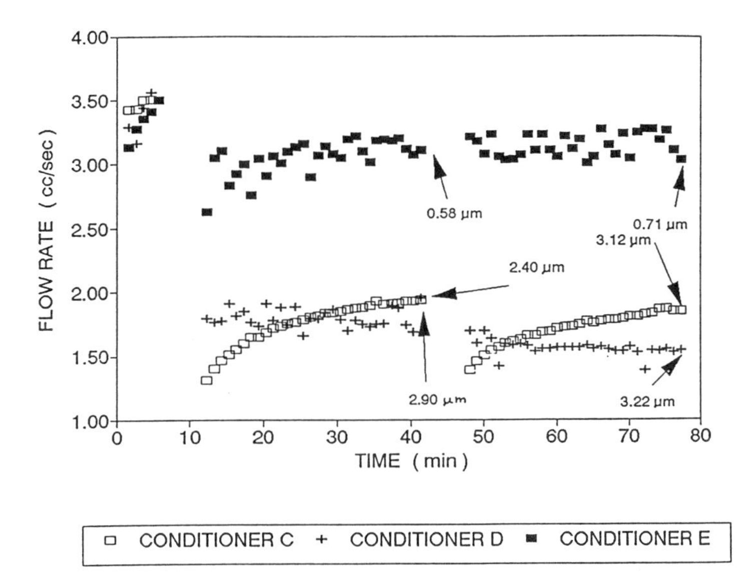

(b)

Figure 7 Experimental traces obtained by using DEPA measurements. (a) Zeta potential, (b) flow rate, and (c) conductivity as a function of time for hair treated with 1% solutions of conditioners C, D, and E (42).

(continued)

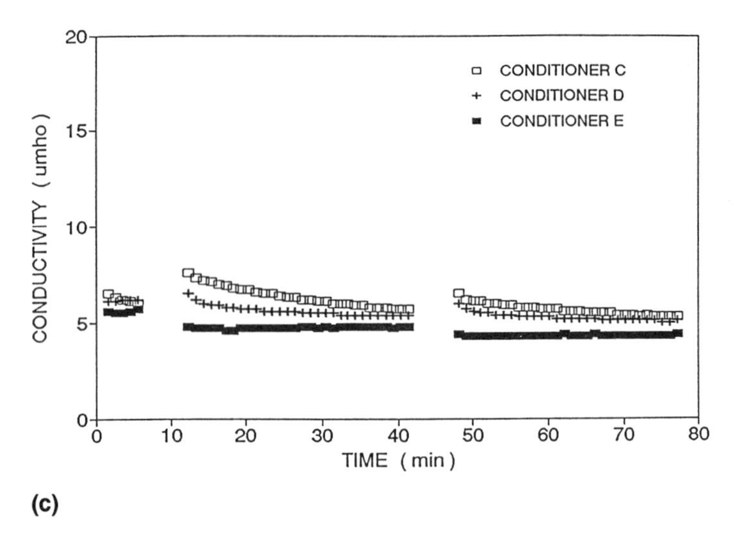

(c)

Figure 7 Continued.

C, D, and E, designed for damaged/dyed, regular, and fine hair (42). The data presented in the figures are consistent with the above-delineated criteria of performance. Conditioner C, based on most substantive ingredients, produced the greatest change in zeta potential of hair and also precipitated the thickest conditioning layer. On the other hand, the use of conditioner E resulted in the least extensive surface modification, as evidenced by a gradual decrease of zeta potentials during rinsing with the test solution and a relatively thin layer of conditioning agents on the hair surface. Conditioner D showed behavior intermediate between that of conditioners C and E.

DEPA experiments with multiple applications of a conditioner followed by shampooing can provide information about the removability of conditioner residues from hair and can be used for the evaluation of buildup parameters. The results of such an experiment are presented in Figure 8 as plots of zeta potential, flow rate, and conductivity as a function of time. The conditioner employed in this study was based on substantive components such as bephenyltrimethylammonium chloride and silicone emulsion DC 929 and resulted in a significant modification of hair as reflected by high positive values of zeta potential and a thick, 3.56-μm layer of the conditioning agent. Subsequent shampooing washed away most of the adsorbed species, lowering the zeta potentials and reducing the thickness of the conditioning layer to 0.71 μm. The data from this experiment can be used to calculate the extent of removability of conditioning residues in a shampooing step and their buildup as a result of two

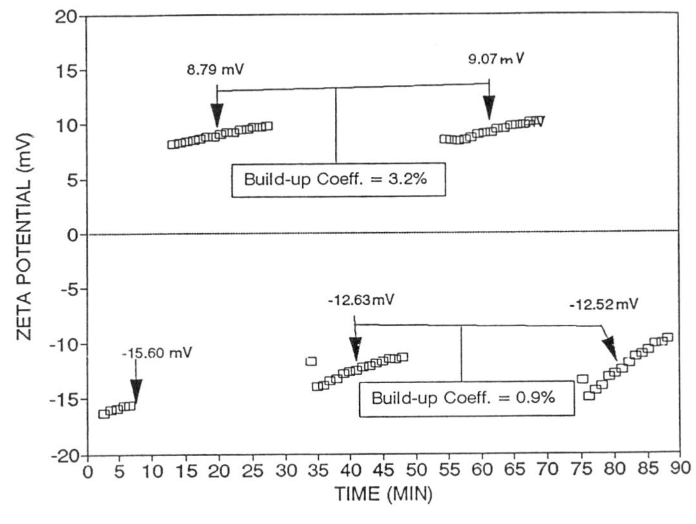

(a)

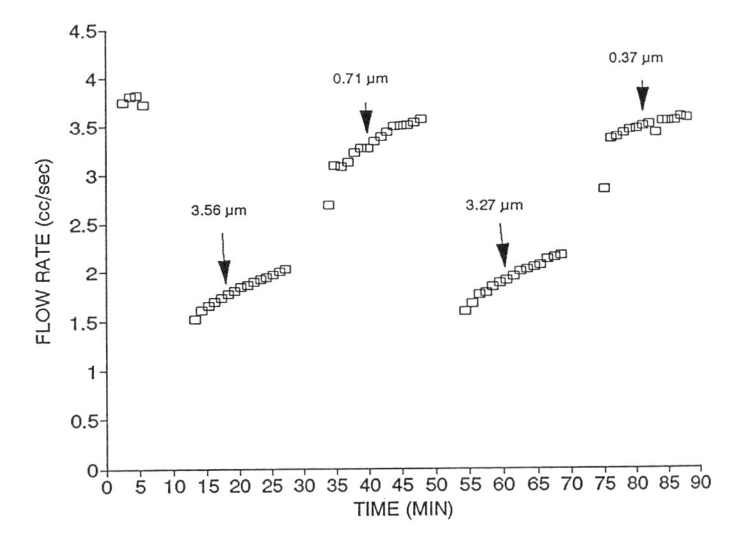

(b)

Figure 8 Experimental traces obtained by using DEPA measurements. (a) Zeta potential, (b) flow rate, and (c) conductivity as a function of time for hair treated in a sequence conditioner-shampoo-conditioner-shampoo.

(continued)

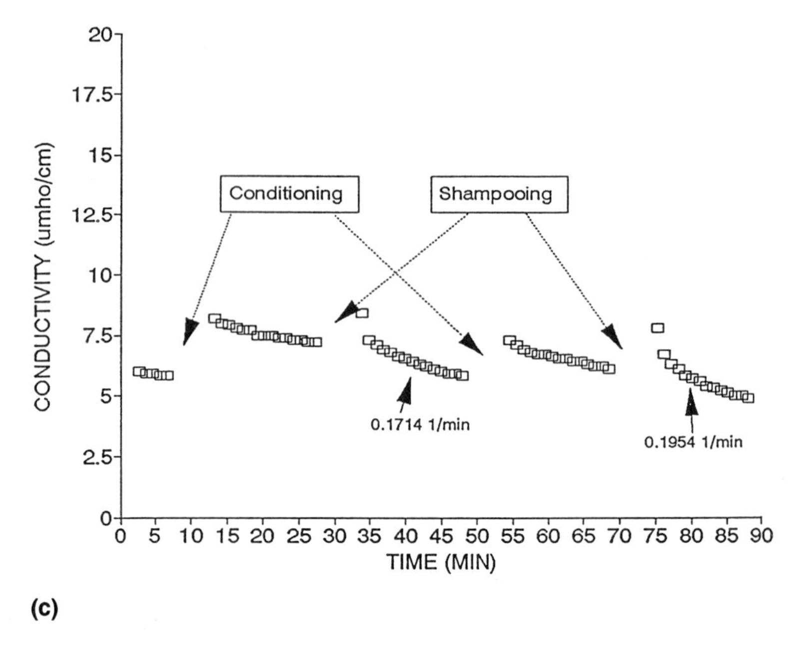

Figure 8 Continued.

conditioner/shampoo applications. The results of such calculations based on zeta potentials are shown in Figure 8a and indicate that the buildup of conditioning agents for this system is small. In addition, the kinetic conductivity data provide rate coefficients of desorption of shampoo surfactants during rinsing with the test solution. This parameter can probably be correated with the ease of rinsing off residual amounts of surfactants from hair, and the results presented in Figure 8c suggest faster desorption rates after the second shampoo application.

Finally, DEPA can be employed to substantiate the "sealing" effects on hair. For example, the "sealing" effect produced by a microemulsion was demonstrated on hair colored with semipermanent dyes (48). On contact with water, such hair exhibits increased plug conductivity resulting from a diffusion of the dyes out of the fiber structure into the test solution. A treatment of dyed hair with cationic silicone emulsions resulted in an immediate reduction in plug conductivity, suggesting that the deposited layer of silicone oil hindered the diffusion of the dyes and surfactants out of the fiber. Similar results were also obtained for other cationic conditioning agents used as treatments for oxidatively dyed or bleached hair (44).

7. Wettability Measurements

Tactile perception of hair and its interactions with hair care formulations can frequently be correlated with the wetting behavior of hair surface. Therefore, the wettability of hair was studied to determine relevant parameters such as contact angles in various liquids, wettability forces, wettability as function of surface modification, etc. (45–50). All of these studies were carried out using the Wilhelmi balance method.

A fundamental surface parameter calculated from wettability measurements is the contact angle of a liquid with the fiber surface, as defined by the Young-Dupre equation:

$$\cos\theta = \frac{\gamma_{SV} - \gamma_{SL}}{\gamma_{LV}}$$

where γ_{SV}, γ_{SL}, and γ_{LV} are solid–vapor, solid–liquid, and liquid–vapor interfacial tensions and θ is the solid–liquid contact angle.

In the case of Wilheimi balance measurements, wettability W, defined as the wetting force w per unit length of the wetted perimeter P, is given by

$$W = \frac{w}{P} = \gamma_{LV} \cos\theta$$

The apparatus for the wetting force measurements consists of an electrobalance and microscope stage for raising and lowering the liquid level. The device can measure both receding and advancing wetting forces, from which parameters such as wettability angles, wettability forces, work of adhesion, and wetting hysteresis can be calculated. Fiber perimeter can be estimated by determining the lengths of the major and the minor axes from microtomed cross sections of the fiber by optical microscopy, or by using a laser diffraction method.

Based on fiber wettability measurements, intact hair (root section) was found to be hydrophobic, with the advancing and receding contact angles equal to 103° and 89°, respectively. The critical surface tension of intact hair was found to be 26.8 ± 1.4 dyne/cm, with the dispersion and polar components equal to 24.8 ± 2.2 and 2.6 ± 1.3 dyne/cm, respectively. These data are based on the measurements of advancing contact angles in water and methylene iodide. Weathering and chemical treatments typically make hair more hydrophilic, resulting in a decrease in contact angle and an increase in wettability forces. For weathered hair, the advancing contact angle was shown to be reduced to 72°, and similar decreases were observed for bleached or permed hair. Deposition of cationic polymers and surfactants, actives frequently employed on conditioning formulations, can modify the surface characteristics of hair. Adsorption of cationic polymers or surfactants on intact hair was found to produce more hydrophilic hair. For example, for hair treated with polyquaternium 10, advancing wettability increased from –8 ± 3 mN/m for untreated hair to 35 ± 16 mN/m for hair with an adsorbed layer of the polymer. This technique was also shown to yield information about the distribution of surface deposits, the effect of temperature on the interactions between hair and surface-modifying agents, and the degree of surface coverage by adsorbed polymers or surfactants (47,49). Multicomponent systems containing surfactants and polymers were also employed to simulate the performance of commercial formulations such as shampoos and conditioners. In the case of shampoos, a commonly occurring phenomenon is the formation of complexes between anionic surfactant and cationic polymer, which were shown to produce hydrophobic or hydrophilic surface deposits depending on the ratio of polymer/surfactant concentrations (49,50). Cationic surfactants, on the other hand, were shown not to interact with the cationic polymers, due to charge repulsion, but to compete with them for the adsorption sites, thus affecting their relative surface coverage.

8. Analysis of Triboelectric Charging (Static Electricity)

Triboelectric charging of hair is frequently observed, especially in low-humidity conditions, giving rise to studies concentrating on the mechanism of this phenomenon. One of the earliest reports was Martin's observation of a directional triboelectric effect for wool fibers (51). He reported that a fiber pulled out of a bundle by the root carried a positive charge, while a fiber pulled out by the tip carried a negative charge. Martin attributed this sign-reversal phenomenon to the piezoelectric effect. The directional triboelectric effect in hair was further confirmed by Jachowicz et al. (52) while performing rubbing electrification experiments using materials characterized by various work functions such as Teflon (reported work function range: 4.26–6.71 eV), aluminum (3.38–4.25 eV), nylon (4.08–4.5 eV), polycarbonate (3.85–4.8 eV), and poly(methyl methacrylate) (4.1–4.7 eV) as rubbing materials. It was shown that the sign and magnitude of static charge on hair depends on both the direction of rubbing (root to tip or tip to root), as well as the nature of the rubbing material.

The instrumentation employed for these studies is presented in Figure 9. A hair tress was mounted in clamps in such a way that it formed a smooth layer with the fibers positioned so that the cuticle edges pointed either downward or upward. Static charge was produced by contact between a rubbing element, in the form of a half-cylinder attached to an adjustable arm rotated by an electrical motor, and the hair fibers. The generated tribocharge was measured as a function of time by means of a static detector probe connected to an electrometer. The rubbing element was exchangeable, which permitted the examination of hair electrification by a variety of materials such as Teflon, aluminum, gold, stainless steel, nylon, etc. Charge decay measurements were

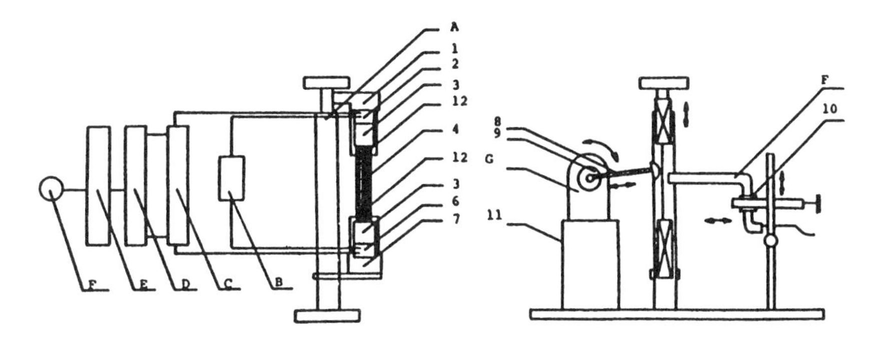

Figure 9 A scheme of an instrument to measure static charge generated by rubbing hair: (A) mechanism for changing fiber tension; (B) power supply; (C) operational amplifiers (Analog Devices, Model 2B31J); (D) computer and A/D converter (Model DT 2801, Data Translation, Inc.); (E) electrometer (Keithley Model 616); (F) static detector probe (Keithley, Model 2503); (G) motor; (1) and (7) load cell holding elements; (2) and (6) load cells (Sensotronics, Model 60036); (3) clamps; (4) fiber tress; (8) adjustable arm with rubbing element; (9) mechanism for adjusting the length of the arm; (10) mechanism for positioning the static detector probe; (11) table for motor; (12) mechanism for positioning the fibers (57).

performed with the same setup by following the changes in generated charge density as a function of time.

A detailed study of triboelectric charging was performed by using several rubbing materials and for two directions of rubbing from root to tip (rt, cuticles pointing downward) and from tip to root (tr). These data suggest that polycarbonate and chitosan acetate are characterized by work functions very close to that of hair (for rt and tr rubbing acting as electron acceptors and electron donors, respectively), that poly(methyl methacrylate) has a lower work function, and that the other materials tested (nylon, Teflon, stainless steel) lie above hair in the triboelectric series. It was also demonstrated that both the kinetics and extent of tribocharge generation may be dependent on the mechanical or electrostatic history of hair. This was ascribed to the fact that triboelectric properties of hair surface might be influenced by its state of strain (permanent deformation and opposite polarization of the cuticles induced by stretching or compressing during rubbing in the rt or tr mode, respectively), or by nonequilibrium distribution of electrons. Further information about the mechanism of triboelectric charging and fiber conductivity was derived from charge decay measurements. It was concluded that, at low humidity, the kinetics of charge decay are nonexponential and the charge carriers can become trapped indefinitely on the hair surface, giving permanent tribocharges and resulting in a flyaway phenomenon. For example, for initial charge density of 2.09, 3.62, and

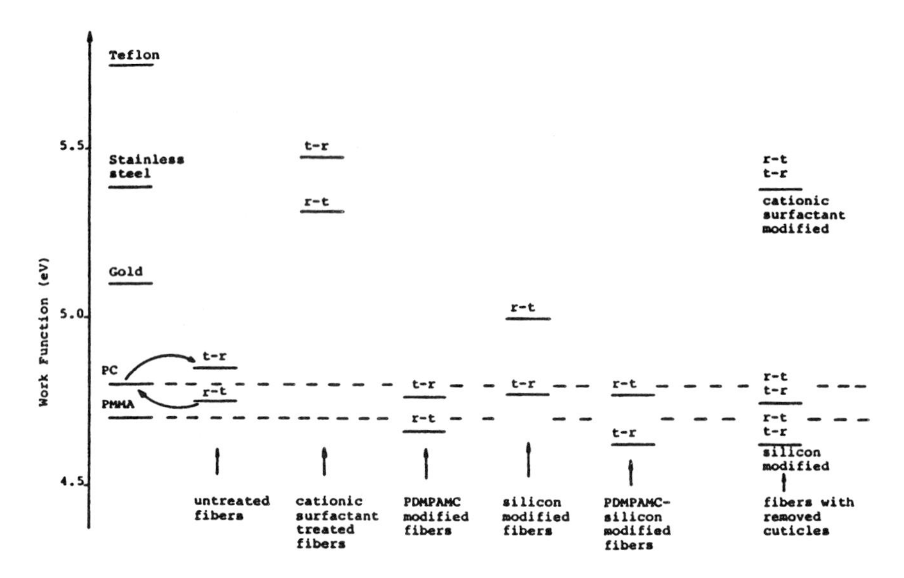

Figure 10 Energy diagram illustrating the effect of surface treatments on the work function of hair.

5.09 C/cm^2, the half-time of the decay was found to be infinity, 50.8 min, and 226.0 min, respectively.

Tribocharging characteristics of hair can be modified by ad(b)sorbed surfactants, polymers, complexes, silicones, acids, bases, and other materials used in cosmetic formulations, including conditioners (53–57). The results of triboelectric charging of surface-modified hair fibers are summarized on the energy diagram presented in Figure 10. Long-chain quaternary ammonium salts increase both the value of the effective work function of the fiber surface and the fiber conductivity. Depending on the length of the alkyl group in the quaternary ammonium salts, the half-times for the charge decays varied in the range from 1.2 to 11.7 min. Longer-chain alkyl quats exhibit much higher ability to increase conductivity than their short-alkyl-chain analogs. This might be related to the fact that the longer-chain alkyl ammonium salts resist rinsing and consequently are deposited on the surface in much larger quantities. The effect of cationic polymers was not as pronounced and clearly defined as that of quaternary alkyl salts. Adsorption of the cationic polymer poly(1,1-dimethyl-piperidinium-3,5-diallyl methylene chloride PDMPAMC or Polyquaternium 6) results in a lower work function of modified fibers, with the effect of the modifying layer superimposed on the directional triboelectric effect. On the other hand, the triboelectric properties of poly(methacrylamidopropyltrimethyl

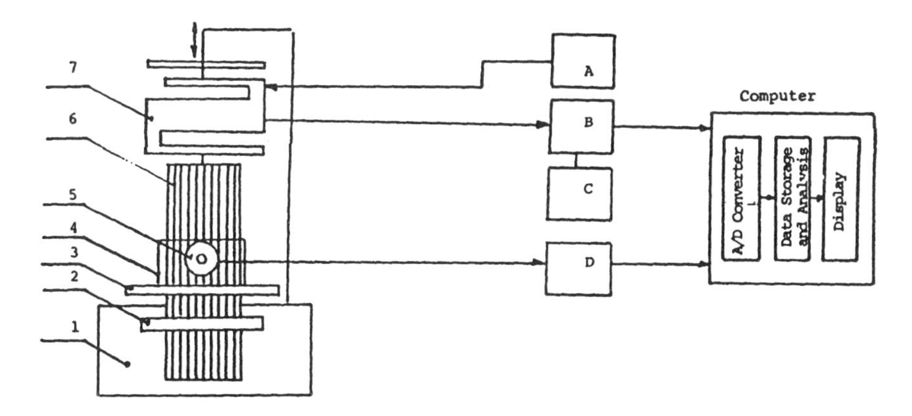

Figure 11 An instrument to measure combing forces and distribution of triboelectric charges on hair: (1) motor; (2) discharging element; (3) comb (four teeth, 1.5 mm thick); (4) holding frame; (5) static detector probe (Keithley, Model 2503); (6) hair tress; (7) load cell (Sensotronics, Model 60036); (A) and (C) power supply (+10V, +15V); (B) operational amplifier (Analog Devices, Model 2B31J); (D) electrometer (Keithley, Model 616).

ammonium) chloride were found to be strongly dependent on the polymer concentration in the treatment solution (57). An even more complicated picture emerged from the analysis of triboelectric charged hair modified by silicone oil, which showed a reversed directional triboelectric effect (the charge became more negative in the root-to-tip than in the tip-to-root mode of rubbing). Silicone oils also lower the conductivity of hair reflected by very low decay rates, and result in a permanent charge on the surface of hair. Data were also obtained for fibers treated with both the cationic polymer, polyquaternium-6, and silicone oil, which illustrated an additive effect of both layers on the kinetics of tribocharge generation (57).

More real-life combing electrification experiments, correlating combing curves with triboelectric charge distributions, can be performed by using an instrument comprising of a load cell and a static detector probe interfaced with a computer (53,58). The instrument, presented in Figure 11, allows measurements of single and multiple combing characteristics, providing plots of combing force or charge density as a function of position in the tress. The data can be analyzed by comparing the charge or force distributions, or by taking an integrated value of these parameters (average linear charged density or combing work expressed in C/cm^2 or G-cm, respectively) over the length of the tress. The measurements performed using insulator combs indicated that the typical charge distribution for clean hair fibers consists of two or three distinct peaks.

The peak corresponding to the upper section of the tress was usually the most pronounced, and the one due to disentanglement of the fiber tips (corresponding to a peak in the combing curve) contributed to the total generated charge density to only a small extent. In the case of metal combs, the charge distributions were different and paralleled more closely the combing curves in which the most prominent peak was generated at the fiber tips. The technique could also be used to study the effect of surface modification with conditioning agents on triboelectric charging. It has confirmed the results of rubbing experiments by showing that the electrochemical surface potential gap between the comb and keratin is a decisive factor in determining the magnitude and sign of the generated charge.

Finally, a simplified analysis of triboelectric charging can be performed by measuring the voltage of the charged surface of hair, generated by manual combing with a static detector probe and an electrometer (56). All such measurements must be carried out at controlled relative humidity in the range of 25–40% RH, and the measurements have to be repeated several times on at least two different tresses to obtain a statistically valid average of static voltage. In addition to this, to ensure the consistency and reproducibility of the data, it is advisable to examine concentration dependencies and to report the results in terms of static voltage relative to untreated controls. As for any kind of triboelectric measurements, especially those involving samples modified by surface treatments, reproducibility may be a problem, due to mass transfer, surface contamination, and surface abrasion during the sliding contact between the comb and the fiber surface.

9. *Substantivity Verification by Dye Staining*

Visual evidence of the presence of cationic conditioning agents on the surface of hair can be obtained by staining hair with anionic dyes such as Rubine dye (59).

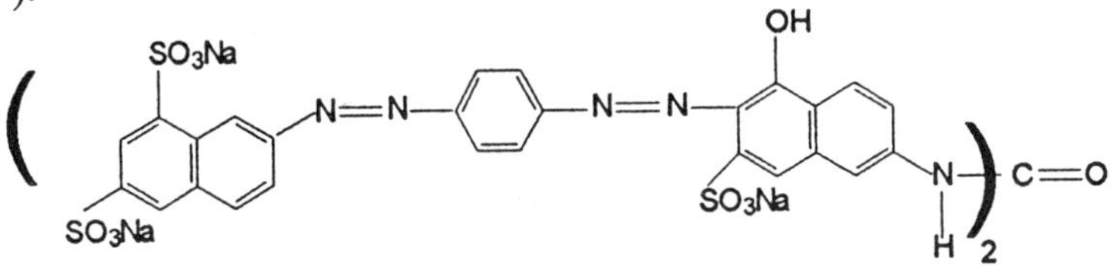

Rubin Red dye

Since this material is no longer available commercially, similar dyes such as Red 80, Red 84, Orange II, and Orange G can be employed. The hair treatment procedure involves the use of 3.4 mM dye aqueous stock solutions containing acetic acid which are diluted 1:5 for a 1-min treatment of hair. The dye solutions produce little change in coloration of intact hair or hair treated with quaternary alkyl compounds with a chain length below C10. For longer-chain cationic surfactants, the extent of hair coloration becomes very high, reaching reflectance values as low as 17% for C16-containing quaternary ammonium compounds. This technique can be employed to detect the presence of cationics on the hair after the application of creme rinse formulations based on cationic surfactants or conditioning shampoos containing cationic polymers.

The staining procedure may also be employed to study the extent of surfactant penetration into the structure of hair. For this, cross sections of quat-treated fibers can be stained with dye solutions and examined under the microscope for the rate and depth of diffusion of a quaternary compound.

10. Microscopy

Both optical and scanning electron microscopy are frequently used to assess fiber integrity and the state of the fiber surface. Major irregularities or damage to the fiber structure, detectable by microscopic methods, include split ends, fibrillation, lifting of cuticles, cuticle abrasion, excessive swelling in water, etc. (60,61). Many of these problems can be avoided by frequent use of conditioners. This can be demonstrated by inducing frictional damage to intact and conditioner-treated hair followed by a statistical analysis of scanning electron microscopy (SEM) micrographs. Other aberrations of the fiber, such as split ends, can be also eliminated by the use of cationic polymers as conditioning agents. SEM, optical microscopy, and fluorescence microscopy may also be employed to analyze the distribution of conditioner on the hair surface (2,50).

B. Noninstrumental Analysis of Hair Tresses

1. Sensory Analysis

A comprehensive review of the whole process of noninstrumental cosmetic product testing is given in several publications (62–65). Moskowitz (62) describes a generic procedure which can be applied to the process of development of any cosmetic product, including hair conditioners. A number of steps are carried out by R&D, marketing, and marketing research, including (1) concept development (marketing), (2) product feasibility analysis (R&D), (3) concept testing (marketing/marketing research), (4) screening of ingredients and prototype development (R&D), (5) instrumental analysis, noninstrumental analysis on hair tresses, (6) prototype screening (small scale by internal expert panel in R&D, salon testing), (7) larger formula optimization (R&D, marketing/marketing research), (8) final in-house optimization (R&D expert panel, small consumer panel), (9) product/concept test (marketing research), (10) confirmatory home-use test (marketing research), and (11) test market (marketing research). A detailed discussion of each element of this comprehensive process, such as the organization and selection of panels, formulation of descriptors, and statistical data analysis, is beyond the scope of this chapter. Also, specific problems such as the segmentation of the product for various hair types (thin, normal, or damaged hair), or the assessment of product fragrancing is not covered by this chapter. In the following paragraphs the discussion is limited to three important steps in the development of a new conditioner: evaluation of a product on hair tresses by an expert panel, salon testing, and home-use consumer testing. The reader may refer to Chapter 12 for additional discussion of these steps.

An important step in the evaluation of a conditioner is testing performed by a panel of experienced evaluators on hair tresses. This test is conducted to compare various attributes of the formulation as they relate to the process of product application on hair and to judge various properties of treated hair. The procedure involves the use of a standardized type of hair in the form of hair swatches. Selection of a proper hair type is critical in the evaluation of hair conditioners because these products have significantly different performance characteristics depending on the type of hair to which they are applied. For example, a product formulated to treat damaged hair will probably be inappropriate for normal or fine hair because of excessive deposition of actives on the hair. A minimum of three swatches is usually used for each product to be evaluated. Hair is shampooed before treatment with a standard surfactant solution (i.e., 3% ammonium lauryl sulfate) and a specified amount of product is applied to hair, usually 1–2 g of formulation per 2 g of hair tress. The formulation is worked into wet hair manually, left on the hair for 1–5 min, and then rinsed off for 30 sec by running water under controlled flow rate and temperature conditions. The key is to develop questionnaires which provide an adequate product description and to properly select panelists to include individuals with varying opinions and perceptions. The rheology, appearance, feel, and texture of the product can be analyzed in terms of the following parameters: (1) viscosity or ease of dispensing the product from a bottle, (2) feel and texture on the palm before application to hair (oily, hard versus soft, waxy, watery, smooth, rich, tacky), (3) color, (4) spreadability or ease of working the formulation into the hair, and (5) ease of rinse from hair and skin after treatment. Hair characteristics are evaluated for (1) ease of wet and dry combing, (2) wet and dry feel, (3) manageability, (4) body, (5) static, (6) luster, and (7) residue on hair. Further analysis of hair condition can be performed after a single or multiple shampooings with a reapplication of the product. The properties of a conditioner which could be judged in this test include (1) removability of a treatment from hair by shampooing and (2) buildup as a result of multiple applications.

Rating systems usually employ a scale from 1 to 10, with 10 being the "top rating" and 1 the "bottom rating." The ratings given by panelists are averaged to give a grade number that can be used to compre various products. Since other scales can also be employed, including individual scales for each participating panelist, any scale can be calibrated by calculating the percentage rating based on the "top rating" characteristic for each test participant. Thus, the calibration facilitates the comparison of the ratings from panelists employing different rating systems (62). Further discussion of various advantages or disadvantages of rating systems, including the use of a bipolar scale for liking/disliking as well as ratio scaling and magnitude estimation, is given in Ref. 62.

2. *Salon Testing*

The next step following the analysis of hair tresses is prototype screening on a small scale by an internal expert panel in R&D or by salon testing (63). Salon testing is usually conducted by product formulators together with salon stylists. Before the test is carried out, a salon operator has to make appointments with panelists possessing a hair type appropriate for a tested product (thin hair, damaged hair, etc.). During the actual test, the most widely used procedure involves the half-head application of a prototype product with the second half-head left untreated or treated with a control formulation. A variety of hair properties can be evaluated on both the prototype– and control-treated side of a head. For conditioners, the key parameters are (1) ease of combing, (2) feel, (3) body, (4) static charge, (5) residue on hair, (6) extent of hair moisturization, (7) buildup upon multiple product application, and (8) removability by shampooing. The test participants also assess the ease of product application, consistency, and color. A detailed questionnaire may be developed including a rating system and questions similar to those presented in Tables 2 and 3. The formulator and the stylist participate in product application, and their inputs, in addition to the opinion of a panelist, are included in the test protocol. Several panelists are usually treated with the same product in order to gather more representative information about the product tested.

3. *Consumer Testing*

Apart from expert panels, guidance for product development and the evaluation of the final product acceptance can be provided by a consumer panel which comprises the final users of the product. Consumer panels can be organized for the analysis of various problems related to the development of a product, including a descriptive analysis panel, a screening panel, or a quality control panel. In addition to this, a home-use consumer test serves as a critical step in the development of a new conditioner. This type of test may include the analysis of a single product or a paired comparison of two different products. An example of the testing protocol as well as the hypothetical test results for a conditioner designed for chemically processed (damaged) hair is given in Table 2. The results demonstrate a very favorble consumer reaction to the

Table 2 Example Protocol for a Home-Use Consumer Test of a Conditioner Designed for Chemically Processed, Damaged Hair: One Prototype Product

Total Respondents	25
Purchase intent	
Definitely/probably would buy	19
Definitely would buy	9
Probably would buy	10
Probably/definitely would not buy	6
Probably would not buy	3
Definitely would not buy	3
Overall opinion	
Excellent	7
Very good	7
Good	6
Fair	3
Poor	2
Product consistency	
Much too thick	2
Somewhat too thick	2
Just about right	16
Somewhat too thin	3
Much too thin	2
Fragrance intensity	
Much too strong	1
Somewhat too strong	4
Just about right	15
Somewhat too weak	3
Much too weak	2
Comparison to usual brand	
Much better	10
Somewhat better	6
About the same	3
Somewhat worse	4
Much worse	2
Product likes	
Mentioned something liked (total)	59
Manageability (total)	50
Leaves wet easy to detangle and comb	10
Leave hair easy to style	7
Gives hair body/fullness	7
Doesn't weigh hair down/no residues or buildup	5
Leaves dry hair feeling soft and silky	8
Adds moisture to hair	7

Table 2 Continued

Does not dull hair	6
Pleasant fragrance/fresh smell/light	5
Light consistency	4
Product dislikes	
Mentioned something disliked	21
Leaves hair flat and limp	6
Don't like the fragrance	5
Had to use too much/more than usual	4
Not viscous enough/Too viscous	6
Attribute ratings	
"Agree completely"	
Hair easy to comb through when wet	19
Rinses out easily and completely	17
Easy to wash off by shampooing	17
Does not leave hair feeling greasy or oily	15
Does not leave hair fly-away or full of static	14
Leaves hair looking and feeling clean	14
Leaves dry hair feeling soft and silky	14
Is good for damaged hair	17
Restores moisture to hair	18
Has a pleasant fragrance	11
Leaves hair healthy looking	14
Does not remove color from hair	20
Leaves hair easy to style	18
Does not remove color from hair	21
Has a nice color	10
Has a good consistency	10
Easy to work into hair	11

Twenty-five white women, ages 18–55, who have chemically processed (oxidatively dyed, permed, or bleached) their hair within the past 6 months and who are users of retail-brand cream conditioners three or more times per week participated in this study. At the company testing center, a product description was red to them, and they were given two 6-ounce bottles of conditioner for damaged hair to use at home for a 3-week period. After 3 weeks they were contacted by phone and probed on product usage, likes/dislikes, purchase interest, overall opinion, and the level of agreement to a list of product attributes.

Table 3 Example Protocol for a Home-Use Consumer Test of a Conditioner Designed for Thin Hair: Paired Comparison of Prototype Products A and B

Total respondents	35		
Overall preference			
Prefer A	21		
A lot	15		
A little	6		
Prefer B	14		
A lot	10		
A little	4		
Preference of preferred test product	Prefer Product versus usual brand		
	A	B	
Prefer test product	14	8	
Prefer usual instant conditioner	3	5	
No preference	3	2	
Total	20	15	
Preference on attributes	Prefer A	Prefer B	No preference
Adding body to hair	17	11	7
Leaving hair manageable	18	12	5
Leaving hair easy to style	19	9	7
Improving the state of damaged hair	17	9	9
Not weighing hair down	15	10	10
Adding fullness to hair	14	10	11
Leaving hair moisturized	13	10	12
Being gentle to hair	15	10	10
Leaving hair shiny	12	12	11
Making hair detangling easy when wet	14	9	12
Leaving dry hair being soft and silky	15	10	10
Not leaving a residue on hair	13	12	10
Not leaving greasy or oily residue in hair	15	10	10
Leaving hair looking and feeling healthy	11	13	11
Having a pleasant fragrance	11	13	11
Rinsing out easily and completely	15	10	10
Not leaving hair flyaway with static electricity	11	14	10
Leaving hair looking and feeling clean	16	12	7
Having an appealing color	9	13	13
Having just the right consistency	3	12	10
Easy to work into hair	14	11	10

Table 3 Continued

Preference on attributes (continued)	Prefer A	Prefer B	No preference
Easy to wash off with a shampoo	15	12	8
Leaving hair feeling stiffer with more bounce	17	10	8

Reasons for preference versus usual brand	Prefer A	Prefer Usual brand	Prefer B	Prefer Usual brand
Hair characteristics				
Imparts more body	12	4	7	3
Easier to comb or detangle	4	4	6	2
Leaves hair with more volume	7	2	5	1
No residue or buildup on hair	8	1	5	1
Leaves dry hair soft and shiny	8	1	4	3
Product characteristics				
Has pleasant scent	7	1	4	—
Washes off easily with a shampoo	5	2	4	3
Rinses out quickly and completely	8	3	6	2
Total	14	3	8	5

Thirty-five white women, ages 18–55, participated in a paired comparison study of products A and B. All panelists described themselves as having hair which lacks body. They also declared themselves users of retail brands of instant conditioner three or more times per week. In order to minimize bias, half of the consumers used formulation A first and half used formulation B first. At the company testing facilities, the participants were given two extra-body conditioners to use at home for a period of 2 weeks. The product to be used during the second week was sealed in a plastic bag and labeled with a date to begin usage. One day before consumers were to switch products they were contacted by phone as a reminder. After 2 weeks consumers were contacted by phone and interviewed to determine an overall preference of test products, preference on a list of attributes, and preference of the preferred test product or their usual conditioner.

product. Purchase intent is high, and the product is judged by the majority of respondents to be better than their usual brand. It also performs all the key functions of a conditioner designed for damaged hair, leaving hair easy to comb, manageable, and remoisturized. Furthermore, the results indicate that the tested conditioner has light consistency, adequate fragrance, and is easy to apply and to rinse off, without leaving an excessive buildup.

The second type of consumer test involves a paired comparison of two different products. This test may include the use of a new prototype conditioner versus an existing, well-established product (for example, a market leader in the analyzed product category), or it may offer two prototypes designed for a new product line. The questionnaire may also explore consumers preference in relation to their usual conditioner brand. Table 3 presents the testing protocol and hypothetical test results for a conditioner designed for women with thin hair who expect an increase in hair body as well as detangling benefits. It can be concluded from these data that both products A and B are strongly preferred over the respondents' usual brand. This can be attributed to superior benefits of the conditioner defined as "giving hair more body or volume," the ability to detangle hair, and the pleasant scent of the formulation. When compared to each other, product A is preferred in terms of a number of hair properties such as hair body or fullness, styling ease, manageability, and combability. Product B appears to be formulated with ingredients characterized by a lower affinity to hair, resulting in less residue on hair. Thus, product B is perceived as less desirable for restoring moisture to hair and for not improving the state of damaged hair. Product B is, however, preferred for rinsing out easily, for not leaving buildup on hair, and for eliminating electrostatic charging. The overall conclusion from the consumer test described in Table 3 is that product A is a good candidate as an extra-body hair conditioner and merits further product development testing.

III. CONCLUSIONS AND FUTURE DEVELOPMENTS

The evaluation of hair conditioners can be performed by both instrumental and noninstrumental methods. At the present time, noninstrumental methods such as panel and consumer testing are predominant, since they are thought to best predict consumer preferences for a given product and ultimately its success in the marketplace. The instrumental methods are usually employed in the early stages of product development in the selection of raw materials, and also for the support of claims when the final formulation is prepared. The use of quantitative techniques is also mandated by new legislation requiring extensive support data demonstrating the effectiveness of cosmetic products and by the widespread use of competitive claims. While the methods of sensory analysis are being constantly refined and improved, there has also been significant progress in physicochemical evaluation techniques. These methods are becoming more automated and computerized, and can frequently provide several physicochemical parameters simultaneously, resulting in a precise and multifaceted description of physical phenomena connected with the process of hair conditioning. Unlike panel or consumer tests, which measure the perception of conditioning effects in terms of a few descriptors well understood by the panelists but not well defined from a physicochemical point of view, the instrumental methods yield precise, quantitative information which is sometimes difficult to relate to human perceptions. However, future development of even more sophisticated quantitative techniques, especially in the area of mechanical measurements (texture analysis, dynamic mechanical analysis), image analysis, and adsorption/desorption techniques should be able to bridge the gap between sensory perception and the physicochemical description of conditioning phenomena.

REFERENCES

1. Jachowicz J. Hair damage and attempts to its repair. J Soc Cosmet Chem 1987; 38:263.
2. Tate ML, Kamath YK, Ruetsch SB, Weigmann H-D. Quantification and prevention of hair damage. J Soc Cosmet Chem 1993; 44(6):347.
3. Jachowicz J, Helioff M, Rocafort C, Alexander A, Chaudhuri RK. Photodegradation of hair and its photoprotection by a substantive photofilter. Drug & Cosmetic Industry, December, 1995.
4. Franbourg A. Synchrotron light: A powerful tool for the analysis of human hair damage. 10th International Hair–Science Symposium, Rostock, 1996.
5. Cooley AJ. The toilet and cosmetic arts in ancient and modern times. Published by Burt Franklin, New York, 1866, reprinted 1970.
6. Davies JT, Rideal EK. Interfacial Phenomena. New York: Academic Press, 1963.
7. Owens MJ. The surface activity of silicones: A short review. Ind Eng Chem Prod Res Dev 1980; 19(1):97.
8. Finkelstein P, Laden K. The mechanism of conditioning of hair with alkyl quaternary ammonium compounds. Appl Polym Symp 1971; 18:673.
9. Scott GV, Robbins CR, Barnhurst JD. Sorption of Quaternary Ammonium Surfactants by Human Hair. J Soc Cosmet Chem 1969; 20:135.
10. Chow CD. Interaction between polyethyleneimine and human hair. Text Res J 1971; 41:444.
11. Woodward J. Aziridine chemistry—applications for cosmetics. J Soc Cosmet Chem 1972; 23:593.
12. Goddard ED, Faucher JA, Scott RJ, Turney ME. J Soc Cosmet Chem 1975; 26:539.
13. Faucher JA, Goddard ED. Influence of surfactants on the sorption of a cationic polymer by keratinous substrate. J Coll Int Sci 1976; 55:313.
14. Jachowicz J, Berthiaume M, Garcia M. The effect of the amphiprotic nature of human hair keratin on the adsorption of high charge density cationic polyelectrolytes. Coll Polym Sci 1985; 263:847.
15. Goddard ED, Harris WC. Substantivity to keratin as measured by X-ray photoelectron spectroscopy (XPS) and electrokinetics (EK). Preprints of the XIVth IFSCC Congress, Barcelona, 1986, Vol. II, p. 1039.
16. Goddard ED, Harris WC. An ESCA study of the substantivity of conditioning polymers on hair substrates. J Soc Cosmet Chem 1987; 38(4):233.
17. Kamath YK, Dansizer CJ, Weigmann H-D. Surface wettability of human hair. I. Effect of deposition of polymers and surfactants. J Appl Polym Sci 1984; 29:1011.
18. Jachowicz J, Williams C, Maxey S. Sorption/Desorption of ions by Dynamic Electrokinetic and Permeability Analysis of fiber plugs. Langmuir 1993; 9(11):3085.
19. Karjala SA, Williamson JE, Karler A. Studies of the substantivity of collagen derived polypeptides to human hair. J Soc Cosmet Chem 1966; 17:513.

20. Turowski A, Adelmann-Grill BC. Substantivity to hair and skin of 125-I-labelled collagen hydrolisates under application simulating conditions. Int J Soc Cosmet Sci 1985; 7:71.
21. Mintz GR, Reinhart GM, Lent B. Relationship between collagen hydrolisate molecular weight and peptide substantivity to hair. J Soc Cosmet Chem 1991; 42:35.
22. Newman W, Cohen GL, Hayes C. A quantitative characterization of combing force. J Soc Cosmet Chem 1973; 24:773.
23. Tolgyesi WS, Cottington E, Fookson A. Mechanics of hair combing. Presented at the Symposium on Mechanics of Fibrous Structures, Fiber Society, Atlanta, May, 1975.
24. Garcia ML, Diaz J. Combability measurements on human hair. J Soc Cosmet Chem 1976; 27:379.
25. Kamath YK, Weigmann H-D. Measurement of combing forces. J Soc Cosmet Chem 1986; 37:11.
26. Jachowicz J, Helioff M. Spatially-resolved combing analysis. Present at SCC Annual Scientific Meeting, New York, December 1995; J Soc Cosmet Chem, in press.
27. Hambidge A, Wolfram L. Effect of comb materials on hair combing. 4th International Hair Science Symposium, Syburg, Germany, 1984.
28. Firstenberg DE, Rigoletto R, Moral L. SCC Annual Scientific Seminar, New York, May 1997, Preprints, p. 36.
29. Robbins CR, Crawford R. Cuticle damage and the tensile properties of human hair. J Soc Cosmet Chem 1991; 42:59.
30. Rushton H, Kingsley P. Treating reduced hair volume in women. Cosmetics & Toiletries 1993; 108(3):59.
31. Hough P, Hey EJ, Tolgyesi WS. Hair body. J Soc Cosmet Chem 1976; 27:571.
32. Yin NE, Kissinger RH, Tolgyesi WS, Cottington EM. The effect of fiber diameter on the cosmetic aspects of hair. J Soc Cosmet Chem 1977; 28:139.
33. Garcia ML, Wolfram LJ. Presented at the 10th IFSCC Congress, Sydney, Australia, 1978.
34. Robbins CR, Crawford RJ. A method to evaluate hair body. J Soc Cosmet Chem 1984; 35:369.
35. Kamath YK, Weigmann HD. Evaluation of hair body. SCC Annual Scientific Meeting, New York, December 1996, Preprints, p. 19.
36. Clarke J, Robbins CR, Reich C. Influences of hair volume and texture on hair body of tresses. J Soc Cosmet Chem 1991; 42:341.
37. Stamm RF, Garcia ML, Fuchs JJ. The optical properties of human hair. Parts I and II. J Soc Cosmet Chem 1977; 28:571.
38. Bustard HK, Smith RW. Studies of factors affecting light scattering by individual human hair fibers. Int J Cosmet Chem 1990; 12:121.
39. Reich C, Robbins CR. Light scattering and shine measurements of human hair: A sensitive probe of the hair surface. J Soc Cosmet Chem 1993; 44:221.
40. Czepluch W, Hohm G, Tolkiehn K. Gloss of hair surfaces: Problems of visual evaluation and possibilities for goniophotometric measurements of treated strands. J Soc Cosmet Chem 1993; 44:299.
41. Maeda T, Hara T, Okada M, Watanabe H. Measurements of hair luster by color image analysis. 16th IFSCC Congress, New York, 1990, Preprints, Vol. I, p. 127.

42. Jachowicz J. Fingerprinting of cosmetic formulations by dynamic electrokinetic and permeability analysis. II. Hair conditioners. J Soc Cosmet Chem 1995; 46.
43. Jachowicz J, Berthiaume M. Microemulsions vs macroemulsions in hair care products. Cosmetics & Toiletries 1993; 108(3):65.
44. The effect of reactive treatments on hair by Dynamic Electrokinetic and Permeability Analysis. 9th International Haire Science Symposium, Prien, 1994.
45. Kamath YK, Dansizer CJ, Weigmann H-D. Wettability of keratin fiber surfaces. J Soc Cosmet Chem 1977; 28:273.
46. Kamath YK, Dansizer CJ, Weigmann H-D. Wetting behavior of human hair fibers. J Appl Polym Sci 1978; 22:2295.
47. Kamath YK, Dansizer CJ, Weigmann H-D. Surface wettability of human hair. II. Effect of temperature on the deposition of polymers and surfactants. J Appl Polym Sci 1985; 30:925.
48. Kamath Y, Dansizer CJ, Weigmann H-D. Marangoni effect in water wetting of surfactant coated human hair fibers. J Coll Int Sci 1984; 102(1):164.
49. Kamath YK, Dansizer CJ, Weigmann H-D. Surface wettability of human hair. III. Role of surfactants in the surface deposition of cationic polymers. J Appl Polym Sci 1985; 30:937.
50. Weigmann H-D, Kamath YK, Reutsch SB, Busch P, Tesmann H. Characterization of surface deposits on human hair fibers. J Soc Cosmet Chem 1990; 41:379.
51. Martin AJP. Triboelectricity in wool and hair. Proc Phys Soc Lond 1940; 53(2):186.
52. Jachowicz J, Wis-Surel G, Wolfram L. Directional triboelectric effect in keratin fibers. Text Res J 1984; 54(7):492.
53. Lunn AC, Evans RE. The electrostatic properties of human hair. J Soc Cosmet Chem 1977; 28:549.
54. Jachowicz J, Wis-Surel G, Garcia ML. Relationship between triboelectric charging and surface modifications of human hair. J Soc Cosmet Chem 1985; 36:189.
55. Patel CV. Antistatic properties of some cationic polymers used in hair care products. Int J Cosmet Sci 1983; 5(5):181.
56. Jachowicz J, Garcia M, Wis-Surel G. Relationship between triboelectric charging and surface modification of human hair: Polymeric versus monomeric long alkyl chain quaternary ammonium salts. Text Res J 1987; 57(9):543.
57. Jachowicz J, Wis-Surel G, Garcia ML. Further observations on triboelectric charging effects in keratin fibers. Polymers for Advanced Technologies, IUPAC International Symposium, Jerusalem 1987, M Lewin, ed., New York: VCH Publishers, 1988, pp. 340–360.
58. Wis-Surel G, Jachowicz J, Garcia ML. Triboelectric charge distributions generated during combing of hair tresses. J Soc Cosmet Chem 1987; 38:341.
59. Crawford RJ, Robbins CR. A replacement for Rubine dye for detecting cationics on keratin. J Soc Cosmet Chem 1980; 31:273.
60. Swift JA, Brown AC. The critical determination of fine changes in the surface architecture of human hair due to cosmetic treatment. J Soc Cosmet Chem 1972; 23:695.
61. Robinson VNE. A study of damaged hair. J Soc Cosmet Chem 1976; 27:155.
62. Moskowitz HR. Cosmetic product testing. New York: Marcel Dekker, Inc., 1984.

63. Cryer PH. Design and analysis of product performance trials in a hairdressing salon. Cosmetic Science, Proceedings of the Second Congress of the IFSCC, AW Middleton, ed., New York: Pergamon Press Book, 1962.
64. Close J-A. The concept of sensory quality. J Soc Cosmet Chem 1994; 45:95.
65. Stone H, Sidel JL. Sensory evaluation practices. San Diego, Academic Press, 1993.

14
Evaluating Performance Benefits of Conditioning Formulations on Human Skin

Ronald L. Rizer and Monya L. Sigler
Thomas J. Stephens & Associates, Inc., Carrollton, Texas

David L. Miller
CuDerm/Bionet, Inc., Dallas, Texas

I. INTRODUCTION

Great strides have been made in our understanding of the mechanism of action of skin care products designed to condition the skin. This progress can be attributed to both an increased knowledge of the function of the skin as well as the development of new technologies used to measure the benefits of "conditioning" products. In this context, the term "conditioning" is defined as an improvement in a definable skin attribute. For instance, an effective moisturizer conditions or improves the attribute dry skin by enabling the stratum corneum to hold optimal levels of water more effectively, by helping to soften the rough, dry skin surface, and by altering the pattern of desquamation, allowing corneocytes to slough as discrete units rather than as clusters of aggregated cells which form dry skin flakes. Other attributes of skin that a conditioning product may benefit or improve include oily skin, large facial pores, skin irritation, and photodamage. Photodamage is a broad term for skin damage that has resulted as a consequence of years of unprotected sun exposure, and is usually manifested by increases in fine lines, wrinkles, mottled pigmentation, pigmented spots, skin looseness, and sallowness. Photodamaged skin is often referred to as photoaged skin, and it is distinguished from intrinsic skin aging, which results from the passage of time. Photodamaged skin is widespread in the population, and thus the market is saturated with skin care products claiming to improve this condition.

The effects of conditioning formulations on the skin are often subtle. Much scientific and clinical research energy has been expended in the latter half of the twentieth century trying to develop techniques for reliably detecting the changes in skin accompanying application and use of conditioning and other therapeutic formulations. There have been "inspirational" achievements that have led the way for development of many of the techniques in use today. The citations below do not necessarily represent the most complete, earliest, or most definitive treatment of the subject, which will always be "work in progress."

For skin dryness, validated grading scales were developed allowing for more reliable clinical assessment by the investigator (1,2). The use of adhesive materials to remove samples of stratum corneum scales for careful ex-vivo assessment was pioneered by Goldschmidt and Kligman (3). Prall introduced the concept of measurement of scattering and reflection of white light by scales held on adhesive substrates, which allowed elegant discrimination of various degrees of scaling (4). Today we have computerized image analysis of scales sampled on commercially available clear adhesive disks designed expressly for this purpose (5).

For examining the texture of the skin surface, technology has progressed along two paths: close-up photography and assessment of silicone replicas of the surface. While many unremarkable photographic techniques have been employed to record changes in the appearance of skin, it was the application of polarized light photography (6) that opened a truly unique way of documenting the skin surface. Viewing the skin under cross-polarized light reveals details of the skin pigmentation and sub-stratum corneum features not readily visible under normal lighting, whereas viewing under parallel-polarized light accentuates the texture of the stratum corneum scales, and surface topography. The analysis of silicone replicas of the skin surface was extensively developed by Cook (7). Cook employed surface characterization tools in use at that time in measuring the smoothness of machined metal surfaces. While these techniques eventually proved much too sensitive to be used in characterizing the rather coarse skin surface (compared to machined metals), adaptation of other methods such as computerized image analysis (8) and laser profilometry (9) has proved most successful in detecting changes in the surface texture condition of the skin.

Electrical measurements of the skin comprise relatively simple-to-apply methods often used to evaluate skin moisturizers. Several instruments based on skin capacitance and impedance are commercially available (10). The difficulty has been in the interpretation of the data produced by these instruments. In 1983, Leveque and deRegal (11) reviewed the various techniques in the literature at that time and provided clear interpretation guidelines. In particular, they pointed out that electrical techniques are most responsive to the water content of the stratum corneum layer; therefore only those conditioning components that actually alter water content will be detected. Conditioning components that address lubricity, for example, are not measured, although they can have a profound influence on the overall effectiveness of a moisturizer. For instance, emollients coat the skin surface and sorb into the upper layers of the stratum corneum, making the skin appear shinier, smoother, and less rough, but typically have little or no effect on the water-holding capacity of the stratum corneum.

A key achievement of the last decade is not attributable to any one person. This has been a decade of development of more than a dozen commercial instruments designed specifically to aid in clinical evaluation of the skin. The technology has moved from the awkward "training wheels" stage of impractical laboratory designs to convenient clinical instruments. A review of these instruments in any depth is beyond the scope of the present chapter. The reader is encouraged to pursue two recent books (12,13) which cover these instruments and many other useful techniques in detail.

In this chapter we describe a five-pronged approach to substantiate product performance claims. This approach includes (a) the importance of panel selection, (b) human perception of product benefit, (c) clinical assessment and photodocumentation of product benefit, (d) objective assessment of product benefit using biophysical methods, and (e) statistically based conclusions. Moreover, we discuss why one must take into account skin complexity, and skin variables such as race, gender, anatomical location, climate, lifestyle, occupation, age, and undefined factors when designing studies. Lastly, we describe selected testing methodologies that are relevant to the success of today's skin care product mix.

II. FIVE-PRONGED APPROACH TO SUBSTANTIATE PRODUCT PERFORMANCE CLAIMS

A. Why Is a Multipronged Approach Desirable (and Necessary) in Substantiating Product Performance Claims?

Product sales are often driven by performance and safety claims. Some examples of product claims are "reduces fine lines," "moisturizes for up to 12 hours," and "nonirritating and safe in normal use." A single clinical or biophysical approach might be used to provide substantiation for any of these examples. However, such an approach is rarely sufficient to characterize a product's benefit, which can lead to support for a false claim. This occurs because a single, *unidimensional* technique does not constitute a method powerful enough to assess the complex interactions that occur when a test product is applied to the skin.

One of the most challenging tasks in supporting a claim is providing credible supporting data. The multipronged approach addresses this issue by employing a series of methods to assess the multifaceted benefits of a product's interaction with the skin. Our method is rooted in an understanding of the anatomy and physiology of the skin and the tools available to measure product effects at the level that they are occurring. Our approach is strengthened by a history of experience in extracting subjective information from panelists about their experiences while using a test product. Lastly, we choose the most appropriate method of analysis and presentation of the data to draw conclusions that provide credible substantiation of product claims.

B. Part I of the Five-Pronged Approach: Panelist Selection

The success of a clinical trial is dependent on the screening of prospective candidates and the selection of qualified subjects. Certain biological factors provide obvious recruiting guidelines. A few such examples are gender, age, and cutaneous conditions or symptoms. For example, consider a study whose goal is to evaluate the efficacy of a facial lotion designed for women to reduce the appearance of fine lines around the eyes. The first goal of recruiting would be to eliminate males, and young women who do not have facial fine lines. In order to provide a successful study, however, recruiting efforts must reach far beyond the basic, well-stated parameters of the study. An example of more subtle (but important) recruiting techniques is illustrated below.

1. Cosmetic Interactions

Individuals should be excluded from a study who have a recent history of using a cosmetic that contains the same active ingredient as the product being tested. Examples include alpha-hydroxy acids, beta-hydroxy acids, and retinoids.

2. Drug Interactions

Certain medications may interact with a product's active ingredient or may mask adverse reactions resulting from product usage. Such drugs include antiinflammatory agents, contraceptives, and androgens. Other medications may induce skin sensitivity to sunlight.

3. *Habits and Practices*

The lifestyle habits and practices of individuals can affect the outcome of a study in certain cases. For instance, individuals who use tanning salons or who are exposed to sunlight as part of their occupation should not participate in a study designed to test a skin lightening product. The selection of habits and practices that are used to screen individuals is dependent on the product being tested and the objective of the study.

4. *Severity of Symptoms*

The selection of a panelist should not be determined solely on an individual possessing the desired cutaneous characteristic(s). Of equal importance is the exclusion of those whose condition could be *too severe* to enable the test material to produce the anticipated effect. In the case of qualifying subjects, more is not always better.

5. *Identifying Specialized Populations*

A specialized population refers to a group of individuals who all possess a well-defined cutaneous attribute such as sensitive skin (14,15), dry or chapped skin, atopic dermatitis, winter itch, rosacea, acne, or cellulite. The screening of individuals for such conditions requires the formulation of a series of questions that accurately defines the condition on a clinical level. At the same time, the questions must be worded so that the volunteer study panelists understand them and respond to them appropriately based on their individual experiences with the condition.

6. *Ethnic Populations*

Companies with global markets can choose to test their products with specific ethnic populations in the United States, or they can perform the tests in countries whose population represents their ethnic choice. Both options provide certain benefits. For instance, skin sensitivity varies in many subtle ways (16), and formulating skin care products to address this issue may determine the success or failure in a foreign market. Premarketing testing in such ethnic populations can be crucial in assuring success in challenging foreign markets.

C. Part II of the Five-Pronged Approach: Human Perception of Product Benefit

The panelists' perception of how a test product performs is possibly the most relevant dimension in product testing. Frequently the subjects' perception of product benefit exceeds that which is captured by clinical grading or biophysical measurements. We obtain information from subjects regarding their perception of test products via questionnaires and poststudy focus group interviews. The following section outlines our experience in capturing subjective information from panelists regarding their experience in using a test product.

1. Questionnaires

Written questionnaires can be administered to subjects at intervals throughout the course of a study and/or at the conclusion of product use. They are not only useful in capturing information about product benefit, they may also be used to inquire about cosmetic preferences, intent to purchase, comparison to their regular brand, or product packaging and delivery. Responses to questionnaires may also provide useful information for selecting the most desirable participants for inclusion in a focus group (see below). Properly designed questionnaires are balanced so that the responder to a given question must choose from one or more positive responses, a neutral response, and one or more negative responses (the number of positive and negative choices must balance). Some questions require a "yes" or "no" answer, or a short narrative description.

2. Focus Group Interviews

Focus groups provide a unique forum for capturing feedback about subjects' thoughts and feelings regarding their experience using a test product. A moderator serves to initiate and guide the topics for discussion. The ensuing interactive group discussion about the subjects' experiences while using the product provides a wealth of dynamic information that cannot be obtained from written questionnaires alone.

One of the most important aspects of conducting a successful focus group is the selection of panelists. The selection process is dependent on the specific information desired about the test products. For instance, if broad feedback is desired, then the selection process might be a random one, where the focus group population is representative of the study population. However, one might prefer to restrict the group to individuals who responded to the postusage questionnaires in a certain way. For instance, if some of the subjects expressed displeasure with some aspect of the product or the way in which it was used or dispensed from the container, then a session with these individuals could be used to learn more about the nature of the specific problem. Focus group panels can also be chosen based on demographics, such as age, skin type, or race.

Each focus group session is designed to last approximately 1 hour and usually takes place with a group of about 10 test subjects. The number of subjects interviewed in a session is kept low to encourage each member's participation in the discussion. Multiple groups can be interviewed in succession if a large number of participants is desired. To optimize interactions between subjects and the moderator, sessions are held in a large conference room, where participants can sit facing one another. The facility is also equipped with a separate, candid viewing room, where sessions can be privately monitored and videotaped. While the information obtained from focus groups is dynamic, it is generally not quantitative, since group sizes are too small. The information obtained is qualitative and therefore not statistically significant. Nevertheless, this information can be some of the most valuable information one can obtain.

D. Part III of the Five-Pronged Approach: Clinical Assessment and Photodocumentation

1. Clinical Assessment

Clinical assessment involves careful inspection of the skin, usually under magnification and blue daylight lighting, by a technician or physician trained to grade the full range of a skin attribute. Clinical grading can be particularly beneficial for assessing skin attributes that are not easily measured with bioinstrumentation. Examples of such attributes are scaling and cupping on the lips, dark circles and puffiness of the eyes, and roughness of calloused skin of the elbows and knees. Clinical grading is especially powerful when combined with high-quality clinical documentation photography.

A product that effectively conditions the skin will produce changes in characteristics that can be perceived and graded by sight or touch. Severity of intensity or degree of improvement can be assessed using a 10-cm analog scale (10-cm line scale) so that changes in an attribute can be quantitated and percent changes from baseline reported. This method of clinical assessment of a product benefit can support claims like "40% improvement in fine lines." Clinical grading can be a simple ranking of an attribute as "mild," "moderate," or "severe." A common application of this method is for assessing erythema (redness), edema (swelling), and dryness/scaling, and subclinical irritation, such as burning, stinging, itching, tightness, or tingling.

a. Photodamage. Most of the attributes listed here are normally graded on the face, although some can also be accurately assessed on the chest and hands, which are the primary areas of concern for most women who have experienced years of unprotected sun exposure. Asterisks indicate attributes best suited for support with clinical documentation photography (see next section).

*Fine lines and wrinkles
Looseness/firmness
*Sallowness
*Mottled pigmentation
*Skin clarity

Tactile roughness
*Skin crepeness

b. Dry/Chapped Lips

*Cupping (small oval depressions on the lips due to dryness)
*Fine lines
*Cracking
*Fissuring

c. Dry Skin and Irritation. These attributes can be assessed on any area of the body including the scalp. Subjective symptoms of irritation (burning, stinging, itching, tightness, and tingling) are assessed by questioning the subject during the examination.

*Scaling
*Cracking
Tactile roughness
*Erythema
Edema

d. Other Skin Characteristics

*Eye puffiness
*Dark circles
*Calloused skin
Cellulite

2. Clinical Documentation Photography

Photography lends credibility to the results obtained from clinical grading of certain skin attributes (see items above marked with an asterisk). Photographs taken prior to product use and at subsequent study visits provide a visual illustration of the gradual benefits that a test product provides. A variety of photographic options are available for capturing the benefits of a test product. The choice of method should best reflect the skin parameter that the product is targeting. Examples of how some photographic methods accentuate different skin characteristics is illustrated in the following section. The photographic methods that are used most often are visible light photography, reflected ultraviolet light photography, and cross-polarized light photography. Descriptions of these methods and demonstrations of their unique contributions to claim substantiation are discussed below.

a. Visible Light Photography. Visible light photography—usually color photos are best—relies heavily on lighting and shadows to document textural facial features such as fine lines, wrinkles, dryness, scaling, and other features (Figures 1 and 2).

b. Reflected Ultraviolet (UV) Light Photography. This type of photography is useful for highlighting pigmentation of the skin that is not apparent with visible light photography. Thus, it is a helpful tool in documenting improvements produced by skin-lightening products or products designed to reduce the appearance of photodamage and aging (Figures 3 and 4).

c. Cross-Polarized Light Photography. Cross-polarized light photography—usually color photos are best—reduces reflection of light from the surface of the skin, thereby eliminating the shiny appearance and/or glare that can accompany visible light photography. This technique is particularly helpful in accentuating the inflammatory nature of acne lesions, and telangiectasia, which is red, finely branching skin capillaries typically found on the nose, cheeks, and chin (Figures 5 and 6).

E. Part IV of the Five-Pronged Approach: Biophysical Methods

An integral part of claim substantiation is the ability to support clinical grading and panelists' perception of product benefit with objective biophysical measurements. Such measurements provide technical support for the more subjectively based results obtained with clinical grading. Full-service clinical research laboratories are equipped with the biophysical instrumentation needed to provide at least one means of support for each of the clinical grading parameters shown in Part I. A brief description of selected biophysical methods and instrumentation is given below.

1. Capacitance Measurements

The NOVA Dermal Phase Meter (DPM) uses an electrical capacitance method to detect changes in the relative moisture content of the stratum corneum. This device is commonly used to test the efficacy of lotions or creams in delivering moisture to the skin or retaining the moisture present in the skin.

The SKICON Skin Surface Hydrometer uses electrical capacitance and conductance to detect the micro water content in the superficial portion of the stratum corneum. It is useful for assessing the therapeutic efficacy of topical agents to deliver and hold water at the skin surface. Like the NOVA Dermal Phase Meter, the SKICON Skin Surface Hydrometer is used to evaluate the effects of moisturizers and other cosmetics on the stratum corneum, and the effects of chemicals such as surfactants on the stratum corneum.

Both instruments can be applied to Tagami's (12) "sorption–desorption" technique to evaluate the water-holding properties of the stratum corneum. Essentially, a droplet of water is applied to the skin, wiped dry after 10 sec, and a reading is taken; thereafter, a reading is taken every 30 sec for 3 min.

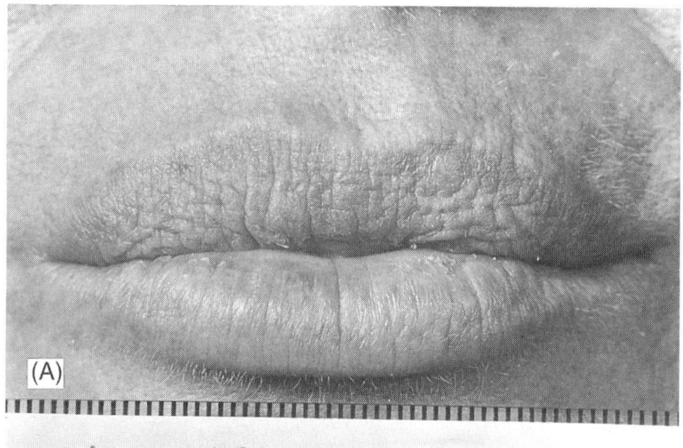

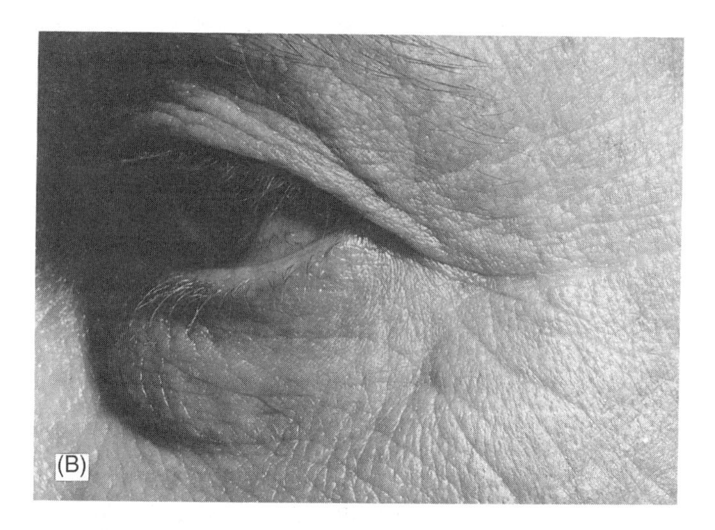

Figure 1 Photographs illustrating the ability to document scaling, fine lines, and wrinkles that accompany (A) dry chapped skin on the lips as well as (B) wrinkles and fine lines of the periocular eye area.

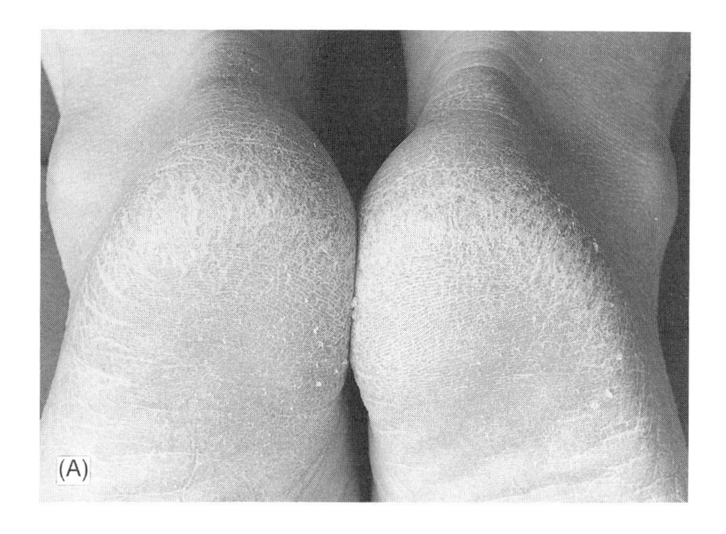

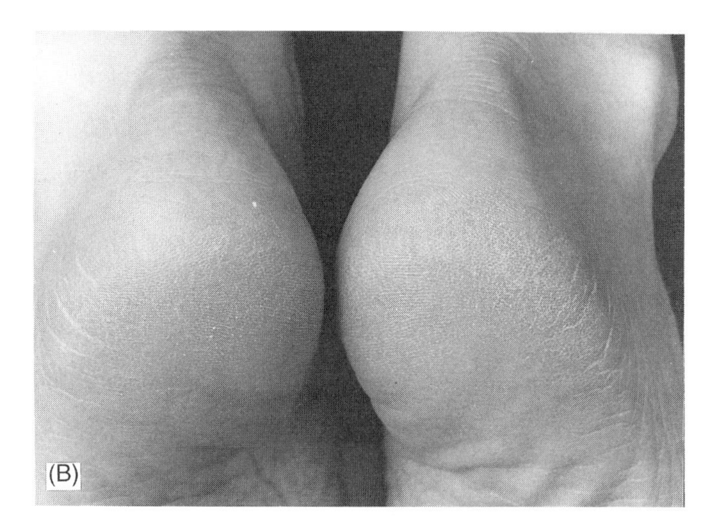

Figure 2 These photographs illustrate the ability of photography to capture dryness, scaling, and calloused skin on the heels. The (A) before and (B) after photos illustrate the ability to show detailed improvements in scaling and dryness that accompanied the use of an alpha-hydroxy acid (AHA) body moisturizer.

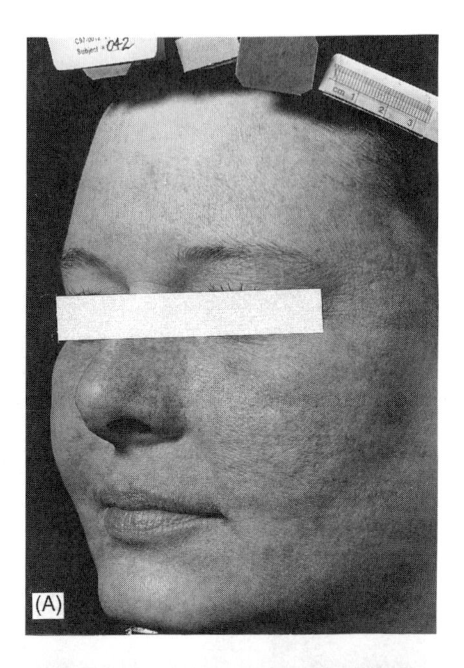

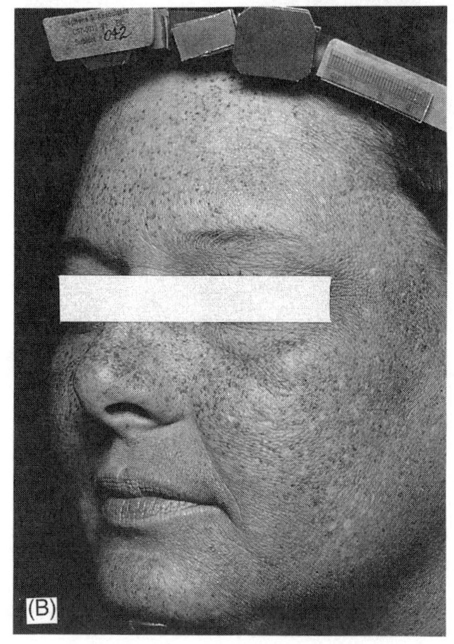

Figure 3 These photographs illustrate the dramatic differences between documenting hyperpigmentation on the face with (A) standard black-and-white photos compared to (B) black-and-white reflected UV-light photographs.

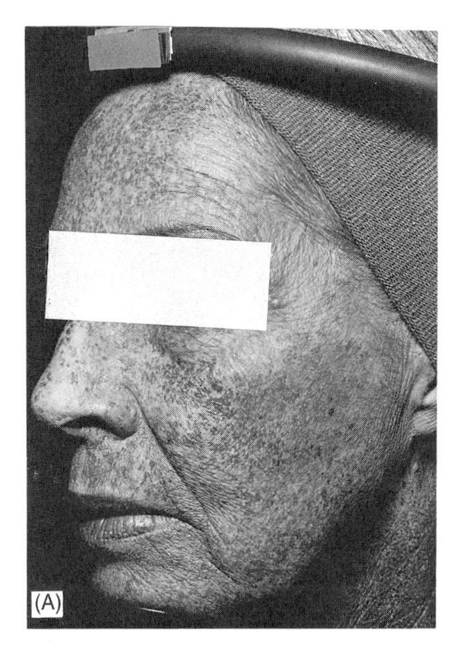

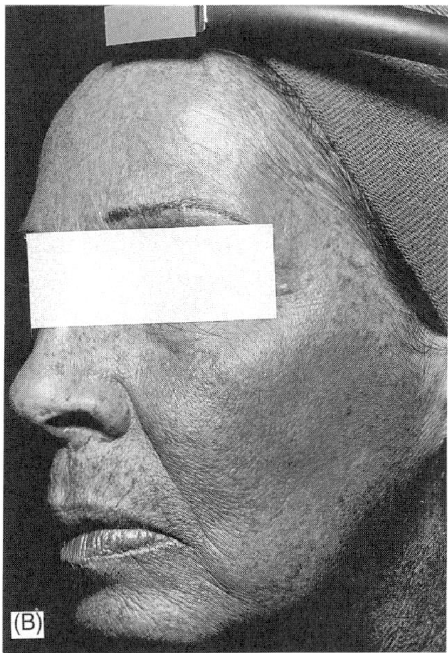

Figure 4 The (A) before and (B) after reflected UV-light photos shown here demonstrate the ability to capture dramatic changes in mottled pigmentation after treatment with an alpha-hydroxy acid (AHA)/retinol facial moisturizer.

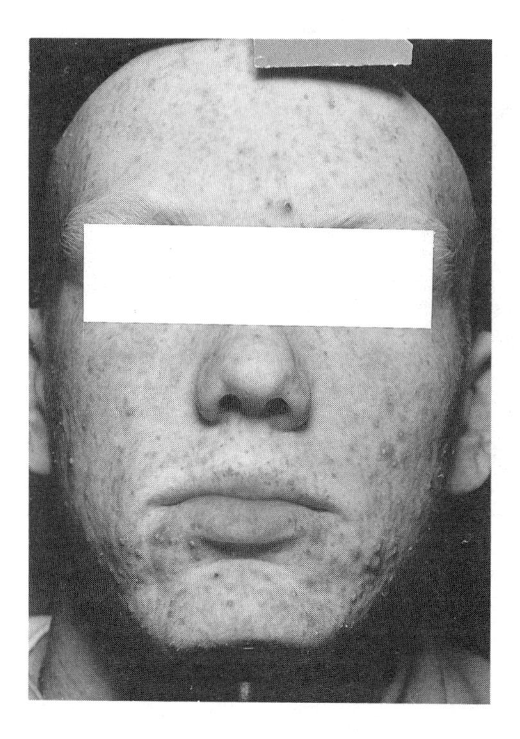

Figure 5 This black-and-white photograph illustrates the effect of using cross-polarization in documenting acne. Notice that the lesions appear pronounced due to the emphasis of inflammation in the papules and the contrast of the inflammation and white within the pustules. Color photographs are superior to black and white, because the cross-polarized filter accentuates the red tones.

The area under the decay curve is the water-holding capacity (WHC) and relates to the capacity of the stratum corneum to retain water.

2. Transepidermal Water Loss Measurements (17)

The ServoMed Evaporimeter is used to measure the transepidermal water loss (TEWL) and provides an estimate of the integrity of the stratum corneum or barrier function. This device is used throughout the course of a study to determine if there are changes in the integrity of the stratum corneum as a result of normal product use or exposure to product under exaggerated conditions.

3. Skin Elasticity Measurements

The Ballistometer (Hargens) (18,19) is a pendulum device used to measure the elastic properties of the skin in vivo. This instrument is used mainly on the

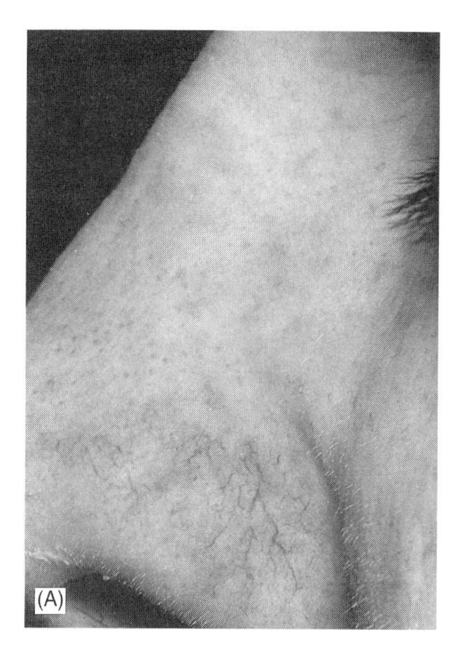

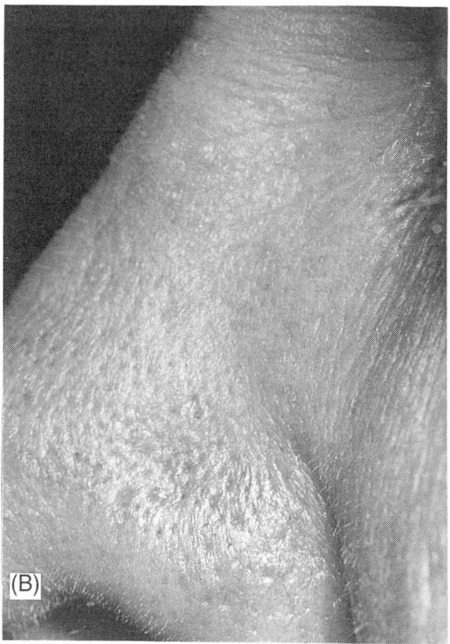

Figure 6 These photographs illustrate the superiority of (A) cross-polarized light over (B) visible light photography in accentuating compacted pores of the nose and telangiectasia. Also note that the lightly pigmented areas (photodamage) are more visible in the cross-polarized photograph.

face and most commonly on the crow's feet area of the eye. Measurements are taken throughout a study to document potential benefits from the use of facial or eye area products designed to firm the skin.

The Cutometer SEM 474 (20) is a device that also provides information about skin elasticity (see Ballistometer above). The Cutometer estimates elastic properties of the skin by applying a precisely controlled vacuum to a small defined area of skin. The measurements provide information about extensibility, resiliency, recoil, and viscous loss. The portability of the Cutometer's probe and its laptop computer interface present few restrictions in choosing locations for measurements.

4. *Parallel and Cross-Polarized Light Video Imaging System (21)*

The Zeiss DermaVision system allows images of skin characteristics to be taken at a magnification that exceeds conventional photographic methods. Magnified images of features such as pores, hairs, and individual pigmented spots can be digitized for image analysis of the size, frequency, and color intensity of cutaneous features.

5. *Skin Color Measurements*

The Minolta Tri-Stimulus Colorimeter measures skin luminosity and chromaticity (22). The instrument records color in a three-dimensional space designated $L^*a^*b^*$. The luminance (L^*) value expresses the relative brightness of the color ranging from black to white, and the a^* and b^* are the two components of chromaticity. The a^* value is the color hue ranging from red (+) to green (–). The b^* value is the color hue ranging from blue (–) to yellow (+).

6. *Squametry (Desquamation Assessment)*

D-Squame adhesive disks (CuDerm, Inc.) (23) are small adhesive disks placed onto the skin that are used to sample the outermost layer of the stratum corneum. The patterns and thickness of the layers of corneocytes that remain on the disk upon removal are used to calculate the coarse and fine flakes and the desquamation index. These parameters, when measured over the course of a study, are sensitive to treatment effects and will increase when the skin becomes dry or irritated.

7. *Sebum Measurements*

Sebutape patches (24) are devices used to measure the rate of sebum excretion and/or the distribution of active sebaceous glands. Estimates of these parameters are derived from image analysis of the transparent patterns left by facial oils deposited into the pores of the microporous, opaque tape. The sebum-filled pores in the tape are no longer capable of scattering light, so they appear transparent compared to the air-filled pores of the surrounding tape.

8. *Measurement of Fine Lines and Wrinkles*

Silicone replicas provide a sturdy model (a negative impression) of the fine lines and wrinkles present on an area of the skin, such as the face. The replicas are routinely evaluated using optical profilometry and image analysis to count and estimate the relative depth of the wrinkle and fine line features. Replicas are commonly taken on the crow's feet area of the eye to monitor the treatment effects of facial products designed to reduce the signs of photodamage or aging.

Examples of how we use some of the bioinstrumentation techniques discussed above are illustrated in Figures 7, 8, and 9.

F. Part V of the Five-Pronged Approach: Statistically Based Conclusions

Data analysis is as important as experimental design and execution in determining the success of a study. Analysis methods are chosen that extract the most information from the data and therefore provide the most general inferences. The choice of analytical techniques depends on (a) the information goals of the study sponsor and (b) the fundamental characteristics of the collected data.

A frequently used design specifies that observations from clinical grading and other measurements are collected at an initial assessment visit, called the baseline visit, and at subsequent assessment visits. This design introduces a time dimension into the study. Post-baseline visit data is compared with baseline data in order to determine whether changes have occurred that are attributable to product use rather than random chance. Therefore the time dimension is used to investigate product effects.

Another experimental design aspect provides a method to compare multiple products by establishing multiple treatment groups of subjects. This allows observations to be collected for each product over time, including at baseline. Postbaseline data is normalized with baseline data for each product, and then products may be compared against one another. Comparisons are frequently made between treatments and controls, test products and standard products, and between different product formulations.

1. *Information Goals*

Sponsors' information goals help determine data management and analysis methods. This idea is perhaps best illustrated with an example. We assume that two products are to be distinguished in terms of performance and that the interaction between ethnicity and performance is to be evaluated. This is initially achieved by employing two subsamples of subjects from the two ethnic groups of interest—say, groups A and B. Both groups use both products—say, X and Y.

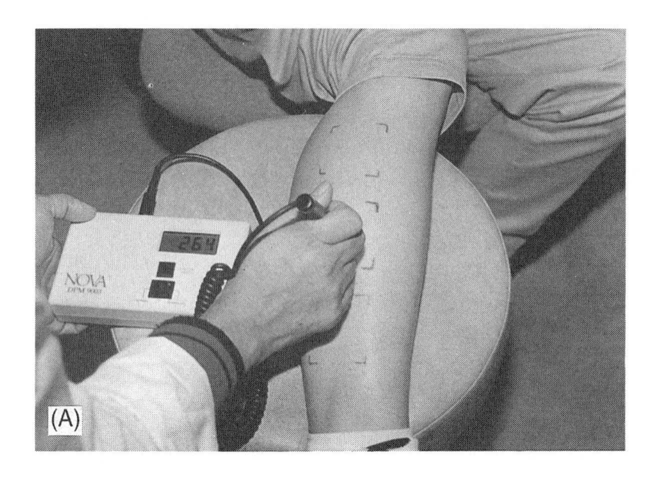

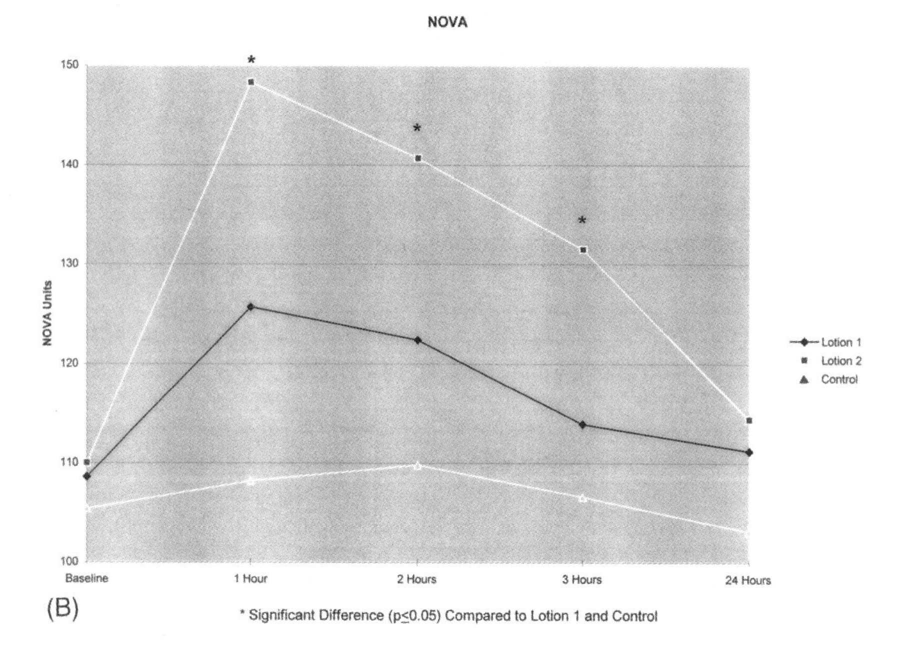

Figure 7 (A) The NOVA Dermal Phase Meter being used in a kinetic dry skin study on the lower leg. NOVA measurements were recorded at baseline and at 1, 2, 3, and 24 hr after two different test lotions were applied to treatment sites. (B) The data for this study. Analysis of the data showed significant differences ($p \leq .05$) between Lotion 2 and Lotion 1 in its ability to moisturize the stratum corneum.

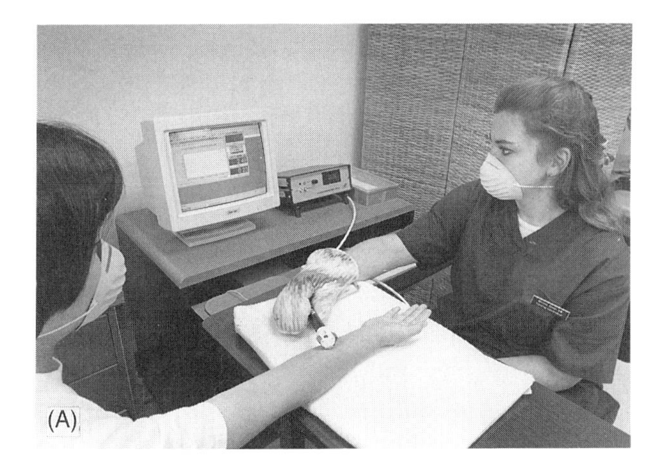

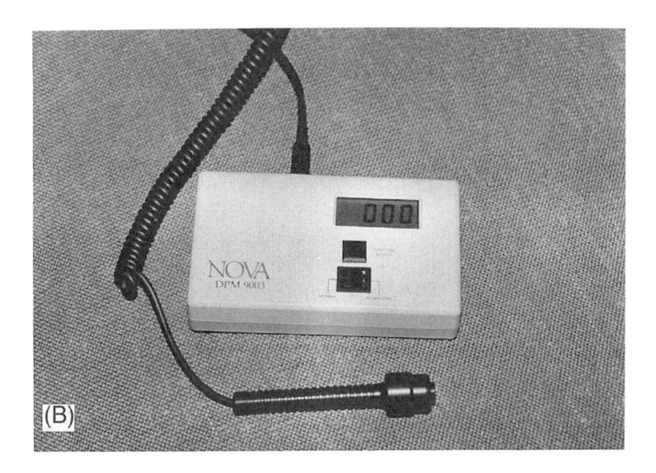

Figure 8 (A) The ServoMed Evaporimeter and (B) the NOVA Dermal Phase Meter. These instruments can be used in conjunction to provide substantiation for claims of moisturization and improvement in barrier function (transepidermal water loss). (C) The graph shows the barrier enhancement effect of Lotion B compared to Lotion A and control. With moisturization studies, transepidermal water loss (TEWL) and relative moisturization (NOVA) measurements usually exhibit an inverse relationship in response to treatment.

(continued)

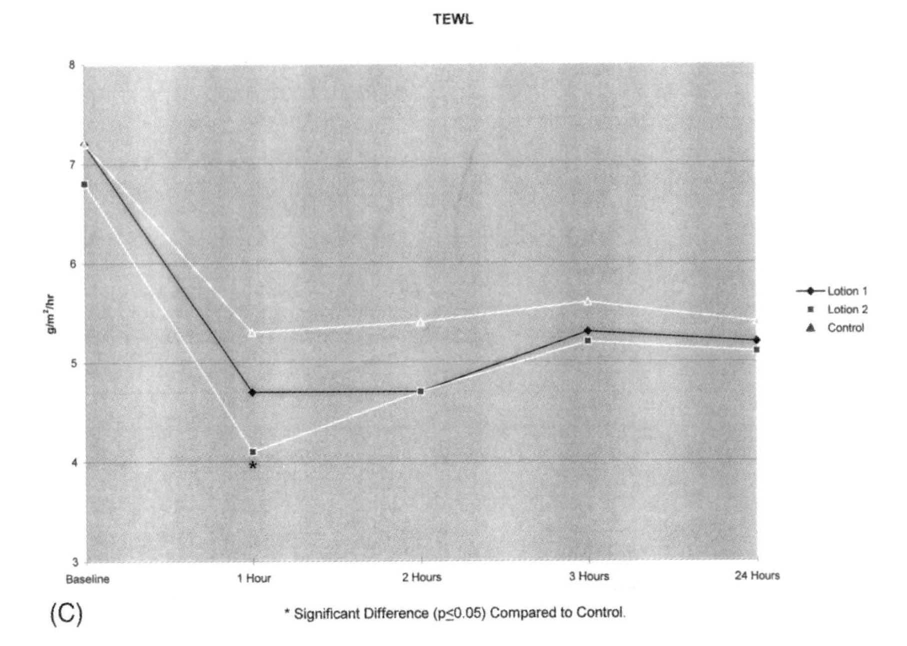

Figure 8 Continued.

Therefore four treatment groups are in operation: AX, AY, BX, and BY. Performance measurements are taken at baseline and at subsequent assessment visits. The data from the postbaseline visits are treated by controlling for baseline variable values, leaving us with baseline-normalized data for each postbaseline visit for each of the four treatment groups. At each postbaseline visit a comparison may be made between the treatment groups. This allows us to investigate simultaneously any potentially subtle ethnic–product interactions that may exist, as well as to compare the products X and Y. However, the analysis is not complete, since more information may be extracted from the data. When the AX and BX groups and the AY and BY groups are combined to produce just two groups—X and Y—we are able to generate a pure product comparison at each post-baseline time point that is more powerful than the comparison involving four treatment groups. In this example, multiple sponsor information goals are achieved by grouping the data.

2. *Fundamental Data Characteristics*

Analysis techniques are selected to match the properties of the study data. Determinants of data properties include experimental design and the fundamental

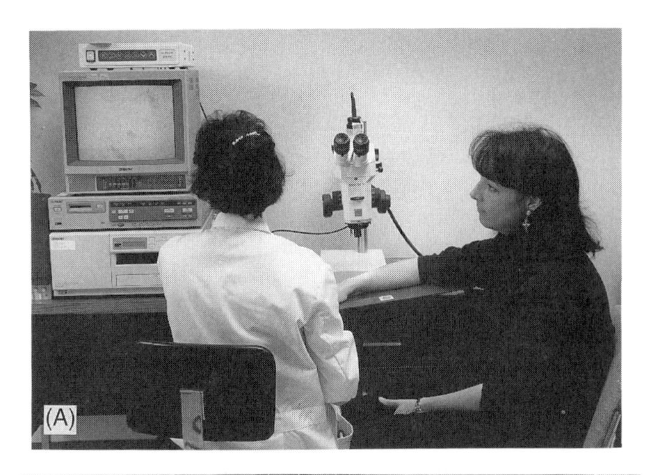

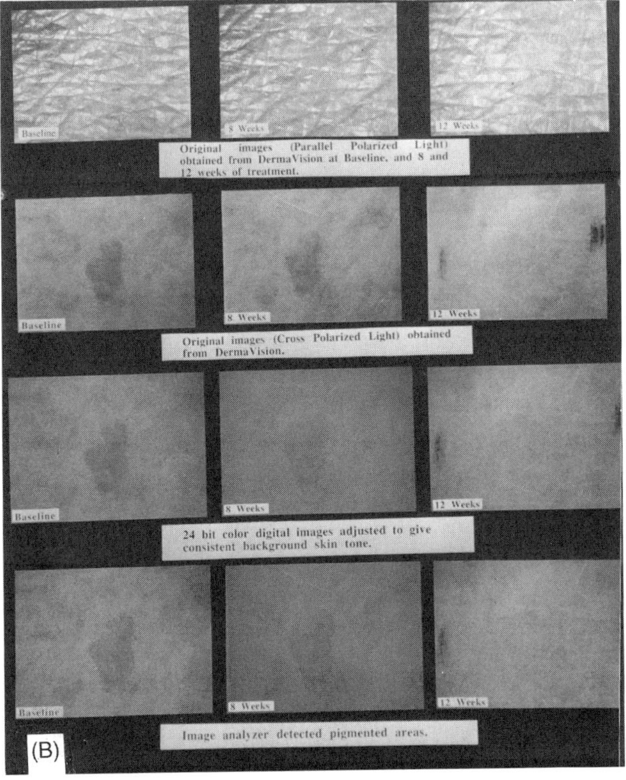

Figure 9 (A) The Zeiss DermaVision system. (B) Photographs taken with this device illustrate the ability of cross-polarized light video to reduce surface features and enhance the visibility of a single pigmented spot, in this case an actinic lentigines lesion. Moving horizontally, notice the gradual fading of the spot at consecutive time points after treatment with the alpha-hydroxy acid (AHA)/retinoid test product.

nature of the measurements made. For example, a test subject who uses a product during the entire course of a study generates paired observations that may be examined for departures from zero in order to determine the existence of product effects. In another situation a test subject may use a product until a certain threshold is reached, at which time the subject's participation in the study ends. The data collected in this case would be survival observations of the time taken to reach the threshold. In the first example, standard univariate techniques (such as *t*-tests) could be used to establish product effects, and multivariable techniques (such as ANOVA) may be implemented to compare products. In the second example, survival analysis methods could be used to examine a single product or to compare multiple products. The preceding examples can also be used to illustrate the differences between data distributions: measurements taken over time for performance variables may well be approximately normally distributed, whereas measurements of the time elapsed until some event occurs are clearly not.

III. SELECTED TESTING METHODOLOGIES/CONTROLLED USE TESTS (25)

A. Controlled Dry Skin Kinetic Studies

The objective in these studies is to demonstrate efficacy of a topically applied product over a short-term period, and may include comparing that efficacy with competitive standards. Examples of claims that positive results from these studies may support are "Moisturizes instantly," "Moisturizes for up to 12 hours," or "Superior to the leading moisturizers." This is usually accomplished using biophysical methods. Twenty to 30 people are typically recruited to participate as volunteer subjects, but the number of subjects required in any clinical trial is determined by the power of a study to detect clinically significant differences in treatments (26). Power is an expression of the ability of a study to detect differences in treatments if one exists, and is dependent on the response rates of the treatments, the significance level desired, and the number of subjects treated (26). Therefore, knowledge of the anticipated differences among treatments is important so that a sufficient number of subjects can be enrolled to ensure with reasonable certainty that a statistically significant difference will be obtained if the anticipated differences between treatments exist. The error of believing that there is no difference between treatments when in fact there is, is referred to as a type II or β error. Many clients unfortunately ignore this aspect of clinical trials, because of the higher costs associated with larger panels.

Treatment sites are usually the volar forearm or the lateral aspect of the lower leg. Three to four sites including a no-treatment site can be marked on each arm or leg with a marker and the aid of a template. Usually 2 mg of test material are applied per square centimeter for treatment sites. This has been found to be an effective quantity of a lotion or a cream to sufficiently cover the skin surface (27). Bioinstrumentation measurements are taken at baseline, and at intervals after product application up to 48 hr. The intervals selected depend on effectiveness times required for a claim. For instance, if one wishes to claim that a moisturizer is effective for up to 6 hr, then the results must show a significant difference in moisturization between the 6-hr time point and baseline, and the untreated control site. An untreated control is important to include, since even several hours can affect skin condition, especially if the environmental conditions are changing rapidly. The effectiveness of a moisturizer or a skin barrier repair agent can be assessed using an evaporimeter to measure transepidermal water loss (TEWL), and a conductance meter to measure the relative hydration of the stratum corneum. Both techniques have been shown to be effective in quantifying moisturization treatment effects on skin (28), and typically show an inverse relationship. However, the opposite is true for irritants, which damage the stratum corneum and thus compromise its integrity (Figure 10).

Barrier repair studies require that the starting skin condition be slightly to moderately compromised. We usually specify that TEWL should be ≥ 7 g/m^2 for the forearm and ≥ 12 g/m^2 for the lower leg. Likewise for moisturization studies, the skin should show moderately dry symptoms.

B. Controlled Modified Dry Skin Regression Studies (1)

The objective of these studies is to evaluate the efficacy of a moisturizer, usually against a competitive standard, in resolving moderate to severe dry skin. This is accomplished using clinical and bioinstrumentation methods. Approximately 20 to 30 subjects (see Section III.A) with mild to moderate skin dryness are employed per treatment cell. Treatment sites are typically the hands or the legs, but could be other areas of the body such as the face if that were the target of the marketing claim or the category of product being tested. The lateral aspect of the lower leg offers an advantage in that it is a uniform, relatively flat surface allowing bioinstrumentation probes to be easily placed in contact with the skin, and clinical scaling and cracking are easier to grade on the leg. Test products are applied once or twice per day for up to 3 weeks, followed by a no-treatment period (the regression phase). Clinical and bioinstrumentation assessments are typically done at baseline, 2 days, 4 days, 1 week, 2 weeks, and 3 weeks during the treatment phase, and every day or every other day during the regression phase until the dry skin condition has returned to baseline levels. Therapeutic moisturizers resolve dry skin more quickly, and they will maintain good skin condition longer after the discontinuation of treatment

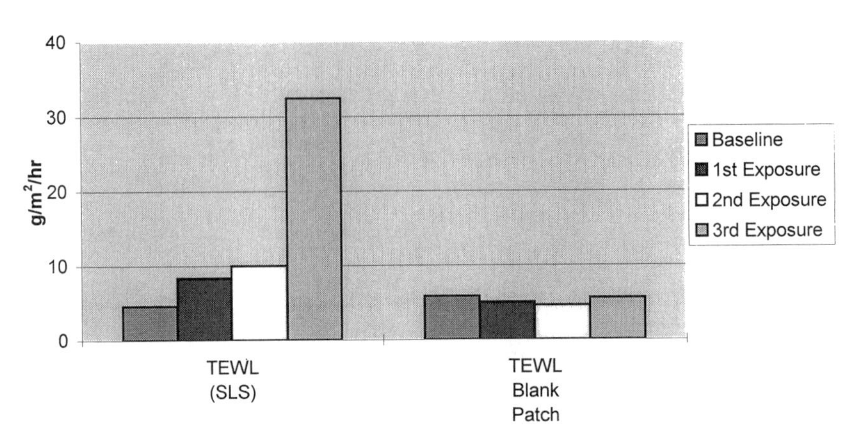

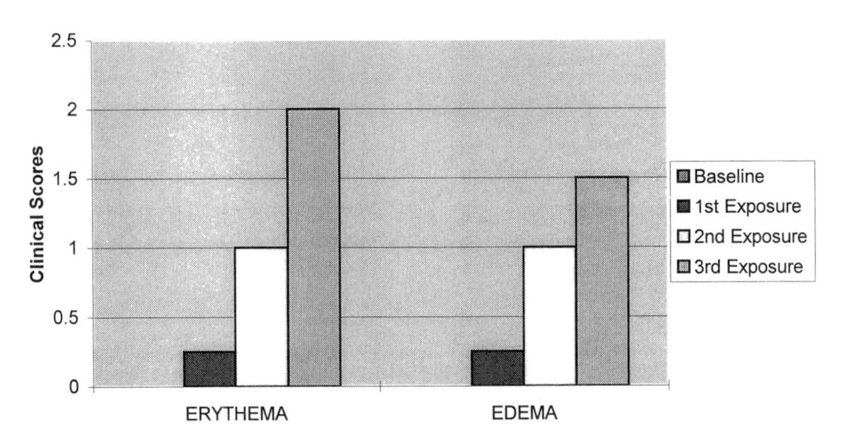

Figure 10 2.0% SLS repeated 6-hr exposures (forearm) chemical irritation. The inverse is usually true when evaluating the effect of a therapeutic moisturizer using evaporimetry to measure TEWL. The lower readings reflect the product's ability to slow water loss from the stratum corneum by improving the barrier properties of the stratum corneum. However, the opposite is true for irritants, as illustrated in this example of repeated exposure to SLS, which damages the stratum corneum and thus compromises its integrity. High TEWL readings result from this damage.

compared to nontherapeutic moisturizers; the regression phase of these studies is thus a good way to differentiate between the two.

C. Controlled Product Usage Studies

1. Photoaging Studies

The objective of these studies is to establish the efficacy and safety of one or more product treatments designed to improve the appearance of photodamage on the face, upper chest, or backs of the hands. These are the most visible areas of the body, and the areas that men or women who have suffered the ravages of years of unprotected sun exposure want improved. Typical efficacy parameters are improvements in fine lines and wrinkles, sallowness, tactile roughness, skin laxity, and mottled pigmentation. Usually 30 to 40 subjects (see Section III.A) are required per cell. Volunteer subjects must meet specific selection (inclusion) criteria, which include having moderately photodamaged skin as determined by clinical examination using a grading scale such as the Glogau classification system (29). After subjects are qualified and enrolled in the clinical trial, they typically undergo a 3- to 7-day "wash-out" period during which they discontinue the use of all moisturizers and moisturizing foundations. They are permitted to use a mild cleanser, and under some circumstances are permitted to use a standard light-duty moisturizer, especially if the environmental conditions are dry and cold. If the skin is allowed to dry out too much, the skin may become compromised and subject to irritation. Many of the treatment actives found in "antiaging" products are irritants in their own right. In our experience, applying all *trans*-retinoic acid or an alpha-hydroxy acid to dry, sensitive skin will illicit skin redness and swelling in approximately 10–15% of the population of a test panel. Even under normal circumstances, approximately 10–15% of subjects may experience a range of symptoms from skin redness and scaling to sensations of burning, stinging, tightness, or itching. During the first 2 weeks of product usage, after their skin accommodates to the irritation, most of the subjects will be able to use the product without further symptoms. However, the skin of approximately 1–2% of the study population will not accommodate to the irritation, and these subjects will need to be dropped from the study.

At baseline, subjects begin using the treatment product assigned, usually twice per day for up to 6 months. Clinical assessment of the effectiveness and irritation of the treatment is usually conducted at monthly intervals. Assessment of irritation includes erythema (skin redness), edema (skin swelling), scaling (skin dryness), acne breakouts, and the subjective sensations of burning, stinging, itching, tightness, and tingling. Bioinstrumentation methods are often employed to substantiate clinical improvement. These include silicone replicas with image analysis for fine lines and wrinkles, D-Squame Disk sampling of surface skin cells (squame) to assess desquamation pattern, conductance meter measurements (e.g., NOVA Dermal Phase Meter or SKICON Skin Surface Hygrometer) to evaluate skin hydration, skin elasticity measurements (e.g., Cutometer SEM 474 or Ballistometer, Hargens), and colorimeter measurements (e.g., Minolta Tri-Stimulus Colorimeter) to evaluate skin clarity changes. In addition, high-quality clinical documentation photography is sometimes employed to document clinical improvement (see Section II.C.2).

2. *Cell Turnover (Stratum Corneum Replacement Method)*

A modest increase in cell turnover is thought to improve the health of the skin by helping to remove dull, skin surface cells (29). Compounds such as retinoids, alpha-hydroxy acids, and beta-hydroxy acids all have keratolytic activity, and they are commonly used in today's antiaging products because of their effects on cell turnover, but also because they are beneficial in improving the photoaging attributes discussed in Section III.C.1. Fluorescent staining of the stratum corneum (SC) with dansyl chloride has been shown to be an effective method for estimating SC replacement time (30). Approximately 18 to 20 subjects are used to do pairwise comparisons on the volar forearm (see Section III.A). This allows one arm to be treated with the test product, and the contralateral arm to be treated with a placebo control or no treatment. Testing multiple test products requires a balanced block design with additional subjects.

In this method forearms are occlusively patched with 5% (w/w) dansyl chloride in petrolatum. Patches are removed after 6 hr and repatched with fresh dansyl chloride. After 24 hr, patches are removed and sites are gently cleansed with moist Webril pads and read in the dark for fluorescence using a Woods light. Subjects then treat their forearms as directed, and return to the clinic after 1 week to begin having their forearm sites graded for the presence of fluorescence. Subjects return every 2 days for grading until both forearms no longer have fluorescence at both patch sites (fluorescence extinction method). A treatment that produces a significant decrease in SC replacement time compared to either an untreated control or a placebo formula as measured by the disappearance of skin fluorescence is regarded as having a beneficial effect. There are variations in this method that involve treatment with product for one or more weeks followed by fluorescence staining of the SC and further treatment. In this case, the skin is preconditioned with the test product.

3. *Pore Size/Sebum Excretion Studies*

These studies are designed to evaluate the effects of topically applied test products to alter either the apparent size of facial pores, or the amount of sebum being delivered to the skin surface. The pores are the outward appearance of the sebaceous follicles, and when they are large and numerous we refer to this condition as "orange peel skin," while too much sebum being delivered to the skin surface results in greasy, oily skin, a condition called seborrhea. The sebaceous follicle is always in dynamic equilibrium, producing sebum within the gland, storing sebum in the follicular reservoir, and delivering sebum from the follicular reservoir onto the skin surface, called sebum excretion. Toners and clarifying lotions help to keep the skin surface sebum-free, and effectively reduce the amount of sebum in the sebaceous follicle. But their action is short-lived, since sebogenesis is a powerful process, and the follicular reservoir can refill and begin spewing sebum onto the skin surface within a few hours.

Studies designed to evaluate either the effects of test products on pore size or skin surface sebum typically incorporate a combination of clinical, photographic, and biophysical methods. At baseline, volunteer subjects are qualified by having their foreheads or cheeks cleansed with a mild liquid cleanser, defatted with hexane, and wear Sebutape patches (31) for 30 or 45 min to collect sebum. Sebutape samples from each individual are graded against a reference standard for low, medium, or high sebum levels. Individuals with medium or high levels are usually selected for enrollment. The Sebutape samples from qualified subjects are carefully placed on view cards, and stored in the freezer under nitrogen gas until analyzed. Subjects are also clinically evaluated for pore size and the degree of impacted material (keratin plug) within the follicle. High-quality color macrophotos are also taken for documentation, and for subsequent image analysis of pore size. Sometimes more than one baseline procedure is done for averaging, in order to reduce the inherent biological variability in sebaceous gland activity within the population. Factors such as age, gender, and endocrine status of prospective subjects may affect the measurement. For instance, oral contraceptives decrease sebum excretion rate (SER), whereas the effect of the menstrual cycle is of minor importance (32). However, a 1°C change in skin temperature produces a 10% change in SER (33), and SER is highest in the morning and lowest during the late evening and early morning hours (34). Consequently, control of these variables will benefit the outcome of these studies.

Test products are administered to the study population for use according to label instructions for varying periods of time, depending on the nature of the product. Some products may have an immediate effect on the appearance of pores, but products designed to affect sebogenesis usually require weeks to months of treatment. A variety of techniques are available for estimating SER. Our method of choice is the Sebutape patch method, which is a lipid-absorbent tape technique (31,35). Other methods include the time-proven gravimetric absorbent paper technique of Strauss and Pochi (36), and the cup technique of Ruggieri et al. (37). A photometric method developed by Schaefer in Europe is based on the reduction in optical density of ground glass in contact with fat (38).

IV. CONCLUSION AND FUTURE DEVELOPMENTS

As Dr. Albert Kligman (39) has suggested in the introductory chapter of the *Handbook of Non-Invasive Methods and the Skin*, "the future is already here" and those at the cutting edge of a discipline that promises to deliver new and improved noninvasive instruments for skin evaluation are like "a rich child in a Viennese pastry shop."

As the personal microcomputer (PC) has become smaller and more powerful, the instruments already in use in the clinical laboratory are becoming "smarter." In the area of clinical documentation photography, the digital era has arrived. Crisp color photos can be archived to CD-ROM disks, providing compact storage accessible to any modern PC with a CD-ROM reader and much more permanent than any print or negative. New smarter instruments will warn the operator of environmental conditions likely to confuse its "sensors." Instruments will "talk" directly to other computers in the laboratory, logging data automatically. Client–server on-line acquisition of instrumentation data in the clinical setting has already been implemented (40). All data including questionnaire and evaluator scores are integrated in a relational database system which ties in subject identification, treatment schedule, etc. The system can automatically apply quality assurance rules to entered data, as well as secure data logging.

More precise and sensitive instruments for visualizing specialized aspects of the skin are on the horizon. These include more advanced spectroscopy techniques that will allow noninvasive determination of the spatial distribution of collagen and keratin fibers as well as other optically reactive substances in various layers of the skin (41).

Magnetic resonance imaging (MRI) of the skin is coming of age with the recent announcement of portable units for obtaining this valuable data (42). Skin MRI promises to noninvasively generate well-resolved images of such things as the actual water profile of the skin, the pilo-sebaceous unit, and other appendages and features of the dermis (39).

All of these emerging technologies will not be able to stand alone, but they will provide much stronger complementary support to clinical and subjective assessments of skin conditioning. The integrated approach, as we have illustrated in this chapter, will stand the test of time, and it will be the best approach for supporting credible marketing claims of product performance benefit. For further information, the reader is referred to Refs. 43–46.

ACKNOWLEDGMENTS

Special thanks are given to Robert Goodman, Professional Photographer, Dallas, TX, for assistance with clinical photodocumentation procedures, and to Andrew Lawson, Ph.D., Lawson and Associates, Austin, TX, for assistance with Section II.F, "Statistically Based Conclusions."

REFERENCES

1. Kligman AM. Regression method for assaying the effectiveness of moisturizers. Cosmet Toilet 1978; 93:27.
2. Seitz JC, Rizer RL, Spencer TS. Photographic standardization of dry skin. J Soc Cosmet Chem 1984; 35:423–437.
3. Goldschmidt H, Kligman AM. Exfoliative cytology of human horney layer. Arch Dermatol 1967; 96:572.
4. Prall JK, Theiler RF, Bowser PA, Walsh M. The effectiveness of cosmetic products in alleviating a range of dryness conditions as determined by clinical and instrumental techniques. Int J Cosmet Sci 1986; 8:159.
5. Schatz H, Kligman AM, Manning C, Stoudemayer T. Quantification of dry (xerotic) skin by image analysis of scales removed by adhesive discs D-SQUAME®. J Soc Cosmet Chem 1993; 44:53–63.
6. Philp NJ, Carter NJ, Lenn CP. Improved optical discrimination of skin with polarized light. J Soc Cosmet Chem 1988; 39:121.
7. Cook TH. Profilometry of the skin—a useful tool for the substantiation of moisturizers. J Soc Cosmet Chem 1980; 31:339.
8. Grove GL, Grove ML, Leyde JJ. Optical profilometry: an objective method for quantification of facial wrinkles. J Am Acad Dermatol 1989; 21:632.
9. Welzel J, Wolff HH. Laser profilometry. In: Berardesca E, et al, eds. Bioengineering of the Skin: Skin Surface Imaging and Analysis. Boca Raton, FL: CRC Press, 1996:part IIB.
10. Distante F, Berardesca E. Hydation. In: Berardesca E, et al, eds. Bioengineering of the Skin: Methods and Instrumentation. Boca Raton, FL: CRC Press, 1995: chap. 2.
11. Leveque JL, DeRegal J. Impedance methods for studying skin moisturization. J Soc Cosmet Chem 1983; 34:419.
12. Serup J, Jemec GBE, eds. Handbook of Non-Invasive Methods and the Skin. Boca Raton, FL: CRC Press, 1995.
13. Berardesca E, et al, eds. Bioengineering of the Skin: Methods and Instrumentation. Boca Raton, FL: CRC Press, 1995.
14. Jackson EM, Stephens TJ, Goldner R. The use of diabetic panels and patients to test the healing properties of a new skin protectant cream and lotion. Cosmet Dermatol 1994; 7:44–48.
15. Rizer RL, Stephens TJ, Jackson EM. A comparison of the irritant contact dermatitis potential of various cloth and plastic wound bandages on sensitive skin individuals. Cosmetic Dermatol 1994; 7:50.
16. Stephens TJ, Oresajo C. Ethnic sensitive skin. Cosmet Toilet 1994; 109:75–80.
17. Pinnagoda J, Tupker RA, Agner T, Serup J. Guidelines for transepidermal water loss (TEWL) measurement. Contact Derm 1990; 22:164.
18. Hargens CW. Ballistometry. In: Serup J, Jemec GBE, eds. Non-Invasive Methods and the Skin. Boca Raton, FL: CRC Press, 1995: chap. 14.8.

19. Fthenakis CG, Maes DH, Smith WP. In vivo assessment of skin elasticity using ballistometry. J Soc Cosmet Chem 1991; 42:211–222.
20. Cua AB, Wilhelm K-P, Maibach HI. Elastic properties of human skin: relation to age, sex, and anatomical region. Arch Dermatol Res 1990; 282:283–288.
21. Dorogi PL, Jackson EM. *In vivo* video microscopy of human skin using polarized light. J Toxicol—Cut & Ocular Toxicol 1994; 13:97–107.
22. Fullerton A, Fischer T, Lahti A, Wilhelm K-P, Takiwaki H, Serup J. Guidelines for the measurement of skin color and erythema. Contact Derm 1996; 35:1–10.
23. Miller DL. D-SQUAME® adhesive disks. In: Wilhelm K-P, et al, eds. Bioengineering of the Skin: Skin Surface Imaging and Analysis. Boca Raton, FL: CRC Press, 1996: part IIA, chap. 2.
24. Miller DL. Application of objective skin type kits in research and marketing. In: Cosmetics and Toiletries Manufacture World Wide (1996 Annual). Herfordshire: Aston Publishing, 1996:233.
25. Jackson Em, Robillard NF. The controlled use test in a cosmetic product safety substantiation program. J Toxicol–Cutaneous & Ocular Toxicol 1982; 1:117–132.
26. Bigby M, Gadenne A-S. Understanding and evaluating clinical trials. J Am Acad Dermatol 1996; 34:555–594.
27. 21 CFR Part 352. Sunscreen drug products for over the counter human use; tentative final monograph. Fed Reg, May 12, 1993; 58:29299.
28. Serup J. A double-blind comparison of two creams containing urea as active ingredient. Acta Derm Venereol (Stockh), suppl 1992; 177:34–38.
29. Glogau RG. Chemical face peels. Dermatol Clin 1991; 9:131–150.
30. Jansen LH, Hojyo-Tomoko MT, Kligman AM. Improved fluorescent staining technique for estimating turnover of the human stratum corneum. Br J Dermatol 1974; 90:9–12.
31. Nordstrum KM, Schmus HG, McGinley KJ, Leyden JJ. Measurement of sebum output using a lipid absorbent tape. J Invest Dermatol 1986; 87:260–263.
32. Cunliffe WJ, Cotterill JA. In: The Acnes: Clinical Features, Pathogenesis and Treatment. London: WB Saunders, 1975.
33. Cunliffe WJ, Burton JL, Shuster S. Effect of local temperature variation on the sebum excretion. Br J Dermatol 1970; 83:650.
34. Burton JL, Cunliffe WJ, Shuster S. Circadian rhythm in sebum excretion. Br J Dermatol 1970; 82:497.
35. Piérard GE, Piérard-Franchimont C, Lê T. Patterns of follicular sebum excretion during lifetime. Arch Dermatol Res 1987; 279:S104–S107.
36. Strauss JS, Pochi PE. The quantitative determination of sebum production. J Invest Dermatol 1961; 36:293.
37. Ruggieri MR, McGinley KJ, Leyden JJ. Reproducibility and precision in the quantitation of skin surface lipid by TLC. In: Touchstone JC, ed. Advances in Thin-Layer Chromatography. New York: Wiley, 1982:249–259.
38. Schaefer H. The quantitative differentiation of the sebum excretion using physical methods. J Soc Cosmet Chem 1973; 24:331.
39. Kligman AM. Perspectives on bioengineering and the skin. In: Serup J, Jemec GBE, eds. Handbook of Non-Invasive Methods and the Skin. Boca Raton, FL: CRC Press, 1995: chap. 1.1.

40. Wilhelm K-P. Client-server based on-line data acquisition for skin bioinstrumentation (presentation abstract). Skin Res Technol 1996; 2:201.
41. Utz SR. *In vivo* human skin spectroscopy: prelude to optical tomography (presentation abstract). Skin Res Technol 1996; 2:201.
42. Zemtsov A. Personal communication (DLM), 1997.
43. Merker PC. Good laboratory practices (GLP's) and good clinical practices (GCP's): beneficial impact on safety testing. J Toxicol 1984; 3:83–92.
44. Larsen WG, Jackson EM, BArker MO, et al. A primer on cosmetics. J Am Acad Dermatol 1992; 27:469–484.
45. Friedel SL. Technical support for advertising claims. J Toxicol—Cut & Ocular Toxicol 1992; 11:199–204.
46. Spilker B. Guide to Clinical Trials. New York: Raven Press, 1991.

Index